G000168557

PHYSICS RESEARCH AND TECHNOLOGY

GENERATION AND APPLICATIONS OF ATMOSPHERIC PRESSURE PLASMAS

Physics Research and Technology

Additional books in this series can be found on Nova's website
under the Series tab.

Additional E-books in this series can be found on Nova's website
under the E-book tab.

Materials Science and Technologies

Additional books in this series can be found on Nova's website
under the Series tab.

Additional E-books in this series can be found on Nova's website
under the E-book tab.

PHYSICS RESEARCH AND TECHNOLOGY

GENERATION AND APPLICATIONS OF ATMOSPHERIC PRESSURE PLASMAS

MASUHIRO KOGOMA
MASAKO KUSANO
AND
YUKIHIRO KUSANO
EDITORS

Nova Science Publishers, Inc.
New York

For permission to use material from this book please contact us:
Telephone 631-231-7269; Fax 631-231-8175
Web Site: http://www.novapublishers.com

Library of Congress Cataloging-in-Publication Data

Generation and application of atmospheric pressure plasmas / [edited by] Masuhiro Kogoma, Yukihiro Kusano, and Masako Kusano.

p. cm.

Includes bibliographical references and index.

ISBN 978-1-61209-717-6 (hardcover)

1. Low temperature plasmas--Industrial applications. 2. Plasma (Ionized gases)--Research. I. Kogoma, Masuhiro. II. Kusano, Yukihiro. III. Kusano, Masako.

TA2020.G46 2011

541'.0424--dc23

2011034609

Published by Nova Science Publishers, Inc. † New York

CONTENTS

PREFACE

Masuhiro Kogoma
Department of Materials and Life Sciences,
Sophia University, Tokyo, Japan

INTRODUCTION

In spite of their physical weakness, human beings have made use of their high intelligence with tools, accelerated their evolution, and have finally reached the highest position among the creatures on the earth. It is observed that chimpanzees and some other animals can also use their tools to take game. However, even for human beings, it probably took several million years to get innovative technology to utilize fire after the appearance of Homo sapiens. If we can see that the impact of IT technology, starting with the emergence of the semiconductor industry, has been leading human beings to a new world with a marvellous speed, we can also say that plasma technology, which fundamentally supports the semiconductor industry, would give an impact to human history in some sense similar to the discovery of fire.

A plasma is an ionized gas at a temperature of several thousand K or in a strong electric field. It contains a variety of high energy species, such as high energy electrons, ions, radicals, electron-excited species, and neutral particles. Plasmas in nature such as lightning and St. Elmo's fire were regarded as supernatural phenomena from God or as signs of misfortune in older days. They were recognized to be physical phenomena owing to development of electro-magnetics and vacuum equipment in the 18 th century. It is said that the term "plasma" was coined by Langmuir in 1929. However, studies of material synthesis using discharge phenomena started much earlier. An experiment on spark discharges was carried out in 1726, and it was reported that an oily substance was synthesized. Eventual industrial application of a plasma was initiated perhaps by Siemens in 1857 for ozone production. Ozone has been used for a variety of applications, such as deodorization for water supplies and swimming pools, sterilization and cleaning of fish products, due to its strong oxidative nature and an environmental safety after dissociation. A huge amount of ozone is produced in the world, and still now nearly 100 % of the ozone is produced using a silent discharge produced by Siemens, since there are no alternative techniques. Here the silent discharge is one example of atmospheric pressure low temperature plasmas. This fact indicates some characteristics and possibilities not only of atmospheric pressure plasmas but also of all plasma technology.

The low temperature plasma technology newly developed after the 2nd World War is one of the indispensable, fundamental technologies for the manufacturing industry of semiconductors in the advanced industrial society. Low temperature plasmas have recently been used for synthesizing various new materials, and for other applications such as exhaust gas treatment and plasma displays. In low temperature plasmas, in which the gas temperature is between room temperature and several hundred K, various chemical reactions can easily occur due to the existence of high energy electrons. Therefore, low temperature plasmas enable deposition of polymeric materials and inorganic compounds on any surfaces, etching and surface modification of materials at low temperatures. Thus an enormous number of studies have been reported for the low temperature plasmas applied for solid surface modification of polymers, metals, and ceramics, and thin film synthesis. However, conventional low temperature plasmas are usually generated at low pressures of less than 1 Torr with a long mean free path of electrons and a long diffusion length of gas molecules in order to ensure the uniformity of the electric field. This kind of plasma requires a vacuum system and high power consumption for reducing the gas pressure. It can be said that low temperature plasmas have mostly been used in high value-added applications, typically in the semiconductor industry. On the other hand, there has been a growing need of the plasma technology in other industries than the semiconductor industry due to its high capacity of adding functionalities. For example, low temperature plasmas often enable treatment of new materials that can not be achieved by the conventional chemical treatment at all. The term "Atmospheric pressure plasma" that appears in the title of this book is a relatively new plasma technique firstly proposed by Kogoma *et al.* [1]. It is characterized by its uniformity, stability and reactivity, which are necessary for the high-technology. This technique has already been or will be applied for large area surface treatment systems of engineering plastic films for adhesion improvement, surface cleaning of glass substrates for liquid crystal displays, treatment of powders, treatment of gas barrier films etc.

1. Characteristics of Plasma Chemical Reactions

The low temperature plasmas discussed in this book are not called by this name simply because their gas temperature is low. In order to understand the meaning of the low temperature plasma, one has to understand the concept of a gas temperature in physical chemistry. If the temperature of a 2-atom-molecule gas increases from room temperature, its kinetic energy, rotational energy and vibrational energy all increase.

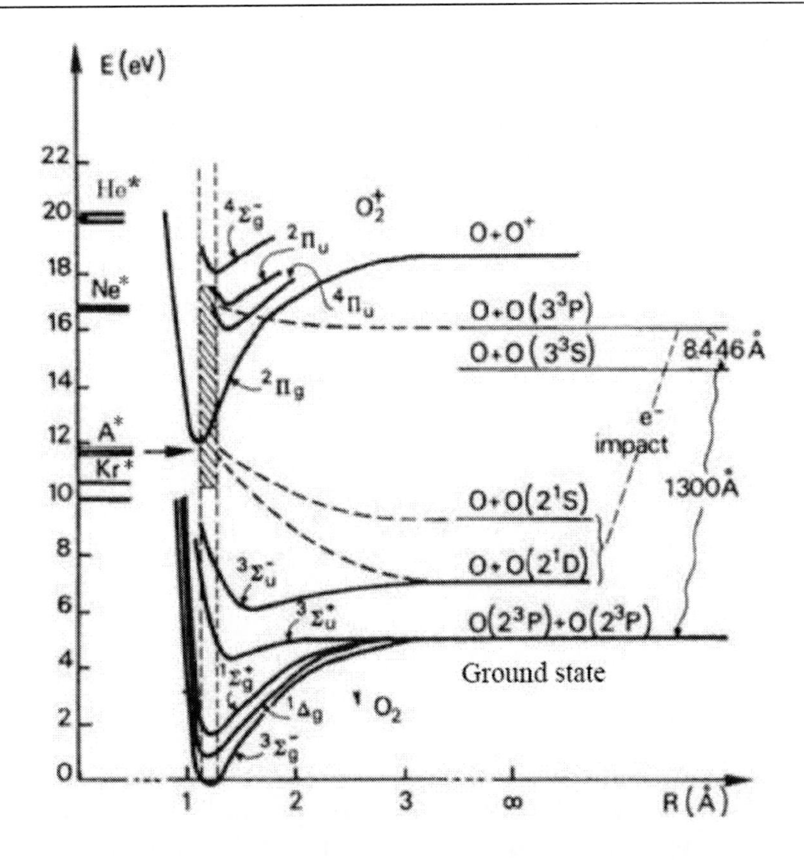

Figure 1. Potential energy curves for oxygen. The short lines at the vertical axis are energy levels of meta stable states of rare gas atoms.

As shown in Figure 1, in the case of oxygen molecules, the dissociation starts when the vibrational energy of the molecules of the ground state exceeds the dissociation energy of the molecules with the temperature rising, namely when the separation of the oxygen atoms in a molecule is the maximum. At this point, the gas temperature T_g exceeds 1000 °C. However, in such molecules, there exist not only the ground state curve of the electron energy but also many higher energy states, which give significant contributions for a plasma state. It is known that, as a core of a molecule is much heavier than an electron, the separation distance of the cores in a molecule would not change even when a high energy electron inelastically collides with a ground state molecule. On the other hand, when the electron energy state in the molecule jumps to a higher state, it is called a transition by the Franck-Condon principle. Collisions of the accelerated electrons in a strong electric field in a plasma induce such energy transitions in a molecule. If a potential energy represented by a curve corresponding to a specific excited energy level decreases as the separation of the cores increases, electron collision dissociation will occur after a vibration. Ionization will be induced when the collisional exciting energy is as high as the ionization energy. These curves usually cross each other; dissociation, ionization and excitation can often occur simultaneously. The electron temperature T_e basically corresponds to electron kinetic energy, in the following relationship:

$$\frac{1}{2}m_e v_e^{\,2} = \frac{3}{2}k_B T_e \tag{1}$$

Here k_B, m_e, v_e, T_e are the Boltzmann constant, the mass of an electron, the velocity of an electron, and electron temperature, respectively. Charged particles such as electrons and ions in a discharge plasma are accelerated by an external electric field, while a discharged plasma itself is thought to be macroscopically neutral. When electrons pass through a gas of molecules, a radius of collisional cross section σ_{eg} is determined by a circle with a radius equal to the sum of a radius of a molecule R_g and that of an electron R_e. Since an electron is much smaller than a molecule, the collisional cross section σ_{eg} is almost equal to πR_g^2. Here the product of the cross section σ_{eg} and an electron's mean free path λ_e is equal to a volume which one molecule dominates:

$$\lambda_e = \frac{1}{\sigma_{eg} n} \tag{2}$$

On the other hand, the density n is proportional to a gas pressure p and is inversely proportional to the temperature T,

$$\lambda_e \propto \frac{T}{p} \tag{3}$$

An electron has on average a kinetic energy obtained during one free path. Due to its small mass, an electron can easily get high kinetic energy from an electric field, and thus it gets a high temperature. At low gas pressures less than 1 Torr the mean free path of an electron is substantially long, and then it can obtain significantly high kinetic energy between its collisions. Therefore high energy electrons can be easily generated in a low pressure plasma. Collisional frequency between high energy electrons and molecules increases at lower pressures, which subsequently results in generation of numerous electrons, ions, excited species and radicals. On the other hand, it is difficult to accelerate ions because an ion is as heavy as a molecule, resulting in lower velocity than electrons. Consequently, ion temperature is only slightly higher than room temperature. As neutral species do not have charges and can not be accelerated by an electric field, the temperature remains at around room temperature at low pressures where the collision with electrons is limited. This is the state of non-equilibrium in which only the electrons are at high temperature while the other species remain at low temperature.

As described already, high energy electrons can easily dissociate or can highly excite molecules by collisions. For better understanding, this process may be compared with a reaction process described by a general chemical technique. In a general chemical process, when reactants (A+B) in a liquid or a gas are externally heated, and when the average temperature of the molecule exceeds the activation energy of the reaction E, a product (C) is produced over the barrier of the activation energy, as shown in Figure 2. The reaction rate constant k of the reaction (A+B -> C) is expressed by the following Arrhenius equation:

$$k = A \exp\left(-\frac{E}{k_B T}\right) \tag{4}$$

where A is a constant. The activation energy in equation (4) is the sum of the enthalpy of formation and the barrier energy. It is normally positive, so the reaction speed increases as the temperature of molecules increases. The thermal dissociation energy of a hydrogen molecule is about 100 kcal mol^{-1}, corresponding to ca. 5 eV. In order for a hydrogen molecule to get this level of dissociation energy, it has to be heated up to at least more than 1000 °C. On the other hand, it can be easily understood how easily a chemical reaction can occur in a low pressure plasma when we consider that an electron can be accelerated to the equivalent energy by a potential difference of 5 volts only. However, when a gas pressure increases, the situation changes as shown in Figure 3.

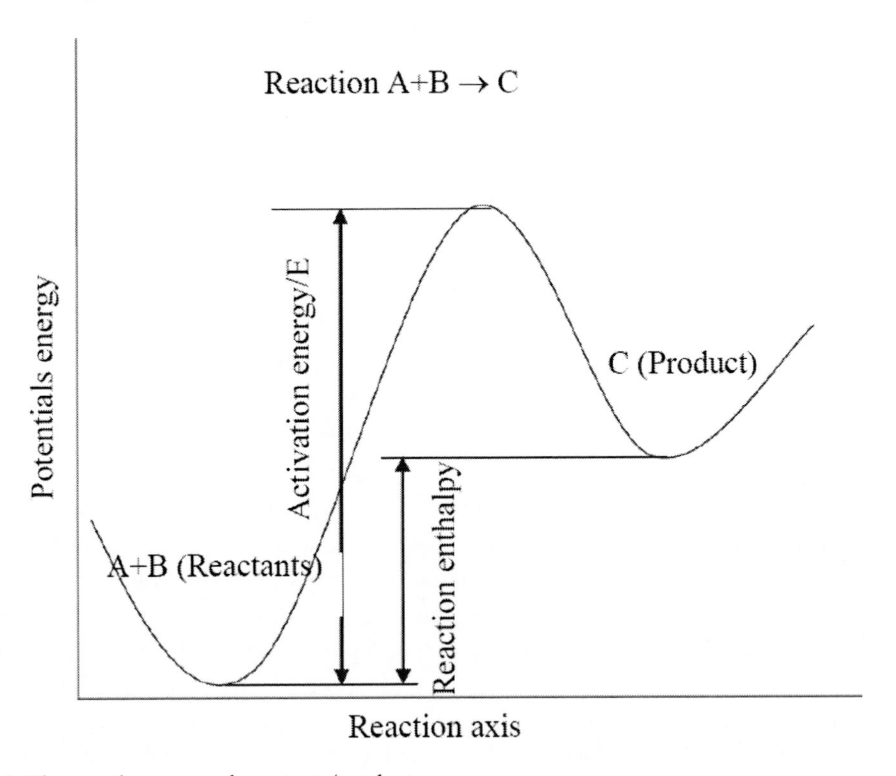

Figure 2. The reaction rate and reactants/product.

As the pressure increases, the frequency of collision with electron increases; subsequently the gas temperature increases by energy exchange with electrons and consequently the electron temperature decreases. As a result, the temperatures of electron and gas become equal when the pressure is higher than 100 Torr, corresponding to the thermal equilibrium state. As gas discharge phenomenon is based on collision ionization by an electron avalanche, in order to sustain a discharge itself, the voltage needs to increase as the gas pressure increases. As electrical input energy increases for sustaining the discharge, the temperature of the net system increases, and the gas temperature will become high. It is the reason why an electrical gas discharge tends easily to be a high temperature plasma or an arc discharge. In an arc discharge at a temperature of higher than 5000 °C, ionization is mainly induced thermally, where electrons are used for generating Joule heat. This discharge is a high temperature thermal plasma, which is completely different from a low temperature plasma.

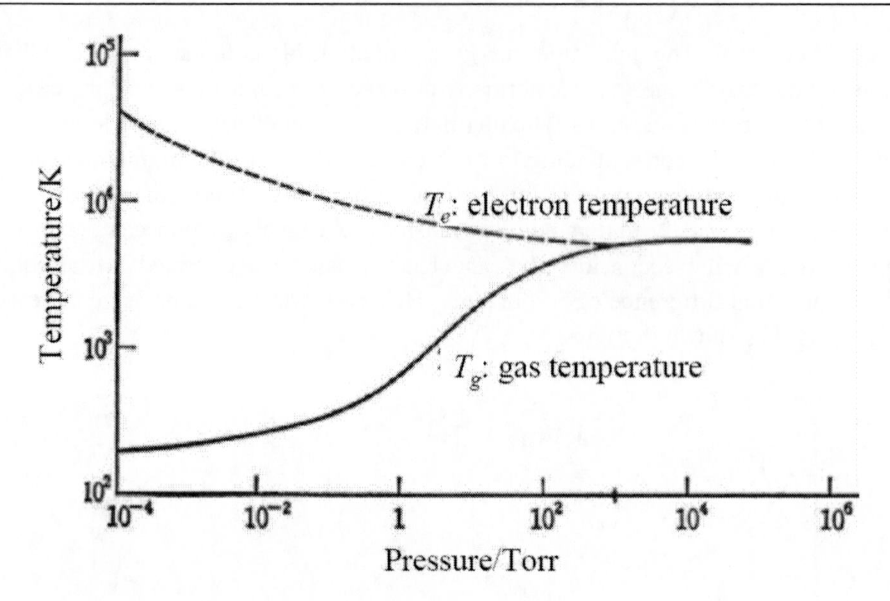

Figure 3. Temperatures in a plasma as a function of gas pressure.

2. Atmospheric Pressure Plasma

As was already discussed in section 1, characteristics of a plasma using a gas discharge phenomenon generally depend on the gas to be used, on the gas pressure, and on the way the energy is delivered to the plasma. Among researchers of plasma chemistry, it has long been common knowledge that a low-temperature plasma means a low pressure plasma, while a high temperature plasma means a high pressure plasma. On the other hand, even at atmospheric pressure, a corona discharge and a silent discharge are regarded as low temperature plasmas. In fact, they have long been applied for treatment of film surfaces. These discharges are inhomogeneous in time and space. They cannot be homogeneous, defusing glow discharges, nor can they have high electron temperatures. So they were thought to be unsuitable for various surface treatments or thin film synthesis which has commonly been performed in low pressure plasmas. However, Kogoma and his group in Sophia University reported at the 8th International Symposium on Plasma Chemistry (ISPC-8 Tokyo, Japan) in 1987 that a defusing glow discharge which is similar to that obtainable at low pressures can be generated even at atmospheric pressure between dielectric layers (dielectric barrier discharge (DBD)) in a noble gas such as helium driven by pulse excitation [1]. After that it has been reported that a variety of treatments performed in low pressure plasmas are also achievable at atmospheric pressure [2-7].

2.1. Research on Atmospheric Pressure Glow Discharge Plasmas

When Kogoma *et al.* firstly reported the processing of atmospheric pressure glow discharge plasma using He DBD in 1987, it did not draw much attention. Apart from their reports, atmospheric pressure glow discharges were seldom reported for some time. In 1989, Horiike *et al.* (Hiroshima University) and Koinuma *et al.* (Tokyo Institute of Technology) reported atmospheric pressure glow discharge plasma torches (after glow). Massines *et al.* in France and Tochikubo *et al.* (Tokyo Metropolitan University) started the research on atmospheric pressure plasmas, and were successful in simulating measured results. In addition Roth *et al.* (Tennessee University) developed atmospheric pressure glow discharge devices with a third electrode. Afterwards generation of a uniform plasma using inert gases other than helium, such as argon or N_2 [13], were reported. A parallel plate type (DBD) was initially used as the discharge system, and then various kinds of systems such as a capillary-tube type [19], a plasma torch (after glow) and a hollow cathode type have been developed.

For readers' references, published journal papers and conference proceedings by Kogoma *et al.* are chronologically listed. They are associated with the development of atmospheric pressure glow (APG) discharge plasmas since the initial stages of the research [13-43].

2.2. Generation of Atmospheric Pressure Glow Discharge

Generating a temporary discharge in a gap is possible by

I. placing a dielectric material at least at one of the electrodes, automatically stopping the development of the arc discharge by the reverse charge effect of the real charges in the discharge. In general, a power supply generating a sinusoidal wave with a frequency ranging between kHz and MHz is used, or
II. utilizing a power supply generating high voltage short pulses.

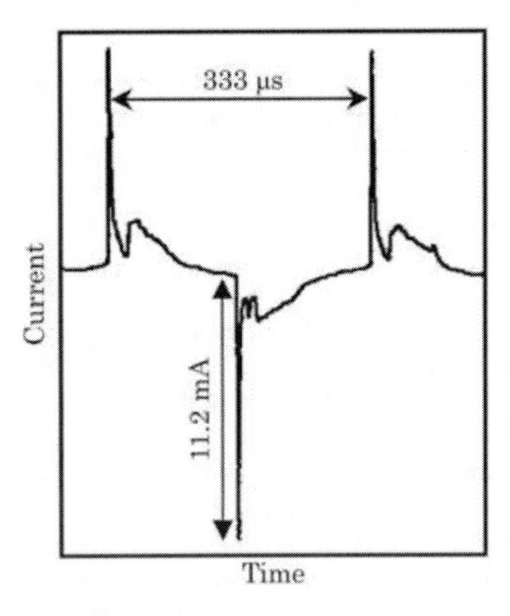

Figure 4. A current wave form of a discharge driven at 3 kHz.

Figure 4 shows the current wave form in the case (I) when 3 kHz sinusoidal high voltage is applied to the electrodes, both of which are covered with glass layers, in atmospheric pressure helium. It is indicated that one pulse discharge of several μs length appears in each half cycle. One may note that Figure 4 reported by Kogoma *et al.* is the current wave form of the helium atmospheric pressure glow discharge observed for the first time in the world. On the other hand, in the case of a conventional ozonizer (DBD), due to its microdischarge, numerous current pulses with extremely short widths appear [8].

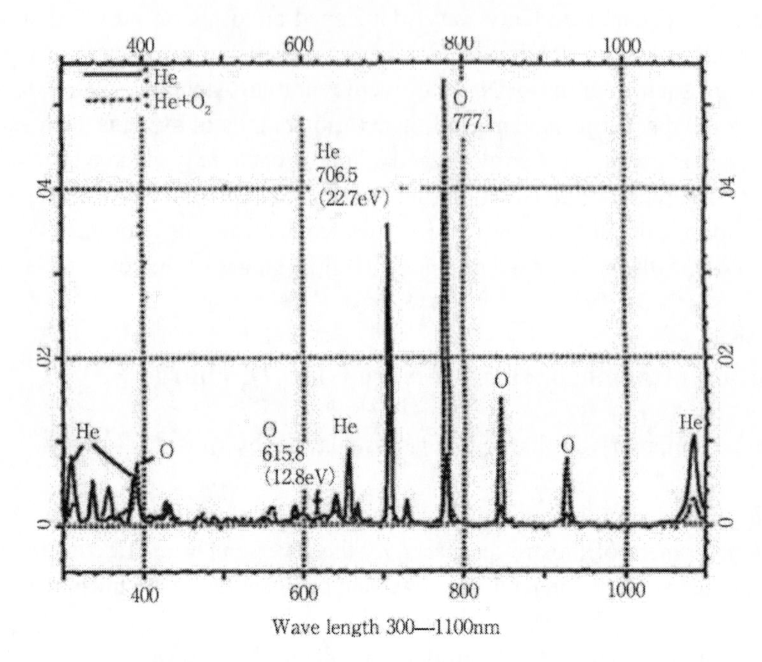

Figure 5. Optical emission spectra emitted from discharges in He (solid line) and He mixed with a small amount of O_2 (dotted line).

These differences are mainly due to whether the discharge generates globally over the electrode surfaces or not. In addition, a pulse width of the current of the glow discharge is approximately 100 times longer than that of the ozonizer in air or O_2 using the same structure of the electrodes. This is because there are some parts of weak strength of the electric field due to the existence of positive columns. It means that using helium gas a glow discharge can be generated, and that this discharge is different from a streamer seen in an air or O_2 discharge of the ozonizer. The positive column of a helium discharge can be eventually observed in a gap of the electrodes. Helium has an ionization energy of 24.5 eV, which is the highest among all gases. At the energy level of 20 eV which is just below the ionization energy, there is a long-lifetime metastable state of helium. In a continuous operation of the discharge, ionization actually requires the difference of these levels, only approximately 4 eV. This is why helium gas shows the lowest ignition voltage of 4 kV/cm among all gases. In addition, the applied electric field in a helium plasma is extremely low, so the transition to the streamer requires a longer time than usual, and thus diffusion of particles parallel to the electrode surface is enhanced. This is the reason why the helium discharge can become a glow even in a DBD. Figure 5 shows optical emission spectra of a helium DBD, containing the strong emissions from helium meta-stable atoms and those from excited states of mixed gas molecules of oxygen atoms dissociated from O_2. In an atmospheric pressure glow discharges, optical emissions are also observed, which originate from OH radicals coming from the water vapours in the discharge, and nitrogen simultaneously. Figures 6 and 7 show the electron energy states of helium and oxygen atoms, respectively.

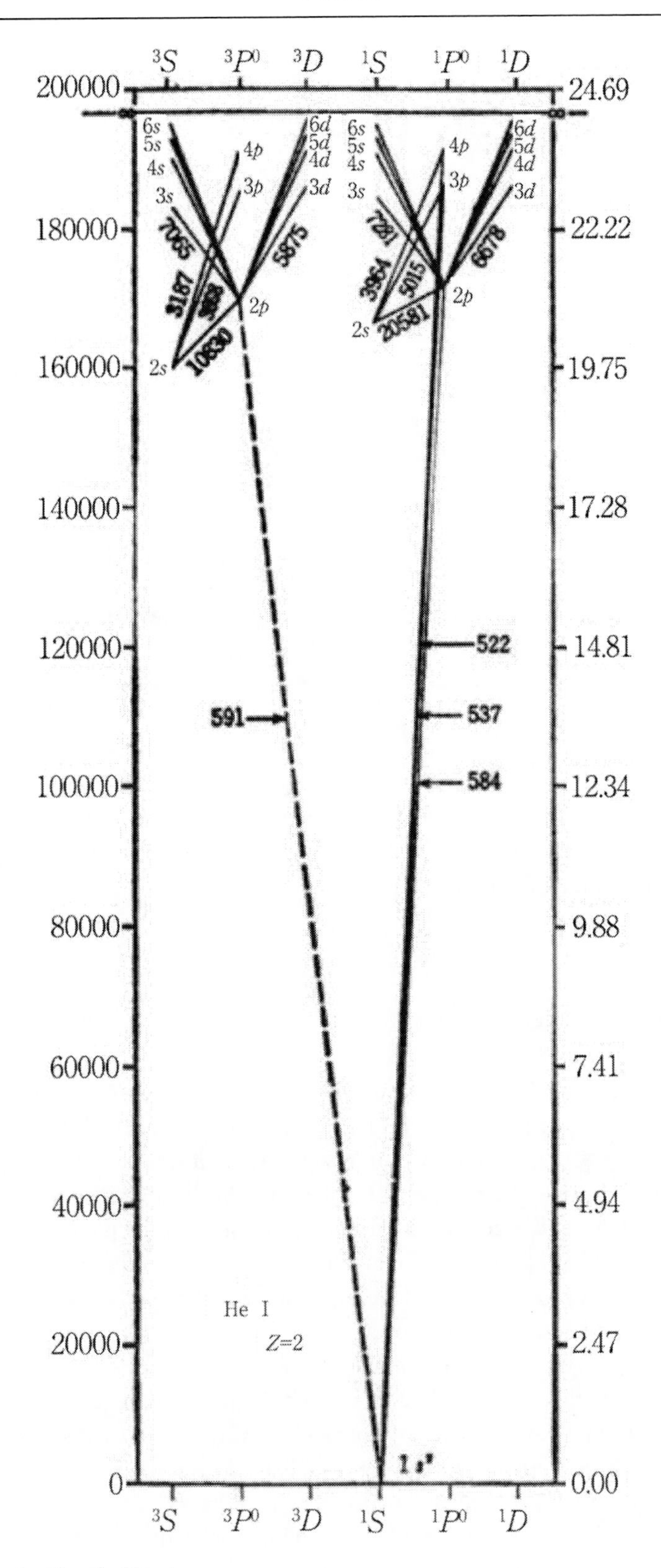

Figure 6. Energy levels of a He atom.

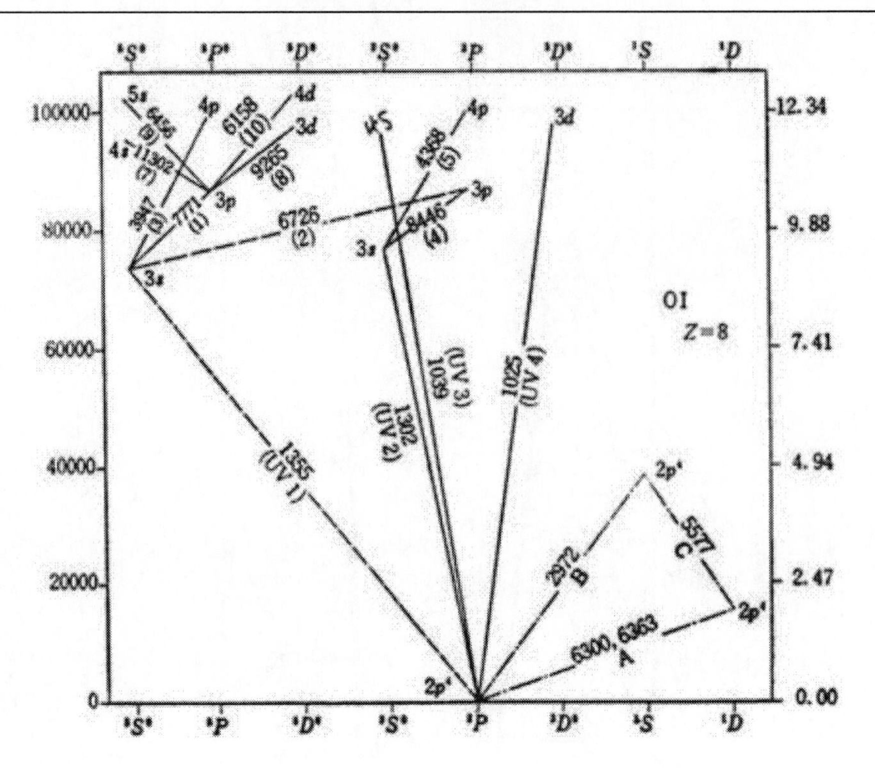

Figure 7. Energy levels of an O atom.

When O_2 gas is mixed in a helium gas, we indicated that the following Penning reactions can occur. Now the optically emitted level of the excited state of oxygen atoms agrees with that of the meta-stable state of helium:

$$He + electron \rightarrow He^*(2^3S, 2^1S) + electron \quad (5)$$

$$He^* + O_2 \rightarrow He + O^* + O^+ + electron \quad (6)$$

It is anticipated that processing of radical reactions and polymerization can occur by adding a small amount of required gas in an atmospheric pressure helium plasma, as is the case for low pressure plasmas. In the case of low pressure plasmas, electrons' collisions with the monomers usually initiate the reactions. On the other hand, in the case of the atmospheric pressure plasmas, helium and argon are used as majority gas. In the plasma, they produce the metastable atoms that have higher electronic excited states. Therefore, such excited rare gas atoms collide with mixed process gas molecules, which are efficiently dissociated. In this way subsequent chemical reactions can easily occur.

In the case of (II), it is recently found that a glow-like diffusing discharge can be generated in N_2 etc. with the steep increase of the applied voltage within several nano-seconds (ns), or between a gap as small as 1 mm. However, as will be discussed in this book, if a normal gas apart from a noble gas is chosen such as N_2 as a majority gas, the discharge is not a glow discharge but is a so-called Townsend-type glow like discharge [9]. It is reported that a homogeneous thin film can be synthesized with such a discharge (HAKONE X). If a noble gas such as helium or argon is used as a carrier gas, a commercially available power supply can be used, and energy states of plasma excited species are high. Consequently high treatment speed and effect are expected. On the other hand, if N_2 [10] or air is used as a carrier gas, the treatment speed and effect will reduce, but the running cost for the gas decreases. Therefore it is important to choose a proper carrier gas depending on the required process.

The atmospheric pressure low temperature plasma, as will be described in the following chapters of the applications, will be a very attractive topic of the research since there are still needs for developing some novel ways of plasma generation.

2.3. Equipment of Atmospheric Pressure Glow Discharge

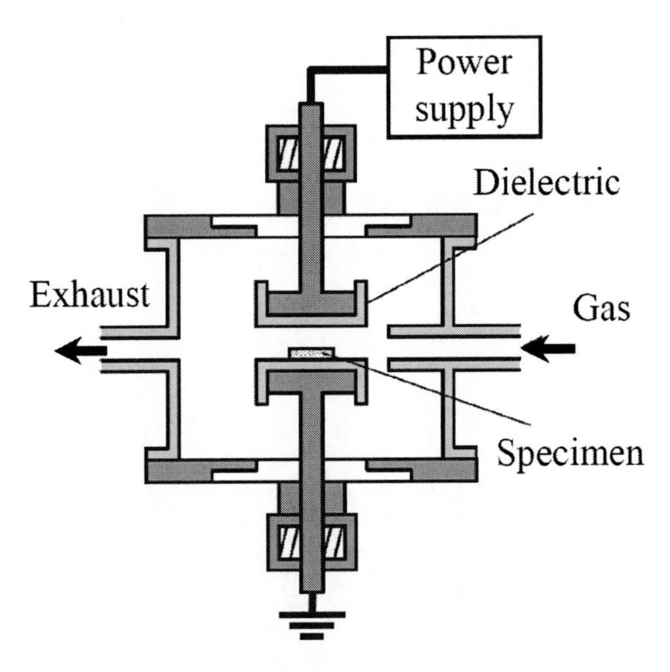

Figure 8. Schematic diagram of a parallel-plate type discharge setup.

Figure 8 shows a typical discharge tube with parallel plate electrodes for atmospheric pressure glow plasma (APG plasma) treatment. One of the two electrodes can be a metal plate. However, if both of them are covered with dielectrics such as glass plates, filamentary discharges like streamers are not easily generated. The gap should generally be thinner than 10 mm so that the uniformity of the plasma can be sustained. When a parallel plate electrode type is used, a substrate is usually placed on either upper or lower electrode surfaces. If a specimen is an insulator such as a polymer, the discharge current will pass through the substrate, even if it is floating.

If a specimen is a conductor, some care has to be taken. Namely when the plasma density is high, a high electric field at the edge of the specimen or at a print pattern can induce locally high density discharge. In the case when a parallel plate electrodes-type is used, the uniformity of feeding a gas in a gap affects the uniformity of the treatment. Apart from the discharge tube with parallel plate electrodes, researchers have developed a remote-type torch like a discharge tube as shown in Figure 9.

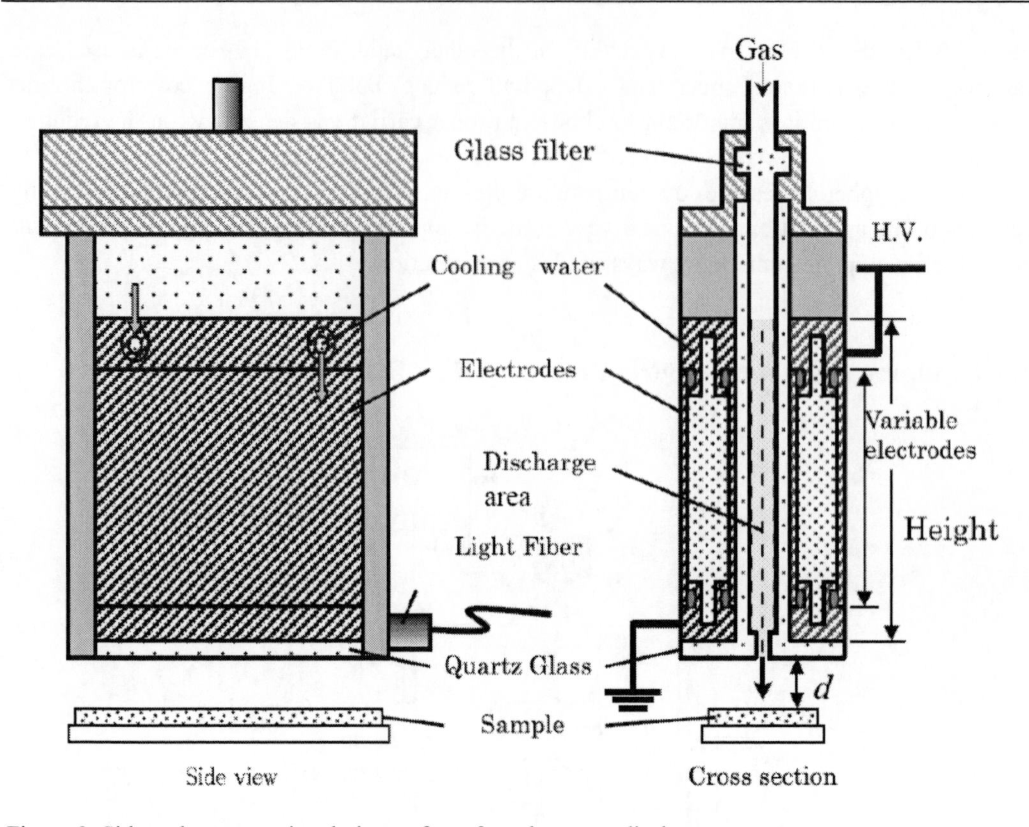

Figure 9. Side and cross-sectional views of an after-glow type discharge setup.

If helium or argon is used as a carrier gas of the remote-type discharge tube, the activated neutral species generated in the discharge can flow outside. The example includes oxygen atoms when O_2 gas is mixed with the carrier gas. By exposing a substrate into the downstream which includes the activated species, one can chemically treat its surface. Utilizing this remote type, Nagata, Horiike *et al.* [3] successfully etched a surface of a silicon substrate in atmospheric pressure air. Recall that the concentration of the activated species decreases as the samples stay in atmospheric pressure air. Using the remote type typically requires higher gas flow rate than using the parallel plate type. However, the remote type is free from direct exposure in a discharge, and thus there is no influence of charged particles such as ions and electrons. On the other hand, the treatment in a parallel plate type is significantly influenced by the charged particles. The remote type is therefore applicable to direct cleaning of substrates on which semiconductor devices and various chips are attached. In addition, it can be easily installed for inline treatment where the plasma device is placed within the existing process. This is one of the marvellous advantages of the atmospheric pressure plasmas, which enables the treatment in atmospheric pressure air. This technique can be applied to the surface cleaning of liquid crystal display panels in the mass production line.

2.4. Condition of Generation of Glow Discharge

The reasons to use a noble gas such as helium and argon are as follows:

a) Low ignition voltages (e.g. helium: 4 kV/cm); difficulty in developing to an arc discharge.
b) High gas thermal conductivity (as high as that of H_2 gas), prompt removal of the generated Joule heat.
c) Long lifetime of meta-stable excited atoms (helium 2^3S, 2^1S), and highest electron energy level among all materials (ca. 20 eV), easy dissociation of mixed gas molecules, and generation of ions and radicals.
d) No formation of compounds with mixed monomers, as a noble gas is inert.

Introductions of nitrogen atoms in the products are often observed when N_2 glow- like discharge is generated.

Applicable frequencies for generating a plasma range from several kHz to more than several hundreds of MHz. It should be noted that gas temperature during the treatment generally increases as the frequency increases. Atmospheric pressure glow discharge is basically a pulse discharge. It will be important to control the plasma density and the increase of temperature, for instance by adding on-off pulse modulated signals on a simple continuous sinusoidal wave. Films can be treated by placing them on the electrode using the parallel plate type or at the after glow using the remote type. However, various types of treatments have been developed in order to fulfil the requirements to treat substrates of different shapes (e.g. tube, continuous film, small sphere, powder, printed circuit boards etc.). They will be introduced in the following chapters of the applications.

REFERENCES

[1] Kanazawa, S.; Kogoma, M.; Moriwaki, T.; Okazaki, S. In *Proc. Int. Symp. Plasma Chem. (ISPC-8 (Tokyo))*. 1987, *3*, 1839-1844.
[2] Kogoma, M.; Prat, R.; Suwa, T.; Okazaki, S.; Inomata, T. *Plasma. Proc. Polym. NATO ASI Series E Appl. Sci.* 1997, *346*, 379-3939.
[3] Nagata, A.; Takehiro, S.; Sumi, H.; Kogoma, M.; Okazaki, S.; Horiike,Y. In *Proc. Jpn. Symp. Plasma Chem.* 1989, *2* ,109-112.
[4] Tanaka, K.; Inomata,T.; Kogoma, M. *Plasma Polym.* 1999, *4*, 269-281.
[5] Mori, T.; Tanaka, K.; Inomata, T.; Takeda, A.; Kogoma, M. *Thin Solid Films* 1998, *316*, 89-92.
[6] Sawada, Y.; Ogawa, S.; Kogoma, M. *J. Phys. D Appl. Phys.* 1995, *28*, 1661-1669.
[7] Babukuty, Y.; Prat, R.; Endo, K.; Kogoma, M.; Okazaki, S.; Kodama, M. *Langmuir* 1999, *15*, 7055-7062.
[8] Austen, A. E.; Whitehead, S. *J IEE.* 1941, *88*, 88-92.
[9] Gherardi N.; Gat, E.; Gouda, G.; Massines, F. In *Proc 6th Int. Symp. High Pressure Low Temp. Plasma Chem. (HAKONE VI)*. 1998, 118-122.

[10] Stahel, P.; Bursikova, V.; Navratil, Z.; Kloc, P.; Brabec, A.; Sira, M.; Janca, J. In *Proc. 10th Int. Symp. High Pressure Low Temp. Plasma Chem. (HAKONE X)*. 2006, 184-187.

[11] Tochikubo, F.; Chiba, T.; Watanabe, T. In Proc. *5th Int. Symp. High Pressure Low Temp. Plasma Chem. (HAKONE V)*. 1996, 225-229.

[12] Gherardi, N.; Gouda, G.; Gat, E.; Ricard, A.; Massines, F. *Plasma Sources Sci. Technol.* 2000, *9(3)*, 340-346.

[13] Kanazawa, S.; Kogoma, M.; Moriwaka, T.; Okazaki, S. *J. Phys D Appl. Phys.* 1988*, 21,* 838-840.

[14] Kanazawa, S.; Kogoma, M.; Moriwaki, T.; Okazaki, S. *Nucl. Inst. Methods, Phys. Res.* 1989, *B37/38*, 842-845.

[15] Yokoyama, H.; Kogoma, M.; Kanazawa, S.; Moriwaki, T.; Okazaki, S. *J. Phys. D Appl. Phys.* 1990*, 23,* 347-377.

[16] Hashimoto, T.; Fukuda, K.; Kogoma, M.; Okazaki, S.; Yoshimoto, M.; Koinuma, H. *Mol. Cryst. Liq. Cryst.* 1990, *184*, 201-205.

[17] Yokoyama, T.; Moriwaki, T.; Kogoma, M.; Okazaki, S. *J. Phys. D Appl. Phys*. 1990, *23,* 1125-1128.

[18] Okazaki, S.; Kogoma, M.; Kawashima, T. In *Proc. 24th Int. Conf. Phenomena in Ionized Gases (ICPIG-24)*, 1999, *vol. I*, 123-124.

[19] Okazaki, S.; Kogoma, M.; Uchiyama, H. In *Proc.* 3rd *Int. Symp. High Pressure Low Temp. Plasma Chem. (HAKONE III)*. 1991, 101-106.

[20] Kosugi, T.; Jinno, T.; Okazaki, S.; Kogoma, M. *Houden-kenkyu*. 1992, *136*, 70-78.

[21] Okazaki, S.; Kogoma, M. *J. Photopolym. Sci. Technol*. 1993, *6,* 339-342.

[22] Okazaki, S.; Kogoma, M.; Uemura, M.; Kimura, Y. *J. Phys. D Appl. Phys.* 1993, *26,* 889-892.

[23] Kogoma, M.; Okazaki, S. *J. Phys. D Appl. Phys.* 1994, *27,* 1985-1994.

[24] Kogoma, M.; Koiwa, K.; Okazaki, S. *J. Photopolym. Sci. Technol*. 1994, *7,* 341-344.

[25] Sawada, Y.; Ogawa, S.; Kogoma, M. *J. Phys. D Appl. Phys.* 1995, *28*, 1661-1669.

[26] Sawada, Y.; Ogawa, S.; Kogoma, M. *J. Adhesion.* 1995, *53*, 173-182.

[27] Sawada, Y.; Kogoma, M.; Tamaru, H.; Kawase, M.; Hashimoto, K. *J. Phys. D Appl. Phys*. 1996, *29*, 2539-2544.

[28] Kogoma, M. *Hyoumen-gijutushi (J. Surf. Finishing Soc. Jpn),* 1996, *47,* 566-570.

[29] Tutumi, K.; Ban, K.; Shibata, K.; Ogawa, S.; Kogoma, M. *J. Adhesion* 1996, *57*, 45-53.

[30] Taniguchi, K.; Tanaka, K.; Inomata, T.; Kogoma, M. *J. Photopolym. Sci. Technol.* 1997, *10,* 113-118.

[31] Sawada, Y.; Kogoma, M. *Powder Technol.* 1997, *90,* 245-250.

[32] Prat, R.; Suwa, T.; Kogoma, M.; Okazaki, S. *J. Adhesion.* 1998, *66,* 163-182.

[33] Mori, T.; Tanaka, K.; Inomata, T.; Takeda, A.; Kogoma, M. *Thin Solid Films* 1998, *316*, 89-92.

[34] Babukutty, Y.; Prat, R.; Endo, K.; Kogoma, M.; Okazaki, S.; Kodama, M. *Langmuir* 1999, *15,* 7055-7062.

[35] Tanaka, K.; Inomata, T.; Kogoma, M. *Plasma Polym.* 1999*, 4*, 269-281.

[36] Prat, R.; Koh, Y. J.; Babukutty, Y.; Kogoma, M.; Okazaki, S.; Kodama, M. *Polymer* 2000, *41,* 7355-7360.

[37] Nakajima, T.; Tanaka, K.; Inomata, T.; Kogoma, M. *Thin Solid Films* 2001, *386,* 208-212.

[38] Ogawa, S.; Takeda, A.; Oguchi, M. Tanaka, K.; Inomata, T.; Kogoma, M. *Thin Solid Films* 2001, *386*, 217-221.
[39] Tanaka, K.; Inomata, T.; Kogoma, M. *Thin Solid Films* 2001, *386*, 213-216.
[40] Tanaka, K.; Kogoma, M. *Plasma Polym.* 2001, *6,* 27-33.
[41] Tanaka, K.; Kogoma, M. *Plasma Polym.* 2003, *8,* 199-208.
[42] Tanaka, K.; Kogoma, M. *Int. J. Adhesion Adhesives* 2003, *23(6)*, 515-519.
[43] Akitu, T.; Ohkawa, H.; Tuji, M.; Kimura, H.; Kogoma, M. *Surf. Coat. Technol.* 2005, *193,* 29-34.

SECTION 1
PROPERTIES OF ATMOSPHERIC PRESSURE PLASMAS

In: Generation and Applications of Atmospheric Pressure Plasmas ISBN: 978-1-61209-717-6
Editors: M. Kogoma, M. Kusano and Y. Kusano

Chapter 1

BASIC CHARACTERISTICS OF ATMOSPHERIC PRESSURE PLASMA

Fumiyoshi Tochikubo
Department of Electrical Engineering,
Tokyo Metropolitan University, Tokyo, Japan

INTRODUCTION

Gas discharge plasmas at atmospheric pressure can be divided in to two; thermal equilibrium plasmas represented by arc discharge, and non-equilibrium plasmas represented by corona and dielectric barrier discharges. Here, the latter will be discussed. In a non-equilibrium plasma, electron temperature T_e is more than several tens thousands K while temperatures of ions (T_i) and gas (T_g) are as low as room temperature. It is also called a low temperature plasma. It is categorized as a weakly ionized plasma. As the ionization ratio (a ratio of the number of charged particles to that of all particles) is as low as between $10^{-7} - 10^{-3}$ in the weakly ionized plasma, the collision of particles in the plasma is dominated by the two-body collisions between electrons and neutrals, and between ions and neutrals. In particular, the collisions between high temperature (high energy) electrons and neutrals determine the characteristics of the plasma, where the materials property of neutral molecules as the source gas strongly affects the characteristics. Figure 1 shows the trajectories of electrons emitted at the cathode (x = 0 cm) in a uniform electric field in Xe gas at pressure of 1 Torr. Each electron is accelerated by the electric field. Because of the stochastic collisions with surrounding neutral molecules it traces complicated trajectory. Although movements of individual electrons and ions are complicated, they move as a group by drift and diffusion which accompany ionization. This movement determines the structure of a plasma. Furthermore, the chemical reaction processes here also determine the plasma process. Figure 2 shows the hierarchical structure of several processes in a plasma. It is probably important to grasp the fundamental phenomena in a plasma in order to understand atmospheric pressure plasma process.

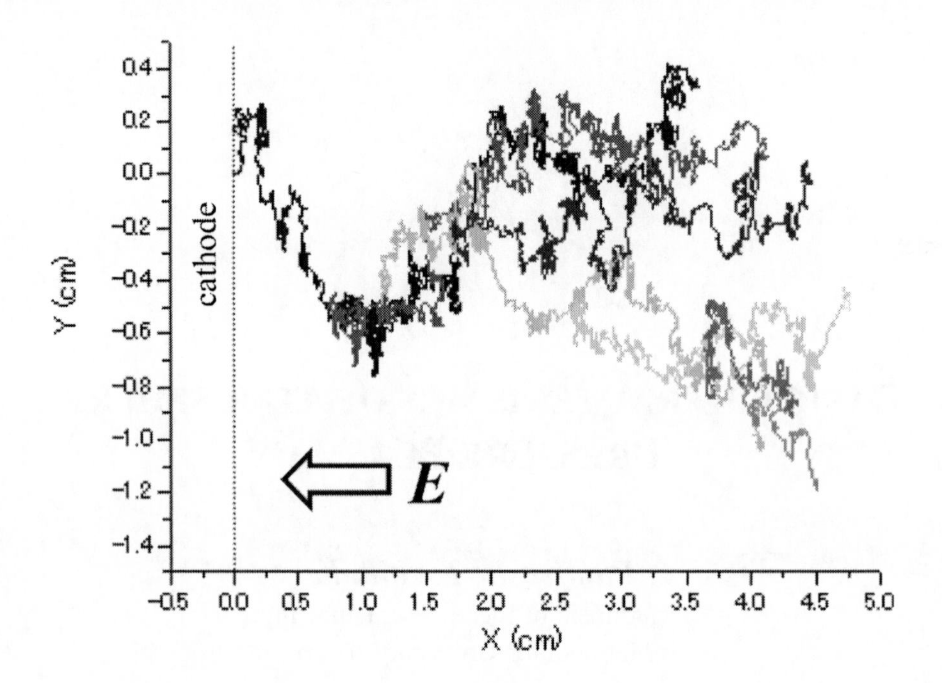

Figure 1. Trajectory of electrons in an electric field emitted from a cathode in Xe gas at the pressure of 1 Torr. E/p = 60 V cm^{-1} $Torr^{-1}$.

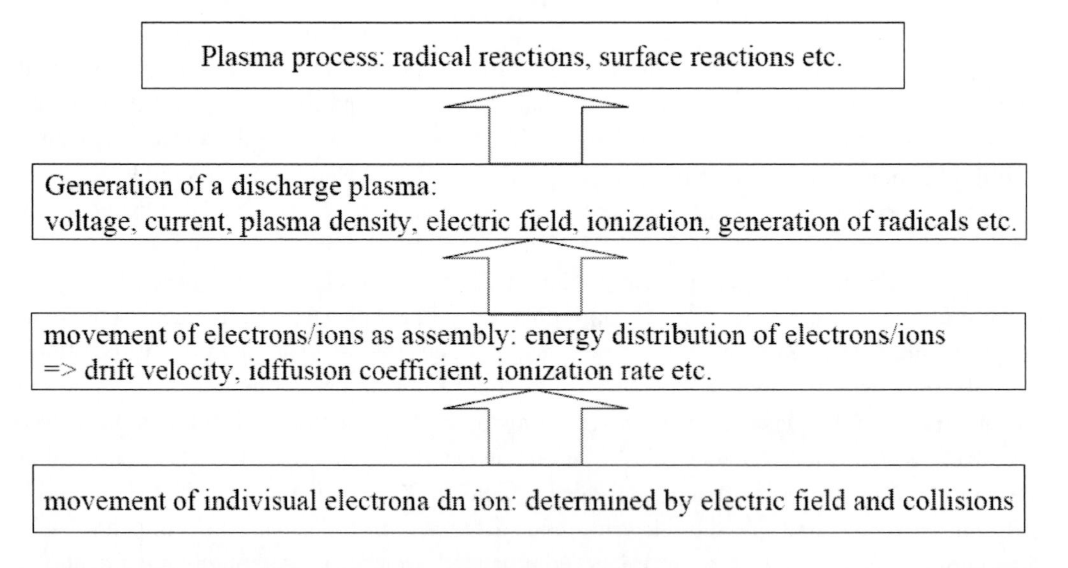

Figure 2. Hierarchical structure of individual processes in a plasma.

In this chapter, basic processes and phenomena of are discussed which are important to consider non-thermal atmospheric pressure plasma. Plasma simulation based on such physical and chemical processes is shown. In addition mechanisms of generation and retaining of non-thermal atmospheric pressure plasma are discussed.

1. BASIC PROCESSES IN A PLASMA

It is high energy electrons that contribute to the important processes such as ionization and dissociation (generation of radicals) in a non-thermal plasma. Therefore basic processes of electrons are mainly described here.

In a discharge plasma, electrons gain energy from the electric field, and loose it by collisions with surrounding neutral particles. It is therefore very important that which collisions how frequently occur. The collision frequency is defined by collision cross section Q [cm^2]. Figure 3 shows the collision cross section of electrons in He, Ar and N_2 gases. The way of electron collision differs in different gases. These differences of collisions reflect the differences of plasma. For example, in molecular gas N_2, there exist energy loss processes of rotational excitation collision and vibrartional excitation collision at a low energy range, and electron energy hardly increases.

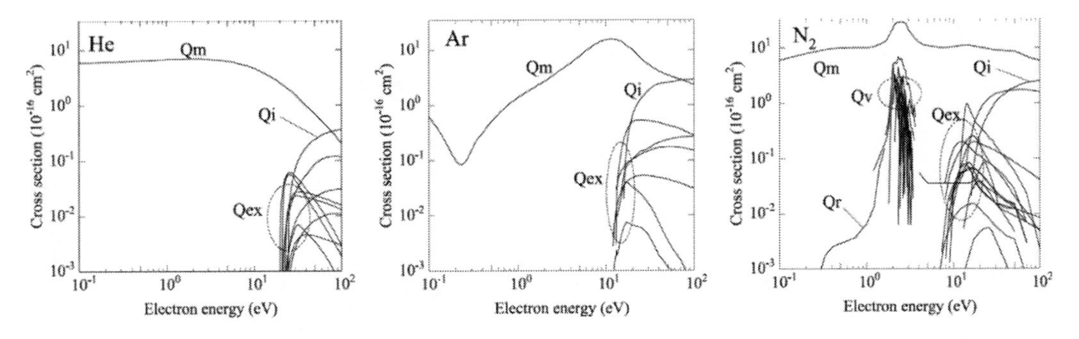

Figure 3. Collision cross sections in real gases.

Q_m: momentum transfer cross section
Q_r: rotational excitation cross section
Q_v: vibrational excitation cross section
Q_{ex}: electron excitation cross section
Q_i: Ionization cross section

Next, lets see Figure 1 again, which monitors movements of individual electrons. Each electron does not move at a single velocity, but it has a distribution. The velocity distribution function *f*(v) at the velocity v is normalized as follows,

$$\int_v f(v)dv = 1 \tag{1}$$

It is noted that *f*(v) represents a probability density function, as is defined so that the integral over electrons with all velocities is 1. Since the velocity and energy are equivalent, using the relation;

$$\frac{1}{2}mv^2 = e\varepsilon \tag{2}$$

energy distribution function of electrons can also be defined. Here e and ε are elementary electric charge [C] and energy [eV], respectively.

$$\int_{v} f(v)dv = \int_{0}^{\infty} F(\varepsilon)d\varepsilon = 1 \tag{3}$$

As the electric field is stronger, the electron energy increases. On the other hand, as the gas density increases, energy loss by collision increases. Therefore, the velocity-distribution function or energy distribution function of electrons is the function of E/N. Here, E and N are the electric field and gas density, respectively. Td is often used as the unit of E/N, where 10^{-17} V cm^2 = 1 Td.

Various important parameters of a discharge plasma can be expressed using the velocity distribution function or energy distribution function of electrons. The electron mean energy $<\varepsilon>$ [eV], the drift velocity v_d [cm/s], and diffusion constant D_e [cm^2/s] are expressed as follows;

$$\langle\varepsilon\rangle = \int_{v} \frac{mv^2}{2e} f(v)dv = \int_{0}^{\infty} \varepsilon F(\varepsilon)d\varepsilon \tag{4}$$

$$v_d = \int_{v} v\cos\theta f(v)dv = -\frac{1}{3}\sqrt{\frac{2e}{m}}E\int_{0}^{\infty} \frac{\varepsilon}{NQ_T(\varepsilon)}\left\{\frac{\partial}{\partial\varepsilon}\frac{F(\varepsilon)}{\sqrt{\varepsilon}}\right\}d\varepsilon \tag{5}$$

$$D_e = \frac{1}{3}\sqrt{\frac{2e}{m}}\int_{0}^{\infty} \frac{\sqrt{\varepsilon}}{NQ_T(\varepsilon)} F(\varepsilon)d\varepsilon \tag{6}$$

Here, m, θ, Qt are mass of a electron, an angle forming the electron velocity and the electric field, and total collision cross section, respectively. It is seen that the mean velocity and the drift velocity are the functions of E/N. On the other hand, as for the diffusion constant, N x D_e is a function of E/N. The diffusion constant is inverse proportional to the gas density (pressure).

The frequencies of various reactions by electron collision are also expressed using the velocity distribution function or the energy distribution function. For example, Equation (7) expresses the ionization frequency v_i [1/s];

$$v_i = \int_{v} NQ_i(v)vf(v)dv = \sqrt{\frac{2e}{m}}\int_{0}^{\infty} \sqrt{\varepsilon}NQ_i(\varepsilon)F(\varepsilon)d\varepsilon \tag{7}$$

This indicates that v_i/N is a function of E/N, and that the ionization frequency markedly increases as the gas density (pressure) increases. It is noted that v_i/N is used as a rate constant k_i [cm^3/s].

$$k_i = \int_v Q_i(v) v f(v) dv = \sqrt{\frac{2e}{m}} \int_0^\infty \sqrt{\varepsilon} Q_i(\varepsilon) F(\varepsilon) d\varepsilon \tag{8}$$

If excitation cross section Q_{ex} and momentum transfer cross section Q_m are used instead of ionization cross section Q_i, one can obtain excitation frequency ν_{ex} and momentum transfer frequency ν_m, respectively. The ionization coefficient α [1/cm] is also frequently used as a parameter, which has a relation to ionization frequency;

$$\alpha = \frac{\nu_i}{\nu_d} \tag{9}$$

Of course, at a constant *E/N*, the ionization coefficient is proportional to of the gas density.

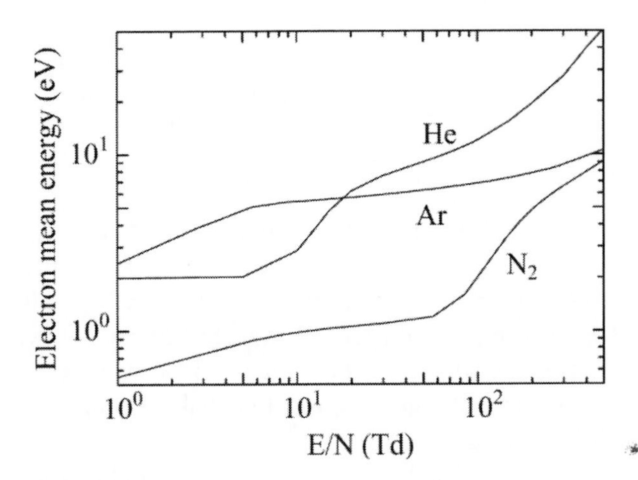

Figure 4. Average electron energy in He, Ar and N_2.

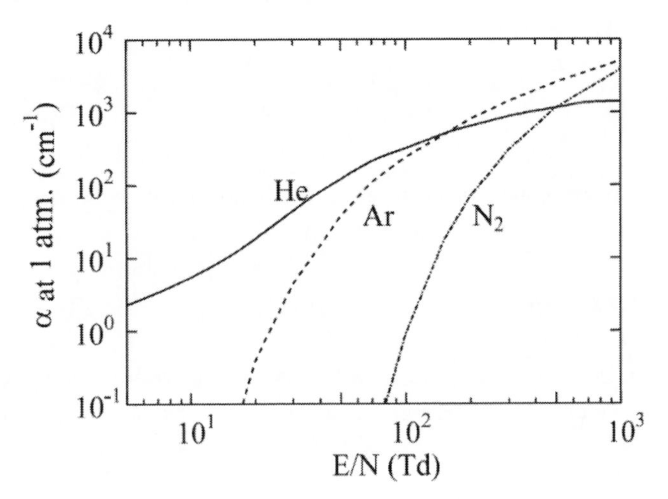

Figure 5.ionization coefficients in He, Ar, and N_2 at 1 atm.

Figures 4 and 5 show the electron mean energy in He, Ar and N_2, and their ionization coefficients at 1 atom. These are obtained by calculating electron energy distribution function $F(\varepsilon)$ from the Boltzmann equation using the collision cross section in Figure 3. Its details are discussed in Refs. [1,2]. However, it is a good idea to know the characteristics of electron collision reactions in a gas which one uses, utilizing a commercially available or free software of energy calculation.

2. Gas Breakdown – Townsend Theory and Streamer Theory

A discharge plasma always starts with electron avalanche initiated by a stochastically existing electron, and results in gas breakdown. Their subsequent discharge modes depend on whether the processes of the gas breakdown follow Townsend theory or the streamer theory.

The electron density distribution $N_e(r,z,t)$ of electron avalanche, departing $(r,z) = (0,0)$ at $t = 0$ in the electric field E_o directing $-z$, is expressed as follows, when the space charge field is negligible,

$$N_e(r,z,t) = \frac{1}{(4\pi D_e t)^{3/2}} \exp\left[-\frac{(z-\mu_e E_o t)^2 + r^2}{4D_e t} + \alpha\mu_e E_o t \right] \tag{10}$$

Here μ_e [cm^2/Vs] is a mobility. Figure 6 compares the electron avalanche in N_2 gas at atmospheric pressure and at low pressure (1 Torr) under the condition that E/N = 180 Td.

Electrons are difficult to diffuse at atmospheric pressure, and ionization occurs abruptly. Consequently the electron avalanche is difficult to extend to the radial direction. On the other hand, at low pressure, the number of electrons does not increase significantly since the ionization coefficient is low. In addition it diffuses to radial direction. It is therefore thought that it is easy to be a glow discharge.

Under Townsend theory, the threshold condition of the gas breakdown is fixed by the ionization coefficient α, secondary electron emission coefficient from a cathode by electron bombardment γ, and the gap of the electrodes d.

$$\gamma\{\exp(\alpha d) - 1\} = 1 \tag{11}$$

The coefficient of secondary electron emission is normally as large as between 0.01 and 0.1,

$$\alpha d = \ln(1+\gamma^{-1}) \approx 2.4 \sim 4.6 \tag{12}$$

Substituting d in Equation (12) gives α which is necessary for the gas breakdown. As α is a function of E/N, the electric field or a gas breakdown voltage can be determined.

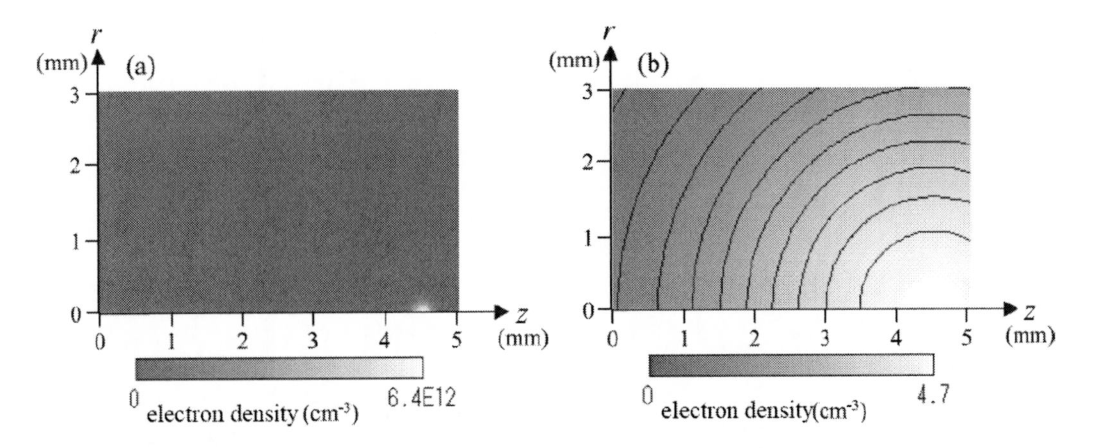

Figure 6. Electron avalanche in N_2 when the reduced electric field is 180 Td. Z = 0 mm and 5 mm correspond to cathode and anode, respectively. (a) atmospheric pressure, (b) low pressure (1 Torr).

As shown in Figure 6, the electron avalanche markedly grows almost without diffusing at atmospheric pressure, and it tends to transform to a streamer. A streamer is a highly conductive discharge column with a diameter of approximately 200 μm, and the electron density inside the streamer can be as high as 10^{14}~10^{15} cm^{-3}. Meeks condition for the transformation from the electron avalanche to a streamer is as follows;

$$E_a = \frac{e}{4\pi\varepsilon_o r_a^2}\exp(\alpha d) \approx E_o \tag{13}$$

Here, E_o, E_a, and r_a are the external electric field, space charge electric field generated at the edge of the electron avalanche, and the edge radius of the electron avalanche, respectively. Equation (13) indicates that when the space charge electric field at the edge of the electron avalanche is comparable to the external electric field, the electric field is highly distorted and a streamer is generated. When $r_a \approx \frac{1}{\alpha}$

$$\alpha d = \ln\frac{4\pi\varepsilon_o E_o}{e\alpha^2} \approx 20 \tag{14}$$

Using this Meeks condition, the gas breakdown voltage accompanying a streamer can be predicted. The ionization coefficient and the electric field given by Meeks condition are obviously larger than those by Equation (12). It means that in order to generate atmospheric pressure glow discharge, it must be driven at the condition no less than Townsends spark condition and no more than Meeks condition. If it is no less than the Meeks condition, the discharge accompanies streamers.

3. Methods of Plasma Simulation

Here, phenomena that the simulations are based on are described, but the details of the simulation methods of discharge plasmas are not described. Just if one knows the meaning of the equations which construct the model, it would certainly be useful to understand simulation results in literature.

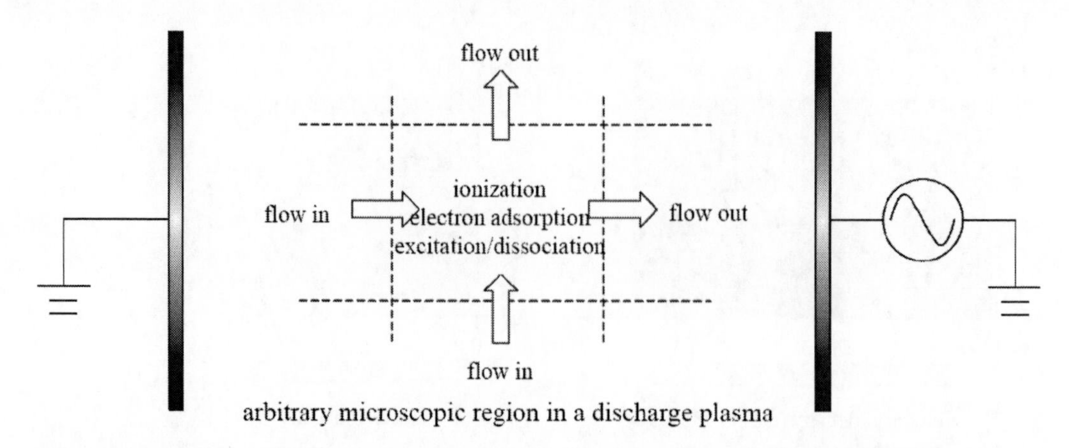

Figure 7. Phenomena occurring at an arbitrary microscale region in a discharge plasma.

The simulation of atmospheric pressure plasma is described using the Pseudo Viscosity Model. This model can approximate movements of electrons and ions by macroscopic flow such as drift and diffusion. An arbitrary microscopic region in a discharge plasma as shown in Figure 7, the continuous density equations for electrons, positive and negative ions are as follows;

$$\frac{\partial N_e(\mathbf{r},t)}{\partial t} + \nabla \cdot \mathbf{J}_e(\mathbf{r},t) = G_e - L_e$$

$$\frac{\partial N_p(\mathbf{r},t)}{\partial t} + \nabla \cdot \mathbf{J}_p(\mathbf{r},t) = G_p - L_p \tag{15}$$

$$\frac{\partial N_n(\mathbf{r},t)}{\partial t} + \nabla \cdot \mathbf{J}_n(\mathbf{r},t) = G_n - L_n$$

Here, N and J are density and flux, respectively. G and L represent generation and loss terms by ionization, electron adsorption, recombination etc. The suffixes of e, p, and n indicate electron, positive ion and negative ion, respectively. Each flux is expressed using a drift velocity and diffusion constant;

$$\mathbf{J}_e(\mathbf{r},t) = N_e(\mathbf{r},t)\nu_{de}(\mathbf{r},t) - D_e(\mathbf{r},t)\nabla N_e(\mathbf{r},t),$$

$$\mathbf{J}_p(\mathbf{r},t) = N_p(\mathbf{r},t)\nu_{dp}(\mathbf{r},t) - D_p(\mathbf{r},t)\nabla N_p(\mathbf{r},t) \tag{16}$$

$$\mathbf{J}_n(\mathbf{r},t) = N_n(\mathbf{r},t)\nu_{dn}(\mathbf{r},t) - D_n(\mathbf{r},t)\nabla N_n(\mathbf{r},t)$$

As the examples of G and L, electron collision ionization and electron adsorptions are expressed as;

Ionization: $\alpha|v_{de}|N_e$ (17)

Electron adsorption: $\eta|v_{de}|N_e$

Here η is electron adsorption coefficient [cm^{-1}]. As was described in section 1, the drift velocity and diffusion constant in Equation (16) and ionization coefficient and electron adsorption coefficient in equation (17) are not constants, but depend on the strength of the electric field. Therefore equation (15) must be simultaneously calculated with Poisson's equation.

$$\nabla\cdot(\varepsilon\mathbf{E}) = e\left(N_p - N_e - N_n\right) \tag{18}$$

Here ε in Equation (18) is a dielectric constant.

This method is called local field approximation, assuming that all the parameters such as drift velocities and diffusion constants of electrons and ions, ionization coefficient by electron collision, and the electron adsorption coefficient functions of the instant reduced electric field *E/N*. It is relatively good approximation at high pressures like atmospheric pressure, and thus in many cases this method is used. In addition, there is the other method assuming that parameters such as the ionization coefficient by electron collision, the electron adsorption coefficient, and the excitation frequency are functions of the mean electron energy, and calculating the following conservation of electron energy;

$$\frac{\partial\left(N_e(\mathbf{r},t)\langle\varepsilon(\mathbf{r},t)\rangle\right)}{\partial t} = -\nabla\cdot\mathbf{q}(\mathbf{r},t) - eE(\mathbf{r},t)\cdot\mathbf{J}_e(\mathbf{r},t) - \frac{2m}{M}v_m\langle\varepsilon(\mathbf{r},t)\rangle N_e(\mathbf{r},t) - \sum_k \varepsilon_k v_k\left(\langle\varepsilon(\mathbf{r},t)\rangle\right)N_e(\mathbf{r},t) \tag{19}$$

$$\mathbf{q}(\mathbf{r},t) = \frac{5}{3}N_e(\mathbf{r},t)\langle\varepsilon(\mathbf{r},t)\rangle v_{de} - \frac{5}{3}D_e\nabla\left(N_e(\mathbf{r},t)\langle\varepsilon(\mathbf{r},t)\rangle\right) \tag{20}$$

Here, $<\varepsilon>$, q, m, M, v_k and ε_k represent mean electron energy, energy flux, mass of electron, mass of neutral particle, inelastic collision frequency and its threshold energy, respectively.

4. Chemical Reactions in Atmospheric Pressure Plasma

When neutral species react in a plasma, these reaction processes can also be estimated using a density continuity equation which is similar to Equation (15).

$$\frac{\partial N_i(\mathbf{r},t)}{\partial t} = D_i\nabla^2\frac{5}{3}N_i(\mathbf{r},t) + \sum_{l,m} k_{l,m}N_l N_m - \sum_j k_{i,j}N_i N_j \tag{21}$$

Here k is a reaction rate constant [cm^3/s], and j, l, and m represent kinds of particles. The second and third terms in the right hand side in Equation (21) represent collision reaction generation by two bodies and annihilation reactions of particle i by two bodies respectively. Three-body collision reaction can also be added in a similar way, though it is not described in Equation (21). The reaction time within which a particle i disappears is particularly important for considering lifetime of radicals in a plasma. The reaction time τ_i of a particle i can be roughly estimated using the following equation.

$$\tau_i = \frac{1}{\sum_j k_{i,j} N_j} \tag{22}$$

It indicates that the reaction time is inverse proportional to the density of reactants and the reaction rate constant. The characteristics of reactions at atmospheric pressure are listed below;

[1] high speed reactions because of an intrinsically large amount of radicals generated due to high pressure,
[2] very high speed reactions which relate to surrounding gas due to the high surrounding gas pressure, and
[3] easy to occur three-body collisions.

Because of the characteristic (1), reactions between radical-radical are easy to occur in a corona discharge accompanying streamers, and a micro-discharges. In addition, because of the characteristics (2), it is often seen that metastable species, which do not show long lifetime, react promptly. Furthermore, there are sometimes the cases that photo-emitting species at excited state electrons react before photo-emission. Because of the characteristic (3) many kinds of species which are hardly produced at low pressure can be generated.

5. Dielectric Barrier Discharge Accompanying Micro-Discharge

A dielectric barrier discharge can be generated by applying alternative or pulsed voltage between electrodes such as parallel plate electrodes and coaxial cylindrical electrodes, at least one of which is covered with dielectrics. The charges carried by the discharge current are accumulated at the dielectrics, voltage applied between the discharge gap decreases, and the discharge is extinguished by itself. Namely the dielectric barrier discharge is the assembly of pulsed discharges. The feature of a dielectric barrier discharge varies by the gas used, the shape of the electrodes, and voltage applying method. The most common feature is a filamentary micro-discharges which appear all over the electrode surfaces. Typical physical quantities of a micro-discharge are the duration time of several tens ~ hundreds ns, the filament diameter of 50~100 μm, the peak current of 0.1 A, the electron density of 10^{14}~10^{15} cm^{-3}, and consumption energy of approximately 5 μJ. Figure 8 shows the example of voltage, current wave forms obtained by experiment. Each current pulse corresponds to individual micro-discharge. Since it is difficult for a simulation to analyze the field in which numerous current pulses exist, phenomena are analyzed by one isolated micro-discharge (single filament) [3-5]. In practice, simulation is performed by applying Equations (15) and (18) to the model of a single filament as shown in Figure 9. Figure 10 shows an example of a current waveform of a single filament obtained by a parallel plate dielectric barrier discharge in N_2/O_2 (15 %) at 1 atom at a temperature of 43 K.

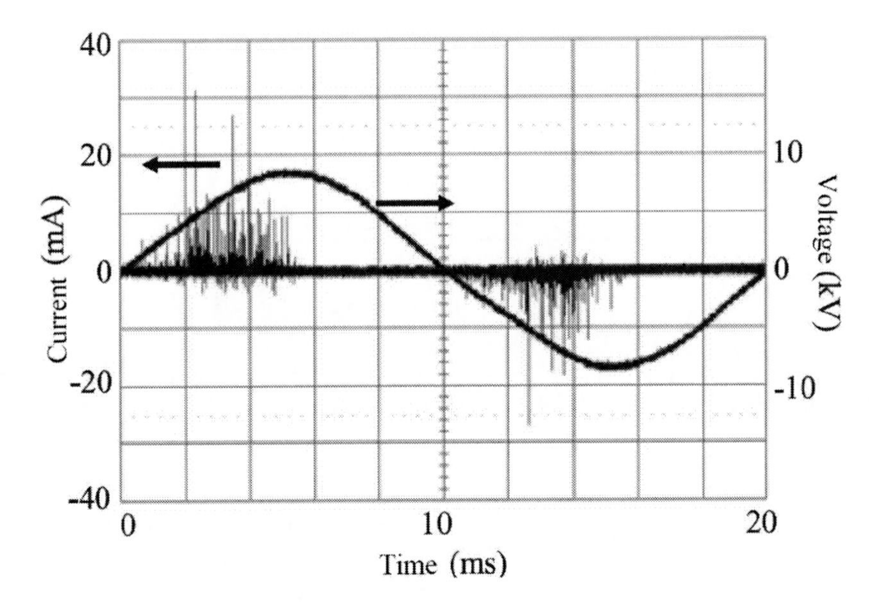

Figure 8. Typical voltage current waveforms of dielectric barrier discharge.

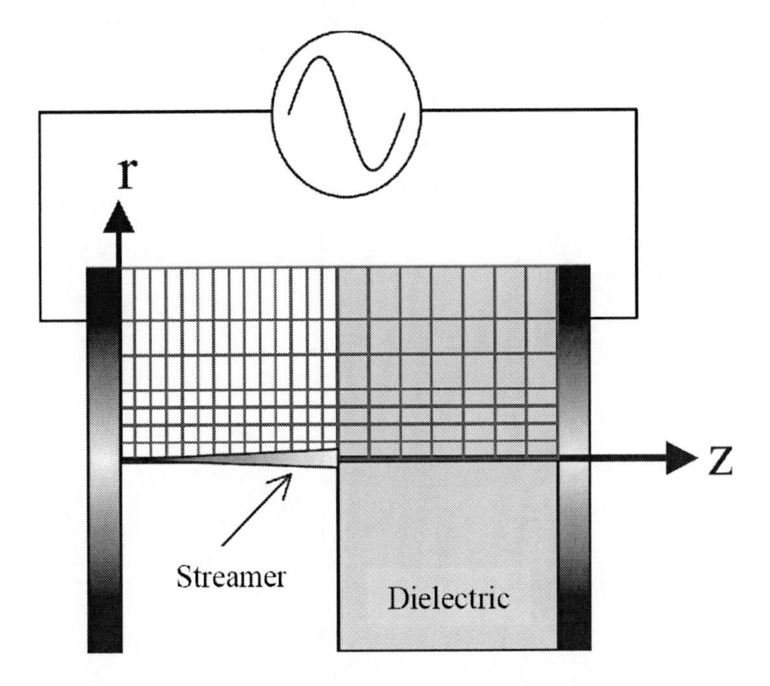

Figure 9. An example of a model for simulation of a single filament.

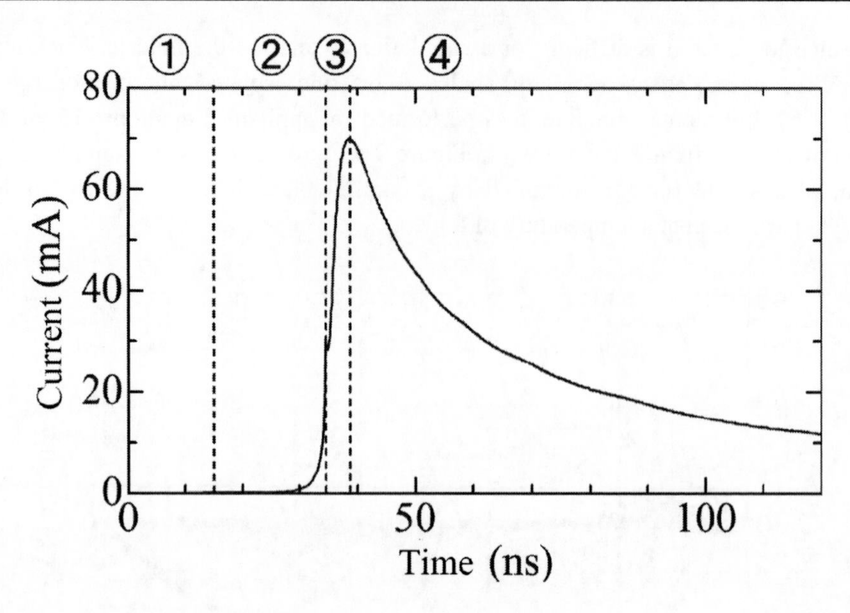

Figure 10. Current waveform of a single filament in a parallel plate type dielectric barrier discharge in N_2/O_2 (15 %) gas mixture at 1 atm at 423 K. The gap is 1.7 mm [5].

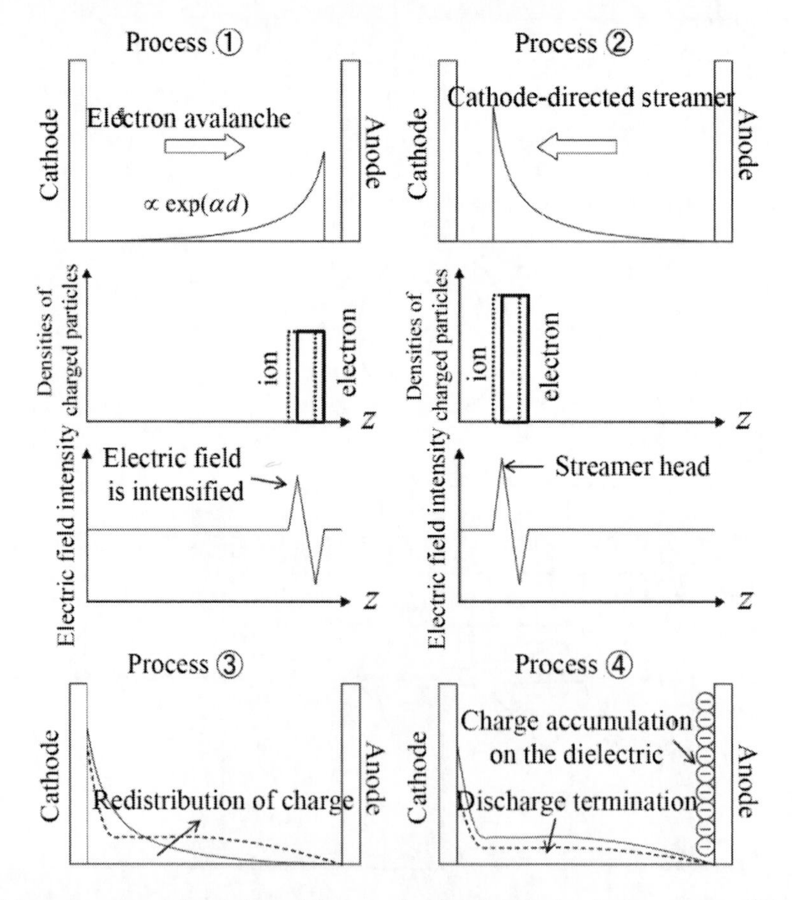

Figure 11. Schematic views of variation of electron density distribution in the process of single filament formation. The numbers (①–④) correspond to those in Figure 10.

Figure 11 shows schematic presentation of the formation process of a single filament. As shown in Figures 10 and 11, the formation process of a single filament can be categorized into four processes. During the process (1), the electron avalanche initiated by the electron existing stochastically grows from the anode surface toward the anode. When the voltage applied by a power supply is sufficiently high, and the electron multiplication rate exp(αd) exceeds 10^8 at the time the electron avalanche arrives at the anode (corresponding to Equation (14). In this case, the charged particle density at the anode surface reaches as high as 10^{12}~10^{13} cm^{-3}), the space charge at the anode surface significantly distorts the electric field. If electrons are supplied by photo-ionization etc., the electron avalanche transforms to a streamer directing the cathode. This is Meeks condition which is already described. At the process (2) cathode directed streamers develop toward the cathode. The edge of the cathode directed streamer (streamer head) is high electric field, where significant ionization occurs. The electron density becomes as high as 10^{14}~10^{15} cm^{-3}. After the cathode directed streamer reaches the cathode, the charge distribution is redistributed, and the electron density near the anode slightly increases (process (3)), the current easily flows all over the volume between the electrodes.

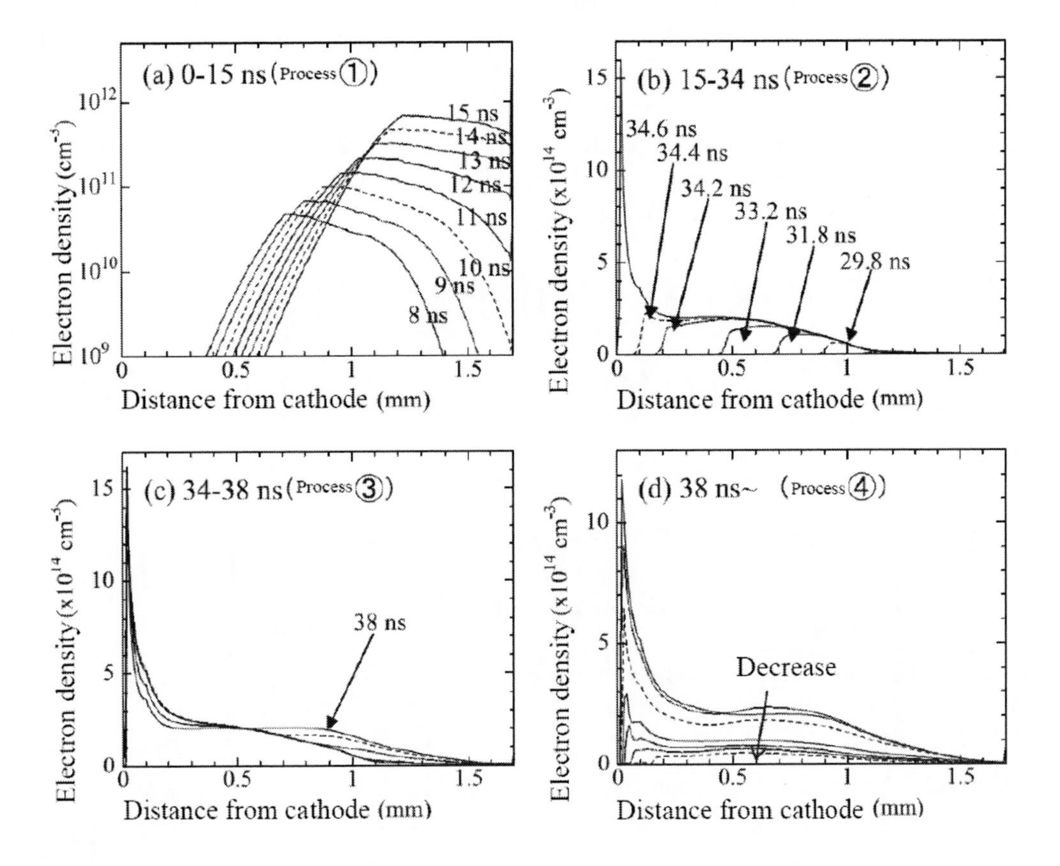

Figure 12.Electron density distribution in the process of single filament formation at the ncentral axis. The condition of the discharge is same in Figure 10.

Therefore the current increases up to maximum. After that, the charge accumulates at the dielectrics by electron current etc. significantly, decreasing the voltage between the discharge gap, and terminating the discharge automatically. Figure 12 shows the electron density distribution at the central axis by actual calculation, reproducing a previously mentioned phenomenon. Here the diameter of the filament was approximately 200 μm. The threshold condition of forming a single filament is to transform the electron avalanche to the cathode directed streamer. The necessary electric field or voltage can be estimated using Equation (14).

Considering density continuity equation for radicals simultaneously, the number of radicals generated during the process of single filament formation can be calculated. Using this, subsequent chemical reaction process (e.g. toxic gas treatment process, radical flux in surface treatment) can also be investigated.

6. Atmospheric Pressure Glow Discharge

Kogoma *et al.* in Sophia University firstly showed that atmospheric glow discharge can be easily generated by introducing atmospheric pressure He in a typical dielectric barrier discharge reactor, and applying AC voltage at the frequency more than several kHz [6]. After that, it is reported that there exist cathode fall region and anode column, and that the discharge is sustained by the secondary electron emission from the cathode by the ion bombardment [7, 8]. It was confirmed that it is certainly a glow discharge. The helium atmospheric pressure glow discharge is ideally uniform in the radial direction, and is different from previously mentioned micro-discharge. Therefore the simulation using a one dimensional model is possible as shown in Figure 13.

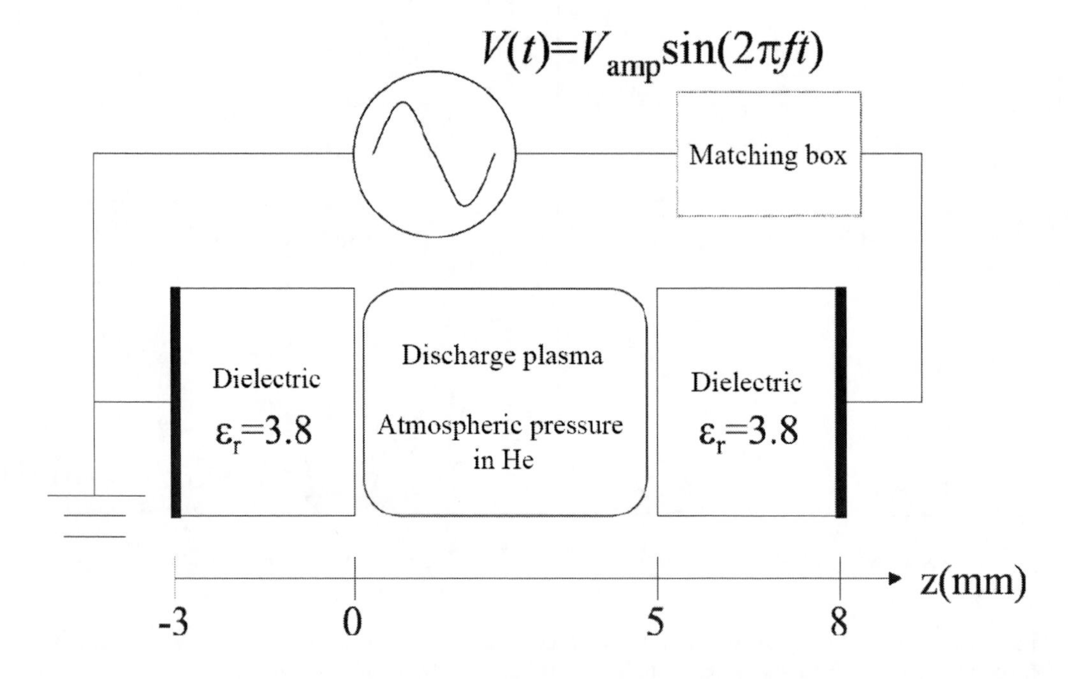

Figure 13. Numerical calculation model of helium glow discharge at atmospheric pressure.

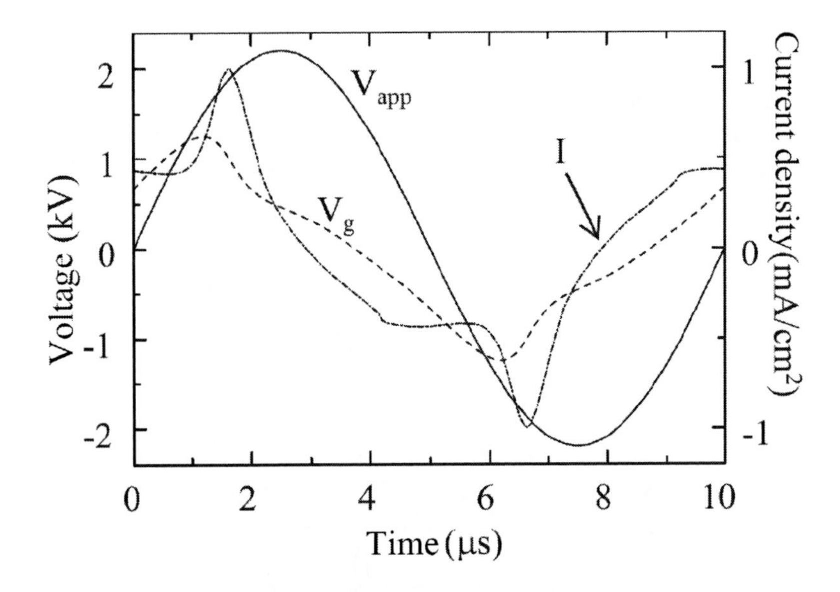

Figure 14. Voltage, current waveforms of helium glow discharge at atmospheric pressure. $V_{appl.}$: applied voltage, V_g: gap voltage, I: current density.

Figure 14 shows the voltage current wave forms obtained from the simulation, applying the voltage of the amplitude of 2.2 kV in a reactor of Figure 13. Here, reaction processes as shown in Table 1 are considered. As it is a dielectric barrier discharge, the discharge is pulse-like. However, random current pulses seen in Figure 8 are not generated. Instead, one current pulse of approximately 1 μs wide appears in half a period. V_g in the figure is the actual voltage applied between the discharge gap which is obtained by subtracting the voltage drop at the dielectrics from the applied voltage. This voltage eventually controls on/off the discharge. It is seen that as the voltage drop at the dielectrics increases by the current, the gap voltage decreases and the discharge is extinguished.

Table 1. Reaction processes used in the simulation of helium glow discharge at atmospheric pressure [8].

Process		Reaction process	Reaction rate
Direct ionization	k_i	$e + He \rightarrow He^+ + 2e$	Function of *E/N*
Direct excitation	k_{ex1}	$e + He \rightarrow He(2^1S) + e$	Function of *E/N*
	k_{ex2}	$e + He \rightarrow He(2^3S) + e$	Function of *E/N*
Production of ionic molecule	k_{He2}	$He^+ + 2He \rightarrow He_2^+ + He$	6.5×10^{-32} cm^6/s
Recombination	k_{rec1}	$He^+ + 2e \rightarrow He^* + e$	7.1×10^{-20} cm^6/s
	k_{rec2}	$He_2^+ + 2e \rightarrow 2He + e$	2.0×10^{-20} cm^6/s
Inactivation	k_{1S}	$He(2^1S) + He \rightarrow 2He$	6.0×10^{-15} cm^3/s
	k_{3S}	$He(2^3S) + 2He \rightarrow He_2 + He$	2.5×10^{-34} cm^6/s
	k_{3Se}	$He(2^3S) + e \rightarrow He + e$	2.9×10^{-9} cm^3/s
Accumulation of ionization	k_{si}	$He(2^3S) + He(2^3S) \rightarrow He^+ + He + e$	2.9×10^{-9} cm^3/s

Figure 15 shows spatial and temporal distribution of electron density, electric field, and ionization rate, corresponding to Figure 14. Electrons move in the gap following the change of the electric field. This movement results in the immediate formation of the ion sheath in front of the cathode. The discharge is sustained by the multiple-ionization at the sheath of secondary electrons emitted from the cathode by ion bombardment. Such a discharge sustaining mechanism is similar to the low pressure glow discharges.

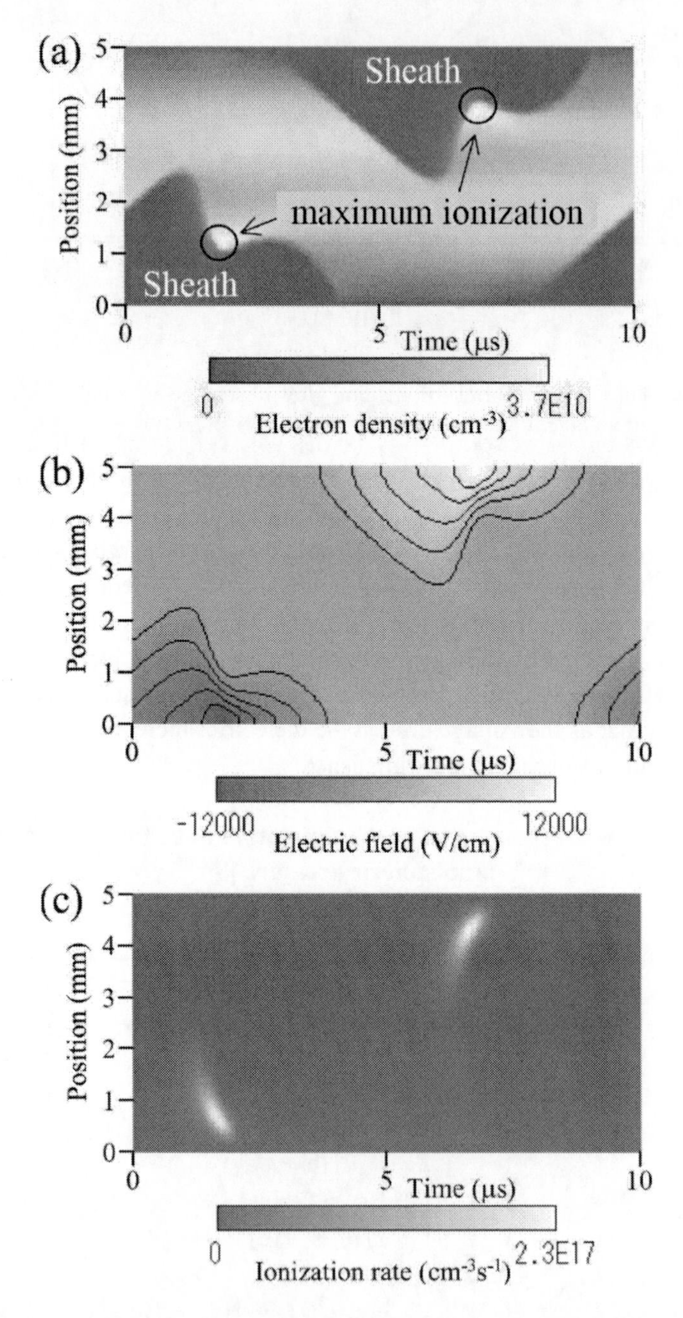

Figure 15. Spatial distribution of electron density, electric field, and ionization ratio in helium glow discharge at atmospheric pressure.

The reason why atmospheric pressure glow discharge in helium is easy to be generated is due to the ionization coefficient as shown in Figure 5. In the atmospheric pressure discharge accompanying streamers like previously mentioned micro-discharges, the reduced electric field at the streamer head is as high as 1000 Td. Under such a high electric field, as electrons have high ionization coefficient and it is also under the high gas density environment of atmospheric pressure, abrupt multiple-ionization tends to induce non-uniform discharge generation. On the other hand, atmospheric pressure glow discharge in helium, the reduced electric field at the ionization region is at most 30 Td or so. In this case, since the multiple-ionization proceeds rather slowly, the discharge will not be non-uniform. It shows that the generation of atmospheric pressure glow plasma depends on how efficiently at low electric field the discharge is sustained. The advantage of helium is its larger ionization rate at the low reduced electric field region than other gases. The ignition voltage of atmospheric pressure glow discharge in helium is nearly equal to the value obtained by the spark condition of Townsend. Namely, it is difficult to generate atmospheric pressure glow discharge unless the voltage is larger than the condition of Townsend spark discharge and lower than the Meeks streamer generation condition. If the ionization coefficients are plotted in Figure 5 corresponding to Townsend's spark condition and Meeks streamer generation condition, it indicates the required condition in various gases which enable generation of atmospheric pressure glow discharge. It is seen that helium needs lower electric filed for generating atmospheric pressure glow discharge, and has wider margin for transforming to streamers than other gases.

By the way, in atmospheric pressure glow discharge in helium, the existence of metastable atoms such as $He(2^1S)$ and $He(2^3S)$ often plays core role in generation of glow. It is discussed that $He(2^1S)$ and $He(2^3S)$ contribute to surface reactions. Table 1 shows the inactivation process and reaction rate constants of $He(2^1S)$ and $He(2^3S)$. If the reaction rates are correct, the lifetimes of $He(2^1S)$ and $He(2^3S)$ in atmospheric pressure helium are estimated by Equation (22) to be 6.8 μs and 6.7 μs, respectively. It is indicated that the densities of these metastable atoms do not become very high. For example, the maximum densities of $He(2^1S)$ and $He(2^3S)$ in Figure 15 are 7.5×10^{10} cm^{-3} and 5.2×10^{11} cm^{-3}, respectively. Therefore, the ionization processes like cumulative ionization via metastable states would not be dominated. As for surface reactions, when the down-flow of helium atmospheric pressure glow discharge is used for surface treatment, it is a little skeptical whether the metastable atoms survive or not. Are the metastable atoms useless at all for formation of atmospheric pressure glow discharges? In fact with the involvement of impurities they occasionally assist the formation very strongly. For example, the reaction rate of metastable He atom and N_2 which is often involved as impurity is

$$He_m + N_2 \rightarrow N_2^+ + He + e \ (k = 5 \times 10^{-11}\ cm^3/s) \quad (23)$$

If N_2 content is as high as 0.01 %, nearly half of the generated metastable atoms contribute as the Penning ionization in Equation (23). This effect increases the effective ionization coefficient at the low reduced electric field region, and thus particularly efficient. The effectiveness of this reaction is experimentally proved as well [9].

As is the case of the capacitively coupled low pressure plasma, in the case of dielectric barrier discharge type atmospheric pressure glow discharge, stable high density plasma can be obtained at low voltage, if radio frequency voltage like 13.56 MHz is used. It is because ions are accumulated between the electrodes since they can not follow the applied alternative electric field, and induce the increase of the electron density. Since the plasma density is high, the current is higher than that with low frequency operation. However, accumulated charge does not become very large as one period is short. In another word, the dielectric does not have the effect of being on/off the discharge plasma. As long as arc discharge can be avoided, it is not necessary to use dielectrics. Comparing with low frequency operation, the current is larger and thus Joule heating of the gas easily occurs. In addition, if it is used for polymerization or CVD, it sometimes dissociates source gas too much because of the high plasma density.

REFERENCES

[1] Makabe, T; *Plasma electronics*; (Japanese) Baifukan: Tokyo, JP, 1999.
[2] *Inst. Electr. Eng. Jpn. Technol. Report*, No. 853, Sept. 2001.
[3] Braun, D; Gibalovt, V; Pietsch, G. *Plasma Sources Sci. Technol.* 1999, *1*, 166-174.
[4] Steinle, G; Neundorf, D; Hiller, W; Pietralla, M. *J. Phys. D Appl. Phys.* 1999, *32*, 1350-1356.
[5] Tochikubo, F; Uchida, S.; Yasui, H;, Satoh, K. *Inst. Electr. Eng. Jpn. Res. Meeting Electr. Discharge,* ED-05-85, 2005.
[6] Kanazawa, S; Kogoma, M; Moriwaki, T; Okazaki, S. *J. Phys. D Appl. Phys.* 1988, *21*, 838-840.
[7] Massines, F; Rabehi, A; Decomps, P; Gadri, RB; Segur, P; Mayoux, C. *J. Appl. Phys.* 1998, *83*, 2950-2957.
[8] Tochikubo, F; Chiba, T; Watanabe, T. *Jpn. J. Appl. Phys.* 1999, *38*, 5244-5250.
[9] Tachibana, K; Kishimoto, Y; Sakai, O. *J. Appl. Phys.* 2005, *97*, 123301.

In: Generation and Applications of Atmospheric Pressure Plasmas ISBN: 978-1-61209-717-6
Editors: M. Kogoma, M. Kusano and Y. Kusano

Chapter 2

GENERATION OF ATMOSPHERIC PRESSURE PLASMA AND ITS PROPERTIES

Koichi Takaki
Department of Electrical and Electronic Engineering,
Iwate University, Morioka, Japan

INTRODUCTION

The purpose of this chapter is to provide the basic knowledge in order to understand the whole content of this book.

An atmospheric pressure glow (APG) discharge plasma was developed in 1988, and has been utilized all over the world due to its various potential advantages, including its simplicity and low cost of the system for plasma processing, its high processing speed [1], its applicability to novel surface treatments such as controlling hydrophilicity and hydrophobicity of polymer surfaces [2,3], material synthesis [4], and its applicability to medical, food, and agricultural industry such as germproof treatment, sterilization and bactericidal [5,6]. It is firstly reported by Okazaki and her group in Sophia University that a pulse glow discharge can be stably generated at atmospheric pressure by 1) inserting dielectric barrier between electrodes, 2) applying AC voltage whose frequency ranges between several kHz and tens kHz, and 3) injecting helium gas or gas mixture diluted with helium gas [7]. Subsequently the effect of the electrode-shapes and the choice of dielectric materials was studied [8, 9], and it now becomes possible to generate APG discharges using various gases including air, N_2 and argon gases [8]. In addition, physical understanding of the APG discharges has progressed through efforts based on experiments and the studies using numerical simulations [10-13]. The reported interpretations of the mechanisms of sustaining APG discharges include Penning effect by excited species [10], and high ionization rates of gas molecules in a weak electric field [11].

It is also possible to generate APG discharges without dielectric barrier 1) using a pulsed power supply [14], 2) using a hollow cathode electrode [15-18], or 3) applying radio frequency or microwave in a narrow gap [19, 20]. Among them the use of a pulsed power supply 1) aims for preventing the discharge from excess heat-up by turning off the high voltage application before it transients to a thermal discharge. On the other hand, 2) and 3) improve spatial uniformity of the APG discharge plasma by reducing electron collision frequency by mainly reducing gap length between the electrodes.

This chapter describes generation of an AGP discharge and its properties, classification of gas discharge plasmas, the generation methods, and the basic properties of the APG discharges.

1. Classification of Gas Discharge Plasmas and Their Properties

Table 1. Classification of discharges and their properties

Type	Ionization mechanism	Electron emission mechanism at the cathode	Properties
Corona	Electron impact ionization (T_e>>T_g, non-equilibrium or locally equilibrium state)	Secondary electron emission by ions etc. (V_c>>V_i)	High electron energy, significant spatial non-uniformity, ion and gas temperatures are almost equal to room temperature.
Glow			Large volume, spatial uniformity, ion and gas temperatures are close to room temperature.
Arc	Thermal ionization T_e=T_g, equilibrium state)	Thermal electron emission (V_c=V_i)	High temperature, high density, temperatures of electrons, ions and gas are almost equal.

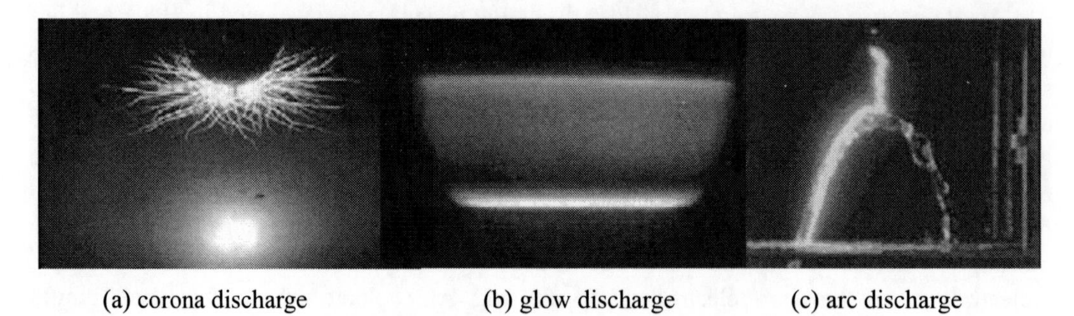

(a) corona discharge (b) glow discharge (c) arc discharge

Figure 1. Light emission from the plasmas of three discharge types.

Discharge plasmas can be generally classified into three; “a corona discharge”, “a glow discharge” and “an arc discharge”. Table 1 summarizes properties of each discharge [21]. Typical photograph of each discharge is shown in Figure 1. The properties of the discharges are distinctively different with each other due to different mechanisms for sustaining discharges.

In a corona discharge there exist high energy electrons with strong spatial non-uniformity. Temperatures of ions and neutral gases are low, similar to room temperature. A most part of energy consumed for the corona discharge is used for acceleration of electrons, while small part of the energy is used for heating ions and neutral gas molecules.

Consequently a corona discharge can be a good choice for material processing and environmental remediation, requiring high energy electrons. For example, many gas treatments require cutting bonding of molecules of the target gases. Otherwise they require cutting bonding of molecules of the carrier gases or changing electron orbits of molecules and/or atoms in order to generate reactive species which can subsequently react with the target gas molecules. These processes need electrons whose energy is higher than the binding energy or excitation energy of the gas molecules. That is the reason why a corona discharge is advantageous because high energy electrons can be efficiently generated by the corona discharge.

The important characteristics of a glow discharge are its spatial uniformity in plasma density and containing many reactive species, with relatively low temperatures of ions and neutral gases. The glow discharge plasma is suitable for sterilization of medical tools and surface modifications. On the other hand, an arc discharge is characterized by its high temperature, more than 6,000 K. It is suitable for material processing requiring thermal chemical reactions, such as thermal spraying.

A direct current (DC) discharge generally develops from a corona discharge to a glow and subsequently an arc discharge, as the discharge current increases as shown in Figure 2 [22]. As the voltage between electrodes increases, a current starts to flow by the movement of space charges by the electric field (non-sustainable discharge). Once the voltage exceeds the breakdown criteria, a continuous discharge is generated. The voltage between the electrodes rapidly decreases down to several hundreds volt. This regime is called Townsend discharge, while the subsequent regime with a constant voltage is called a "normal glow". Further increase of the current induces the increase in the voltage (abnormal glow). After that, the discharge mode transients to an arc when the current exceeds approximately 1 A, and the voltage decreases to several tens V.

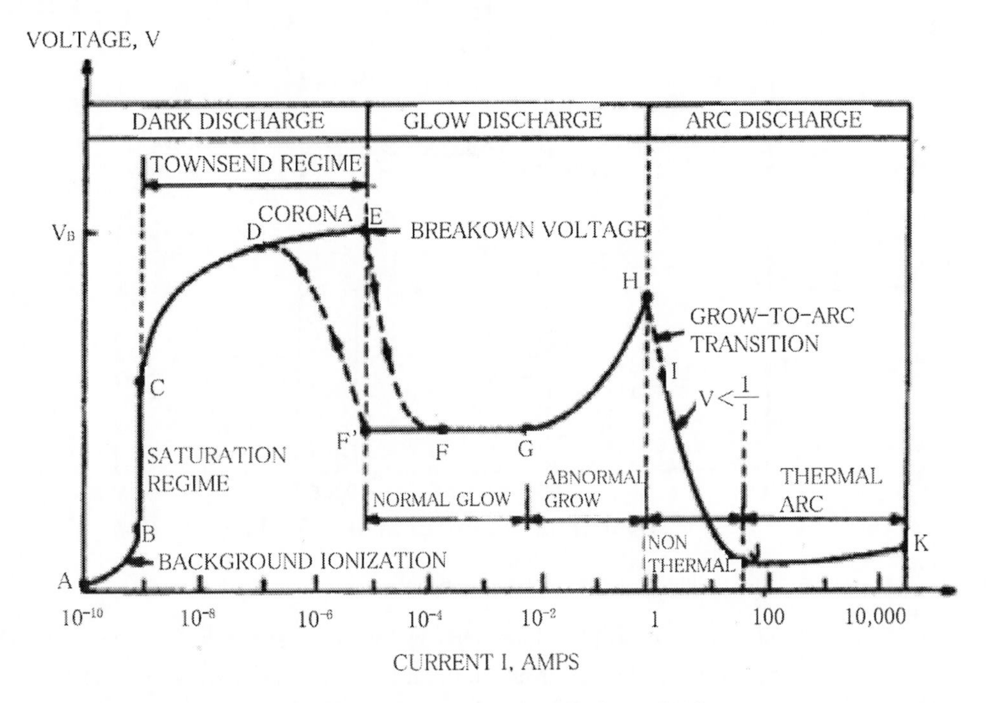

Figure 2. Voltage-current characteristic of a direct current discharge [22].

The difference of the glow and arc discharges is generally characterized by the mechanisms of the electron emission from the cathode and the ionization in the space. In a glow discharge the secondary electrons are emitted by the impact of ions, photons or high energy neutrals to the cathode (γ-effect), while in an arc discharge the thermal electron emission is dominant. The ionization mechanisms are impact ionizations (α-effect) for the glow discharge, and thermal ionizations for the arc discharge. However, a glow discharge is mostly generated at a pressure in the range of mTorr without electrodes using radio frequency or microwave, and thus it becomes difficult to classify the discharge by the mechanism of secondary electron emission from cathodes. In such a case, a discharge can be characterized using ion species and temperatures. For example, if the majority of ions are created in the gas, the discharge can be classified to a glow, while if it is from the electrode materials, it is an arc. If the electron temperature is higher than the ion temperature, it is mostly classified to a glow, while if it is comparable, it is an arc.

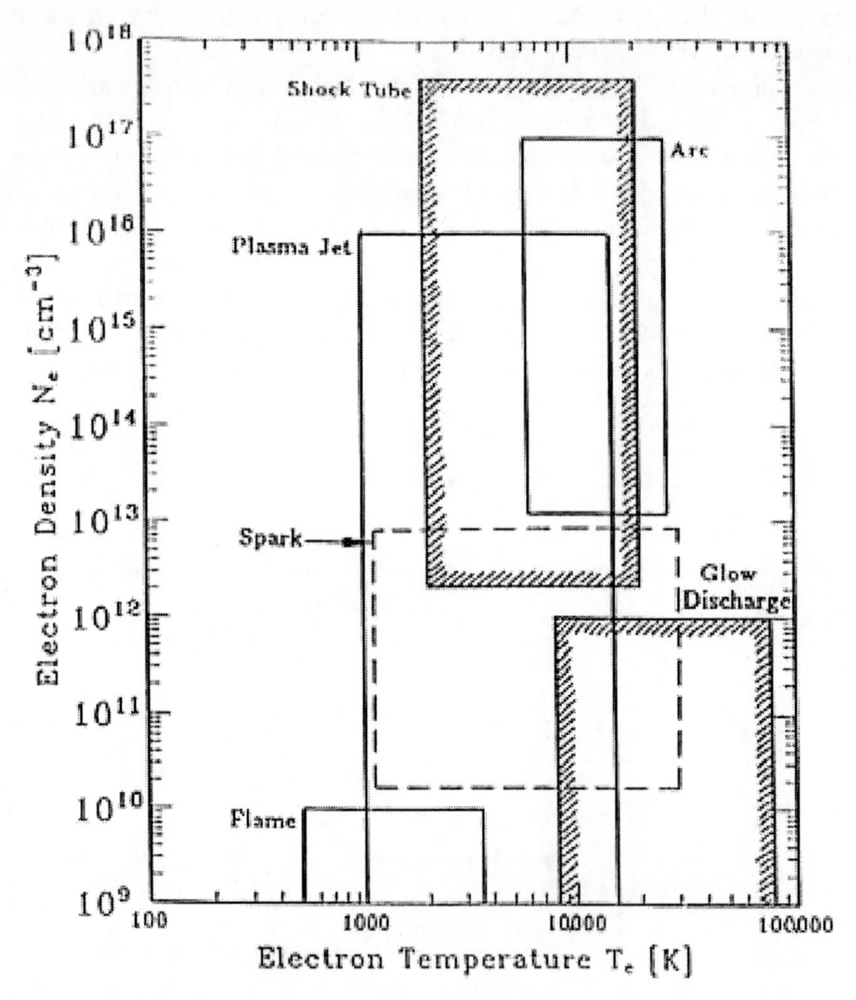

Figure 3. Electron temperature and density of process plasmas [23].

The gas discharge plasma is often characterized by the electron temperature and the plasma density. Figure 3 shows the electron temperature and density of various discharges [23]. The horizontal and vertical axes indicate electron temperature in K (11,600 K =

approximately 1 eV), and electron density in cm^{-3}, respectively. Table 2 summarizes process plasmas which are commonly used for industrial applications [23]. The details of apparatus for the plasma generations and basic properties of these plasmas are presented in reference [24]. The typical value of electron density in a low pressure plasma is in a range between 10^{11} and 10^{12} cm^{-3}. A hot filament is used to generate DC discharge plasmas. Radio frequency and microwave sources can also be used to generate discharge without electrodes. Relatively high density plasmas can be generated using a permanent magnet to produce a magnetic field in the plasma region, whose examples include an electron cyclotron resonance (ECR) discharge. Arc discharges are applied for welding, cutting, material processing using arc jet. Although the electron densities of these plasmas in Table 2 are in a range between 10^{11} and 10^{13} cm^{-3}, the actual electron density and the electron temperature at the cathode spot can be approximately 10^{20} cm^{-3} and 4 eV, respectively, while at the arc channel in a range between 10^{16} and 10^{19} cm^{-3}, and between 0.5 and 1 eV, respectively.

Table 2. Various process plasmas, their properties and applications [24]. (1bar=10kPa=750Pa)

discharge mode	pressure (mbar)	n_e (cm^{-3})	T_e (eV)	applications
DC glow	10^{-3}-100			
Cathode region			100	SP, DE, SE
Negative glow		10^{12}	0.1	CH, RA
Positive column		10^{11}	1-10	RA
Hollow cathode	10^{-2}-800	10^{12}	0.1	RA, CH
Magnetron	10^{-3}			SP
Arc, hot cathode				
External heating	1	10^{11}	0.1	RA
Internal heating	1000	10^{13}	0.1	RA, WE
Rf capacitive	10^{-3}-10^{-1}	10^{11}	1-10	RA
Hollow cathode	1	10^{12}	0.1	PR, SP, DE
Magnetron	10^{-3}			PR, RA
Rf inductive	10^{-3}-10	10^{12}	1	PR, ET
Helicon	10^{-4}-10^{-2}	10^{13}	1	PR
MW				
Closed structure	1000	10^{12}	3	CH
SLAN	1000	10^{11}	5	PR
Open structure				
Surfatron	1000	10^{12}	5	PR
Planar	100	10^{11}	2	PR
ECR	10^{-3}	10^{12}	5	PR
Electron beam				
BPD	10^{-2}-1	10^{12}	1	PR
DBD	1000	10^{14}	5	OZ, PC

MW: Microwave, SLAN: Slot antenna, BPD: Beam product discharge, DBD: Dielectric barrier discharge, SP: Sputtering, DE: Deposition, SE: Surface elementary, CH: Chemistry, RA: Radiation, WE: Welding, PR: Processing, ET: Etching, OZ: Ozone, PC: Processing chemistry

2. IONIZATION AND BREAKDOWN

Table 3 summarizes ionization energies of various gas atoms and molecules. Here the ionization energy is defined as the energy required for removing the high energy outer shell electrons of a neutral species such as an atom and a molecule. Generally ionization is caused by 1) thermally, 2) photo-irradiation, 3) potential energy of excited species, or 4) the kinetic energy of high speed electrons.

Table 3. Excitation voltages and direct ionization voltages of various gases [26].

Gas	V^* (eV)	V_i (eV)	Process
H_2	7.0	15.37	$H_2 \rightarrow H_2^+ + e$
H		13.6	$H \rightarrow H^+ + e$
N_2	6.3	15.57	$N_2 \rightarrow N_2^+ + e$
O_2	7.9	12.5	$O_2 \rightarrow O_2^+ + e$
Ar	11.7	15.7	$Ar \rightarrow Ar^+ + e$
He	21.2	24.5	$He \rightarrow He^+ + e$

In a high temperature gas, the kinetic energy of the gas atoms or molecules, corresponding to their random movement, can be enough to initiate ionization (thermal ionization). The 1 eV corresponds to approximately 11,600 K. Thermal ionization hardly occurs at the temperature around several hundreds K. Thermal ionization can be expressed as following reaction:

$$A + B^{(*)} \rightarrow A^+ + B^{(*)} + e.$$

The Photo-ionization can take place with photons of short wavelengths such as ultraviolet by reaction:

$$A + h\nu \rightarrow A^+ + e, \text{ or}$$

$$A' + h\nu \rightarrow A^+ + e.$$

where $h\nu$ indicates a photon.

Penning ionization by excited species with long lifetimes can occur by reaction:

$$A + B' \rightarrow A^+ + B + e.$$

The direct ionization by fast electron is called impact ionization (α-effect):

$e_{(fast)} + A \rightarrow A^{+} + e_{(slow)} + e_{(slow)}$.

The impact ionization is the probabilistic process, and takes place when an electron can be accelerated long enough without collision so that its kinetic energy exceeds the ionization potential. Its distance is called mean free path for ionization, λ_i.

The ionization probability is expressed as collision cross section σ_i (cm^{-2}) similar manner to an elastic collision. Collision cross sections of various gases are shown in Figure 4. Kinetic energies of species are statistically expressed with the velocity distribution function. As a result, a generation rate of charged particles, which is equal to the ionization frequency ν_i is given by

$$\nu_i = \frac{n_g \int_0^\infty \sigma_i(u) u f(u) du}{\int_0^\infty f(u) du} \tag{1}$$

where $f(u)$, n_g, u are the distribution function of an electron velocity, the density of the gas molecule, and the velocity of the particle, respectively.

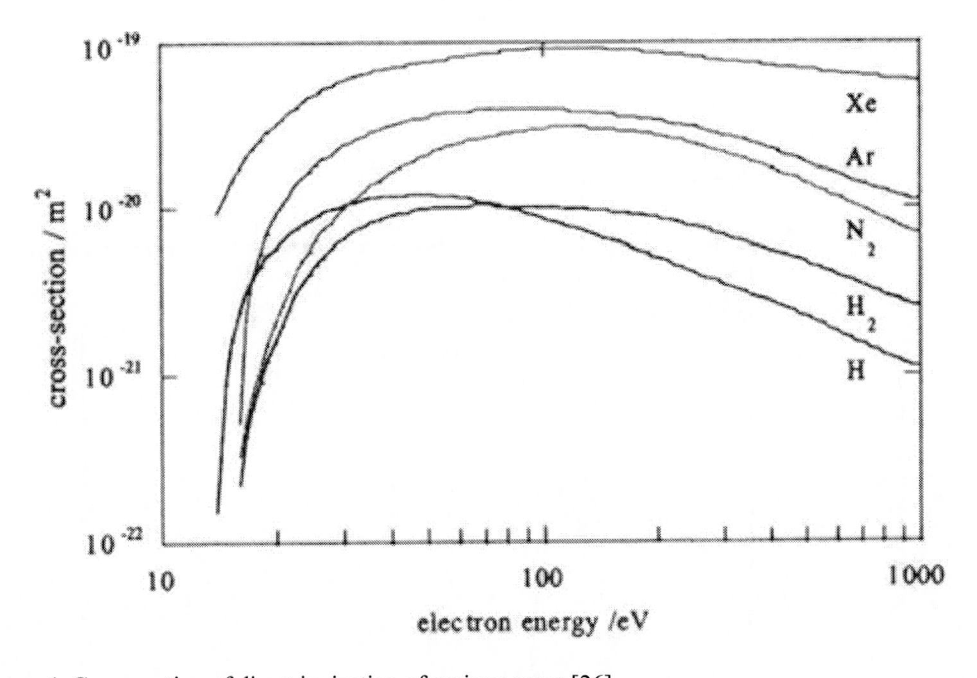

Figure 4. Cross section of direct ionization of various gases [26].

Secondly, the dielectric breakdown condition of the gas is discussed here. When an electron is in an electric field E, the force to the electron $-eE$ induces the change of the momentum, $m_e u_e \nu$, where e, E, m_e, u_e, and ν are the unit charge of an electron, the electric field, the mass of an electron, the electron velocity, and the collision frequency, respectively. Therefore, when the electron mean free path λ (e.g. ca. 1/3 mm in an argon gas at a pressure of 1 Torr) is much shorter than the gap of the electrodes, d, the drift velocity of the electron u_{drift} is given by;

$$u_{drift} = \frac{eE}{m_e u_e} = \mu_e E \tag{2}$$

where μ_e is the electron mobility, which is a function of the gas pressure. Using Einstein's equation,

$$\frac{D}{\mu_e} = \frac{kT}{e} \tag{3}$$

where D and k are the diffusion constant and Boltzmann constant, respectively. On the other hand, the increase of the number of electrons towards the opposite direction to the electric field (x-axis) is given by

$$dN_e = \frac{N_e dx}{\lambda_i}. \tag{4}$$

and thus

$$N_e = N_{e0} \exp\left(\frac{x}{\lambda_i}\right) \tag{5}$$

where N_{e0} is the number of electrons at $x = 0$.

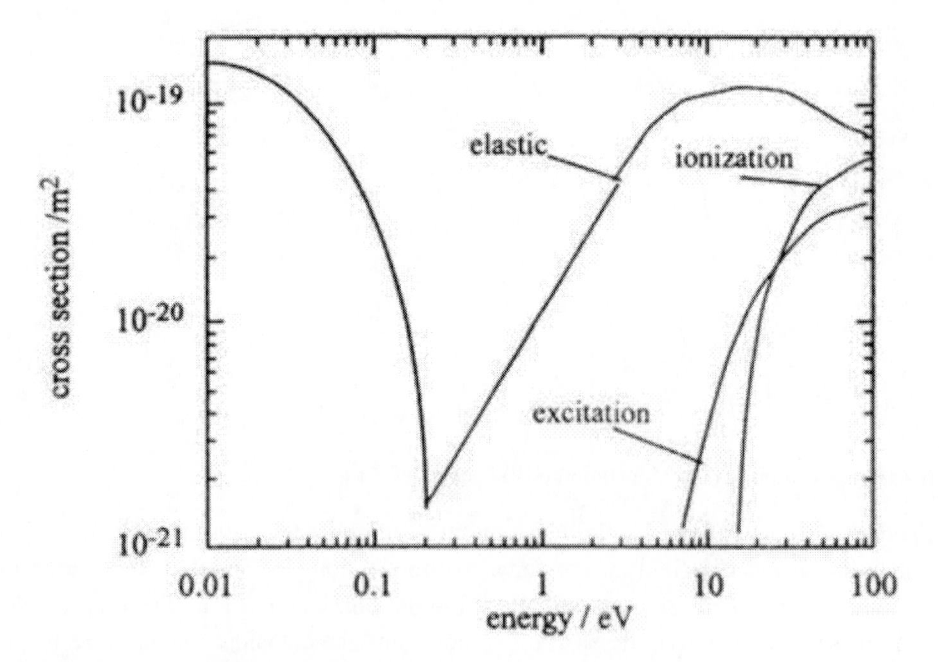

Figure 5. Cross sections of argon [26].

A collision cross sections (probability) of argon gas for different reactions as a function of the electron energy is shown in Figure 5. When the electron energy is low, the elastic collision dominates. However, at higher energies, the probabilities of the excitation and ionization increase. Considering the relation between the electron energy obtained from the electric field ($eE\lambda$) and the energy which is necessary for the ionization (eV_i), Townsend's ionization coefficient α is given by

$$\alpha = \frac{1}{\lambda_i} \propto \frac{1}{\lambda}\exp\left(-\frac{V_i}{E\lambda}\right). \tag{6}$$

As the mean free path is inversely proportional to the gas pressure p,

$$\alpha = Ap\exp\left(-\frac{Bp}{E}\right) \tag{7}$$

where A and B are constants, depending on the gas species. Assuming that the secondary electrons generate from the cathode due to the γ-effect appeared in DC glow discharge, the condition of sustaining the discharge is given by

$$\gamma N_o\left[\exp(\alpha d)-1\right] = N_o \tag{8}$$

or

$$\alpha d = \ln\left(1+\gamma^{-1}\right). \tag{9}$$

where γ represents the number of secondary electrons per positive ion impact, involving one or more of several secondary mechanisms. Substituting equation from (9) into (7) yields the famous Paschen's law in a uniform electric field,

$$V_{br} = \frac{Bpd}{\ln(Apd)-\ln\left[\ln\left(1+\gamma^{-1}\right)\right]} \tag{10}$$

where $V_{br}=E_{br}/d$ is the gas breakdown voltage.

Equation (10) indicates that there is a minimum breakdown voltage (Paschen minimum) as a function of the product pd [26] as shown in Figure 6. Atmospheric pressure discharge generally occurs at the right hand side of the minimum (Paschen right).

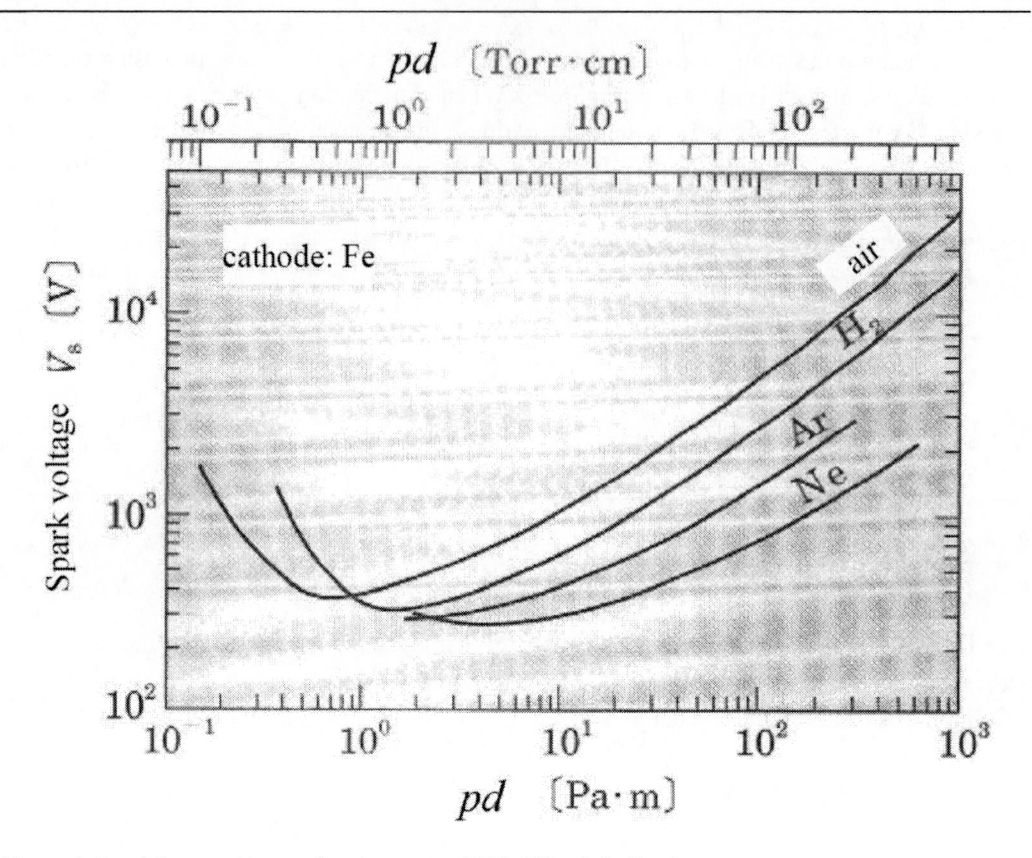

Figure 6. Breakdown voltages of various gases [27]. (1Pa=7.5mTorr)

3. Structure of a Glow Discharge and Arc Transition Criteria

Figure 7 shows a typical structure of a glow discharge [5]. The luminous region is divided into a negative glow near a cathode and a positive column. One of the characteristics of a glow discharge is that the cathode voltage drop between the negative glow and the cathode (several hundreds V) is much higher than that of an arc discharge (several tens V). The thickness of the cathode voltage drop region (cathode sheath) decreases as the gas pressure increases, and is several μm at atmospheric pressure [28]. The electric field at the region of the cathode voltage drop is extremely higher at high gas pressure with high energy consumption density than that in a low pressure. This can cause thermal instability and easily transient to an arc discharge. Another important characteristic of a glow discharge is its uniformity; the positive column and the negative glow almost diffuse uniformly in the radial direction (Figure 8 ①-③).

A glow discharge at high pressures near atmospheric pressure or at high current over 1 A can easily lead to arc transition owing to a thermal instability [29,30]. Under such conditions, a glow discharge tends to be transitory as shown in Figure 8. A glow discharge with ca. 300 A is kept for 1.5 micro seconds, and then luminous spot appears on the cathode and the plasma decreases its volume. Accordingly, the current rapidly increases while the discharge voltage rapidly decreases from approximately 1kV to several tens V.

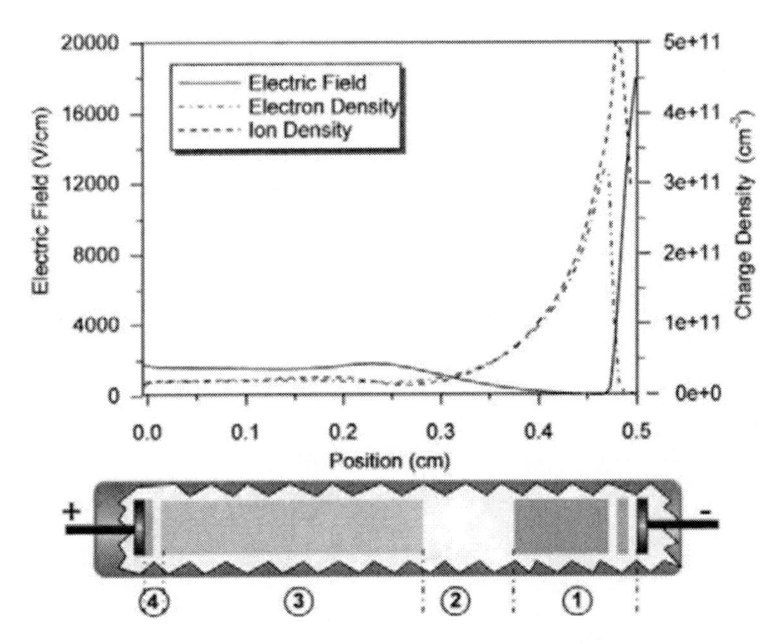

Figure 7. Structure of a DC glow discharge; negative glow ①, Faraday dark ②, positive column ③, anode region ④ [5].

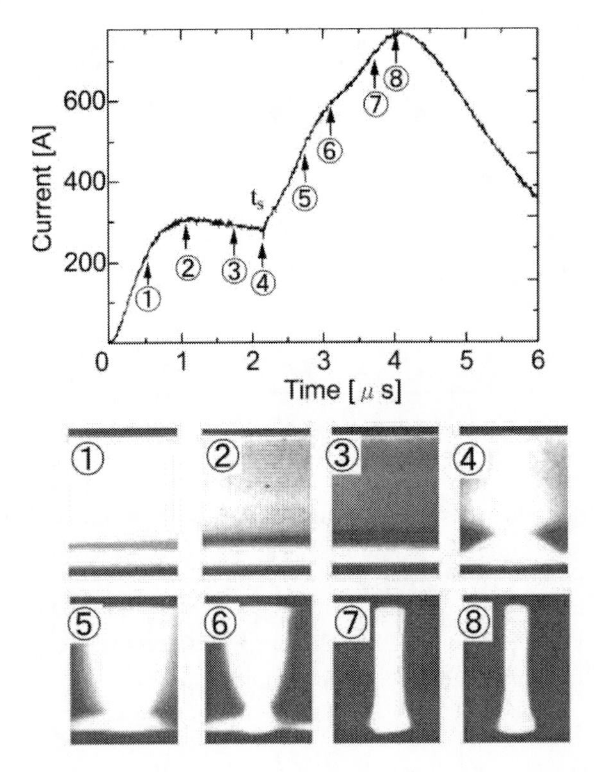

Figure 8. A current waveform and light emission of a high current transient glow discharge (gas: air, pressure: 5.5 Torr, gap: 2 cm, cathode: lower electrode, anode: upper electrode) [28].

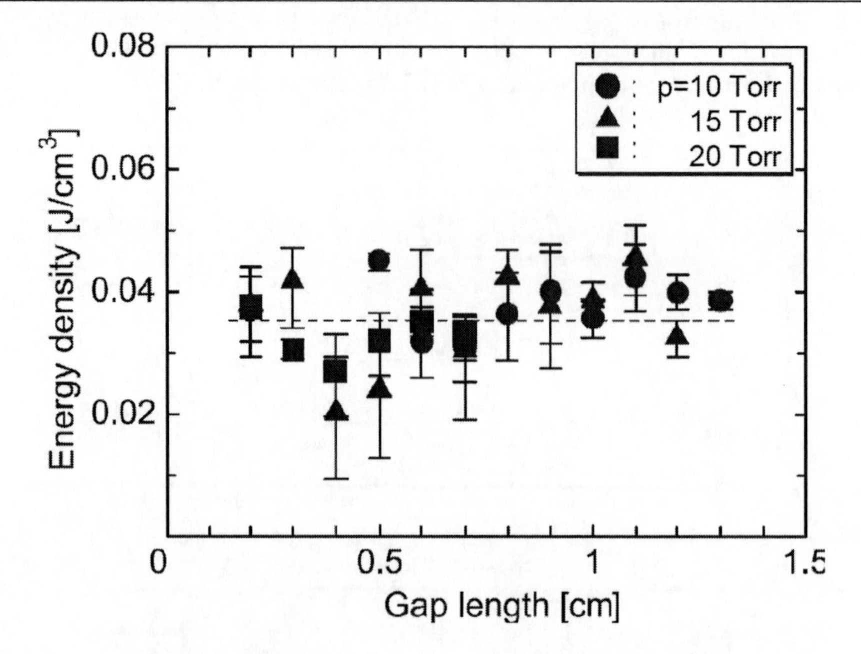

Figure 9. Consumed energy density in the cathode fall region before arc transition in N_2 gas [28].

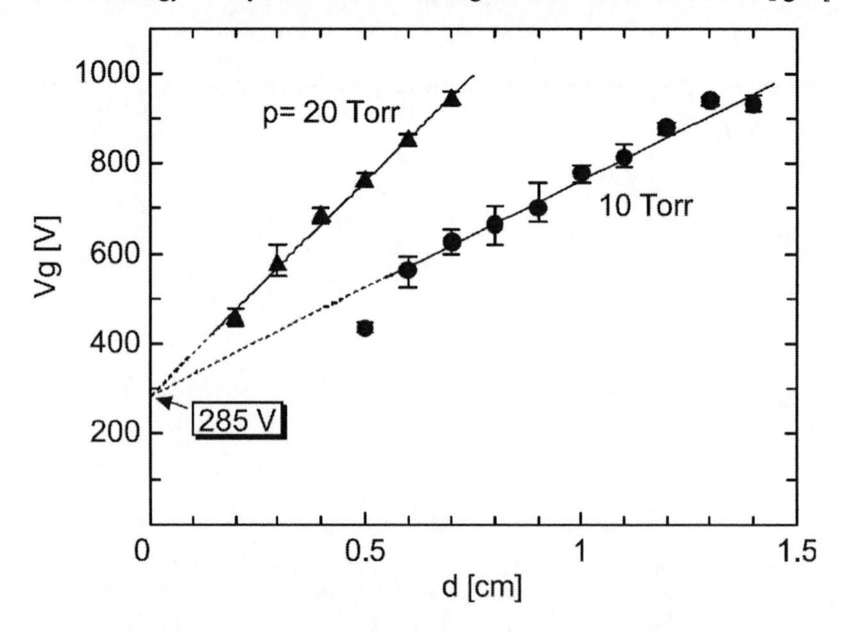

Figure 10. Glow discharge voltage v.s. the gap length in N_2 gas at two different gas pressures [28].

Figure 9 shows the criterional energy density which is estimated by dividing the energy consumption in the cathode sheath by the volume of the sheath until the luminous spot appears under various gas pressures [28]. The horizontal axis is the gap length. The energy consumed in the sheath E_c is obtaeined by the equation:

$$E_c = V_c \int_0^{t_s} i(t)dt \tag{11}$$

where V_c is the cathode fall voltage. As shown in Figure 10, V_c is the "zero gap voltage", determined by extrapolating the plot of the glow discharge voltage V_g as a function of the gap length to the intercept at the zero gap length ($d = 0$) [31]. As shown in Figure 8, t_s is the time from the ignition of a plasma to the arc transition. The volume of the sheath is estimated by the product of the area of the electrode and the following sheath length l_c of the normal glow [28],

$$l_c = 0.23 \cdot p^{-1} \text{ [cm]} \tag{12}$$

where p is a gas pressure in Torr. It is seen in Figure 9 that the criterional energy density is approximately 35 mJ cm^{-3} with some scattering, independent of the gap length and/or the gas pressure. This result revels that the glow-to-arc transition can be evaluated using the criterional energy consumed in the cathode sheath. When the pulse voltage is applied, a glow discharge can be sustained without the arc transition by adjusting the pulse width and the pulse voltage so that the energy consumed into the sheath is lower than the criterional value.

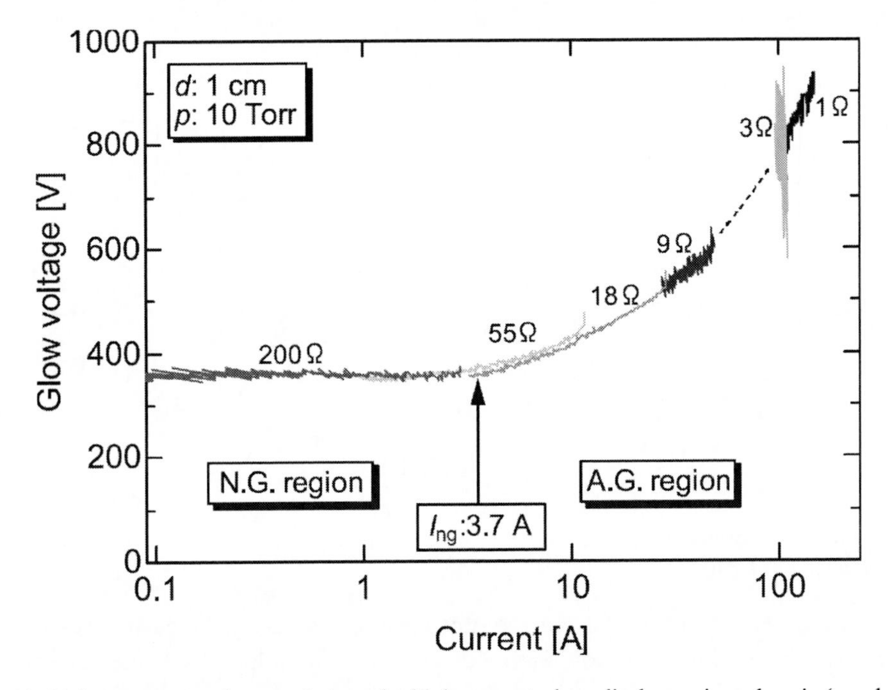

Figure 11. Voltage-current characteristics of a high current glow discharge in a dry air (gap length: 1 cm, pressure: 10 Torr) [32].

Figure 11 shows the voltage-current characteristics of the large-current glow discharge in dry air, adjusting the condition of the input energy lower than the criterional value [32]. Copper disk electrodes were used to generate a discharge in a gap of 1 cm at a pressure of 10 Torr. The current density of a normal glow discharge depends on the cathode material and a gas species. In the case of Figure 11 with the copper electrodes and air, the current density [A cm^{-2}] of the normal glow discharge is given by

$$J_n = 240 \times 10^{-6} p^2 \tag{13}$$

where p is a gas pressure in Torr. For example, at a pressure of 10 Torr, the current density is 0.024 A cm^{-2}. As the surface area of the cathode is ca. 155 cm^2, the maximum current of a normal glow discharge is 3.7 A. The arrow in Figure 11 indicates this value. Namely the lower and higher current regions divided by this current in Figure 11 correspond to a normal glow and abnormal glow, respectively. It is indicated that a glow discharge can be obtained ranging from the normal glow region with a constant voltage to the abnormal glow region with the high current exceeding 100 A. Here the abnormal glow region is represented by the increasing voltage as the increasing current. The plasma density is estimated to be ca. 3 x 10^{17} m^{-3} using the kinetic database of N_2 gas.

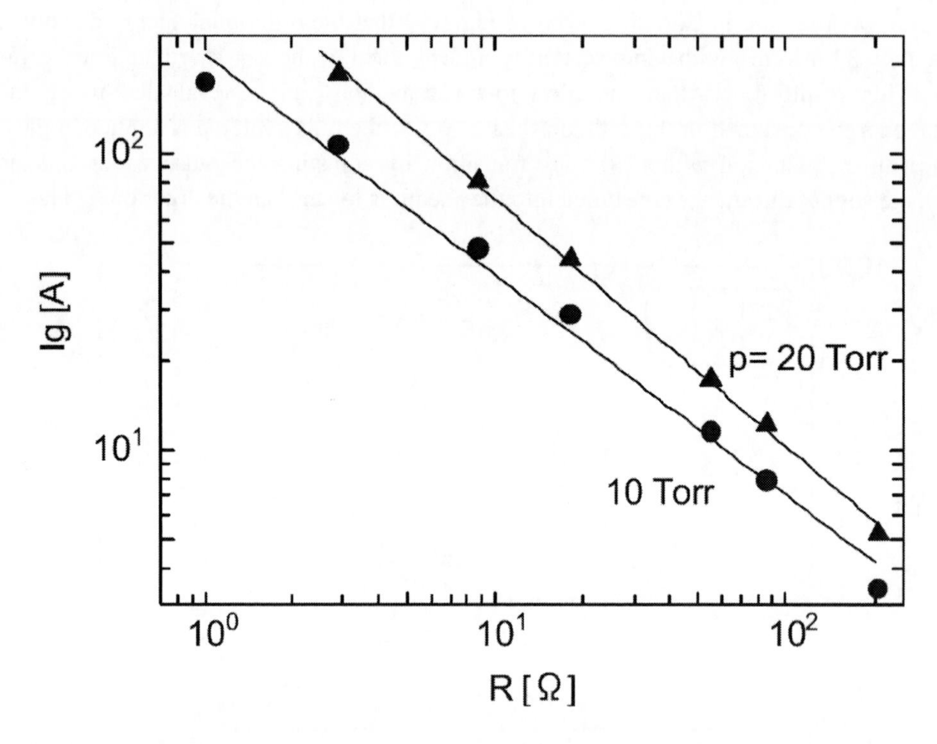

Figure 12. Glow discharge current v.s. resistance of the damping resistor at two different gas pressures (gas: air, gap length: 1 cm) [33].

One of the ways to prevent from the arc transition is, as predicted in equation 11, to extend the sustaining time of a glow discharge by sustaining a low discharge current using a damping resistor. Figure 12 shows the relation between the resistance of the damping resistor, R, and the glow discharge current I_g just before the arc transition [33]. The discharge was generated in air in a gap of 1 cm. It is seen that the glow discharge current decreases as the resistance increases. Figure 13 shows the current density obtained by dividing the current by the luminous area on the cathode. The dot lines indicate the current densities of the normal glow discharge estimated using equation (13). The discharge current decreases as the resistance increases. The current densities decrease as approaching to the values of the currents for the normal glow discharges. After that, the discharge area on the electrodes decreases and thus the current densities can remain almost constant. These results indicate that by reducing the resistance of the damping resistor, the mode of the transient glow discharge can be continuously controlled from the normal to the abnormal glow.

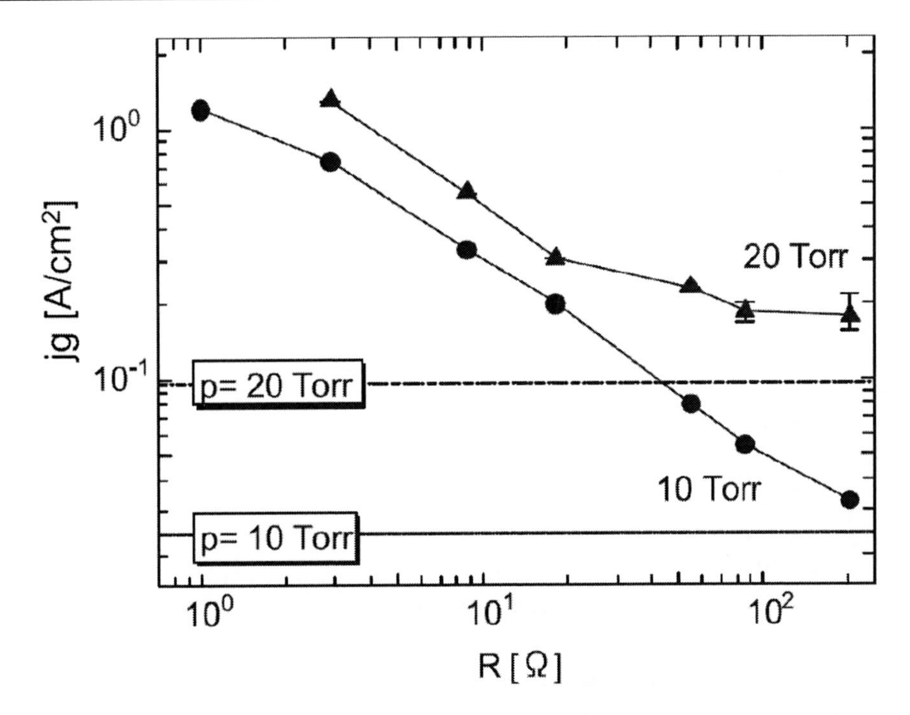

Figure 13. Discharge current density v.s. resistance of the damping resistor at two different gas pressures (gas: air, gap length: 1 cm) [33].

4. Properties of Large Current Glow Discharges

The properties of plasmas are generally characterized by the electron temperature and densities of electrons, ions and excited species [26]. Figure 14 shows the time- and spatial-averaged electron density and temperature in the positive column of the large current glow discharge of N_2 gas at various gas pressures. The gap length is 1cm. The electron density N_e is obtained by;

$$N_e = \frac{j}{eu_{drift}} \tag{14}$$

while the electron temperature is obtained using the Einstein's equation (3) [34]. Here j is the current density. Analyzed region is the positive column of the abnormal glow discharge. It is seen that as the current increases, both the electron density and the electron temperature increase. The electron densities at ca. 100 A are estimated to be 5.5 x 10^{11}, 6.1 x 10^{11}, and 6.6 x 10^{11} cm^{-3} at the gas pressures 1, 5, and 10 Torr, respectively. The electron temperatures with the same conditions are 1.57, 1.52, and 1.48 eV, respectively. It is noted that these values are similar to those obtained using probe measurements [32].

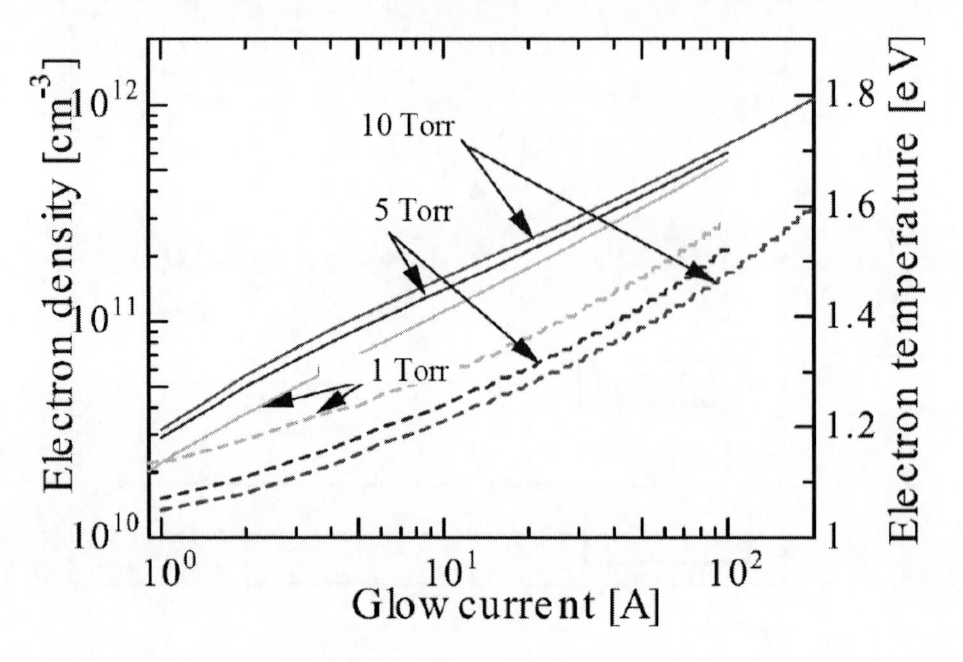

Figure 14. Electron density (solid lines) and electron temperature (dashed lines) v.s. glow discharge current at three different gas pressures (gas: N_2, gap length: 1 cm) [34].

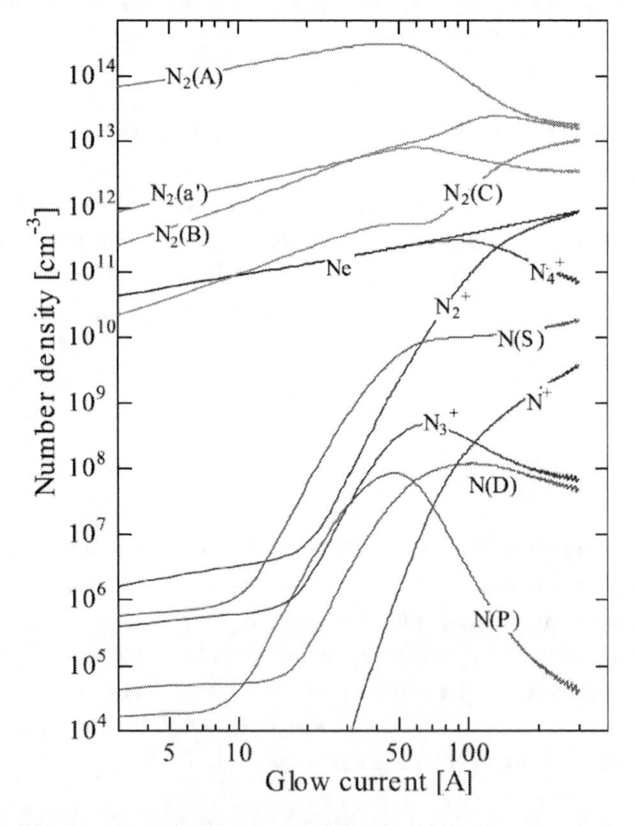

Figure 15. Number densities of ions and excited species as a function of glow current (gas: N_2, pressure: 10 Torr, gap length: 1 cm) [35].

Figure 15 shows the number densities of the ions and the excited species as a function of the glow current in a gap of 1.0 cm in an N_2 plasma at a pressure of 10 Torr [35]. The number density of $N_2(A)$ is 6.7 x 10^{13} cm^{-3} up to the region of the normal glow discharge corresponding to the current of 3 A, which is ca. three-order higher than the electron density of 3.0 x 10^{10} cm^{-3}. The electron density at the region of the abnormal glow discharge, corresponding to the current of around 250 A, is 7.4 x 10^{11} cm^{-3}. The number density of $N^2(A)$ of 10^{13} – 10^{14} cm^{-3} presented here shows a good agreement with the calculated value of 10^{13} cm^{-3} for the DC glow discharge reported by Guerra *et al.* [36] and the measured value of 4.0 x 10^{13} cm^{-3} using Q-mass by Matsuda *et al.* [37].

5. Pulse Generated Atmospheric Pressure Glow Discharge and its Properties

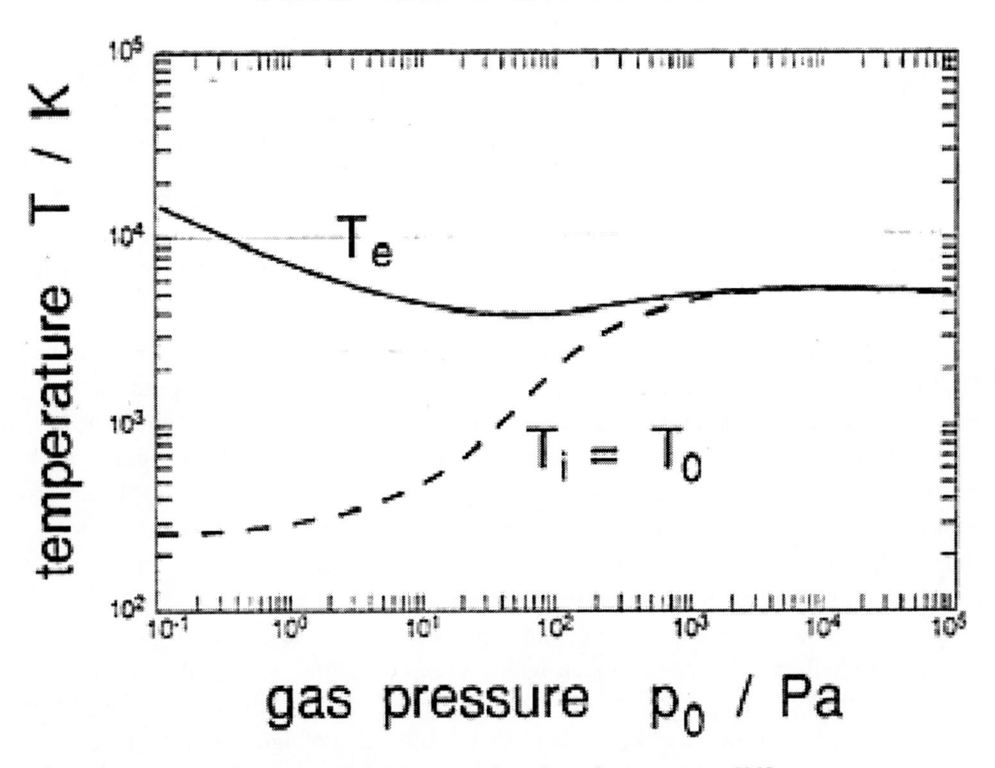

Figure 16. Electron, ion and gas temperatures as a function of gas pressure [38].

Even at low gas pressures, as the current of a glow discharge increases, the thermal instability enhances at the sheath near the cathode surface (cathode fall region), and subsequently transits to an arc discharge as shown in Figure 8. As the gas pressure increases, the gas number density increases and thus the collision frequency increases. Therefore at higher pressures, energy exchange among particles becomes more frequent, and thus the difference of the electron temperature from the temperatures of ions and gas decreases as shown in Figure 16 [38]. It means that the local thermal equilibrium state (non-thermal plasma; low temperature plasma) transients to thermal equilibrium state (thermal plasma; high temperature plasma). In addition the thickness of the sheath is inversely proportional to the gas pressure as expressed as equation (12). Therefore even if the discharge current is the same, the energy deposited per a unit volume increases with the increase of gas pressure, easily inducing thermal instability. The methods to prevent such instability were already discussed in section 1. Here the method of generating APG discharge plasmas using pulse voltage and a needle-array electrode, and the properties of these discharges are presented.

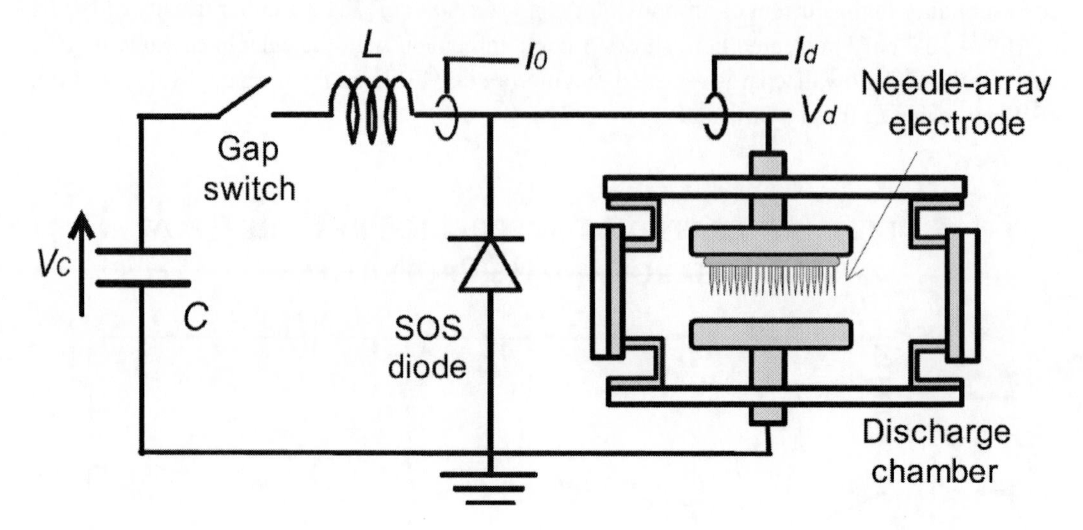

Figure 17. A schematic diagram of an APG discharge setup using the inductive energy storage system pulsed power generator [39].

Figure 17 shows an example of a schematic diagram of the setup for the generation of an APG discharge plasma by inductive energy storage system pulsed power generator [39]. An inductive storage energy is transformed using a semiconductor opening switch (SOS) diode. The SOS diode is commonly used for a pulse generation circuit to excite an excimer for laser source work as a spiker and sustainer with a simple circuit [39, 40]. The pulsed power generator consists of a capacitor C, a gap switch GS, an inductor L, a fast recovery diode (FRD) which is used as the SOS diode. A needle-array electrode is used in order to generate a homogeneous discharge all over the region between the electrodes [39].

Figure 18 shows the photographs of the electrodes and the discharge at pulse repetition rate of 50 pps (pulses per second), taken with the exposure time of 1 s. It is seen from the image that a glow discharge is generated without arc transition all over the needle-array electrode. Figure 19 shows the waveforms of the applied voltage and the current of the discharge. The charge up voltage, the capacitance and the inductance are -10 kV, 4.2 nF, and 12.6 μH, respectively. It is indicated from Figure 19 that the glow discharge is generated as follows; application of a pulse voltage induces the generation of streamers, and the discharge circuit becomes a series-circuit with C, L, and the impedance of the glow discharge Z_G, and then the glow discharge is generated and sustained by discharging from the capacitor. The impedance of the pulse power generator Z after the opening theSOS diode is 55 Ω, while the impedance of the glow discharge Z_G, is ca. 250 Ω, estimated by dividing the voltage by the current. As $2Z < Z_G$, the waveforms of the voltage and the current are decreasing to 0 with over-damping manner.

Figure 18. Images of the needle-array electrodes and the light emission from the atmospheric pressure glow discharge (gas: N_2, gap length: 1 cm).

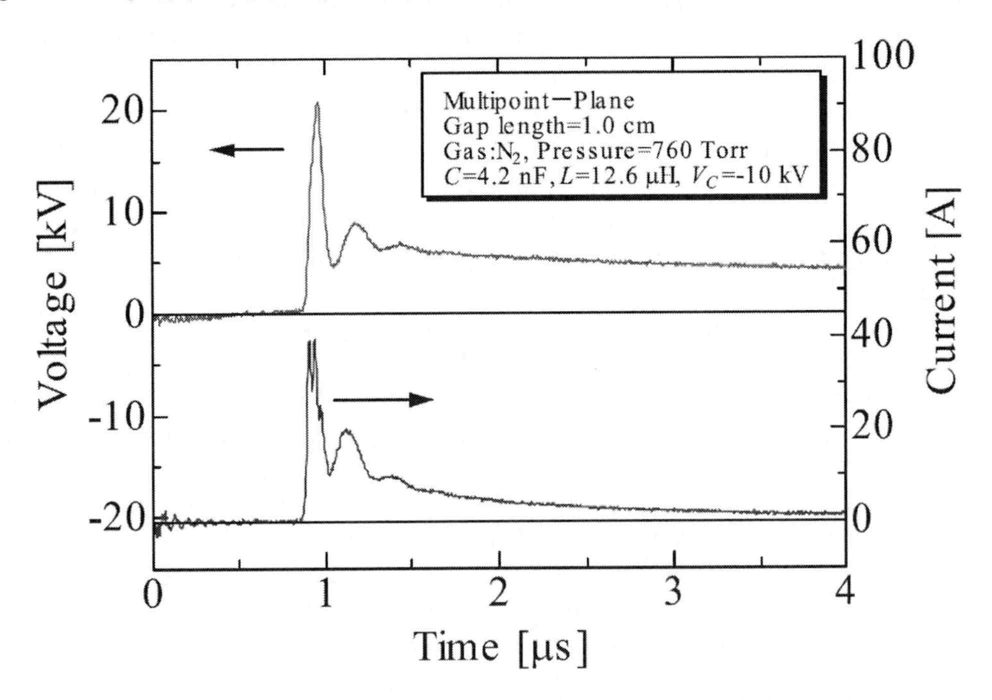

Figure 19. Voltage and current waveforms of atmospheric pressure glow discharge in N_2 gas [39].

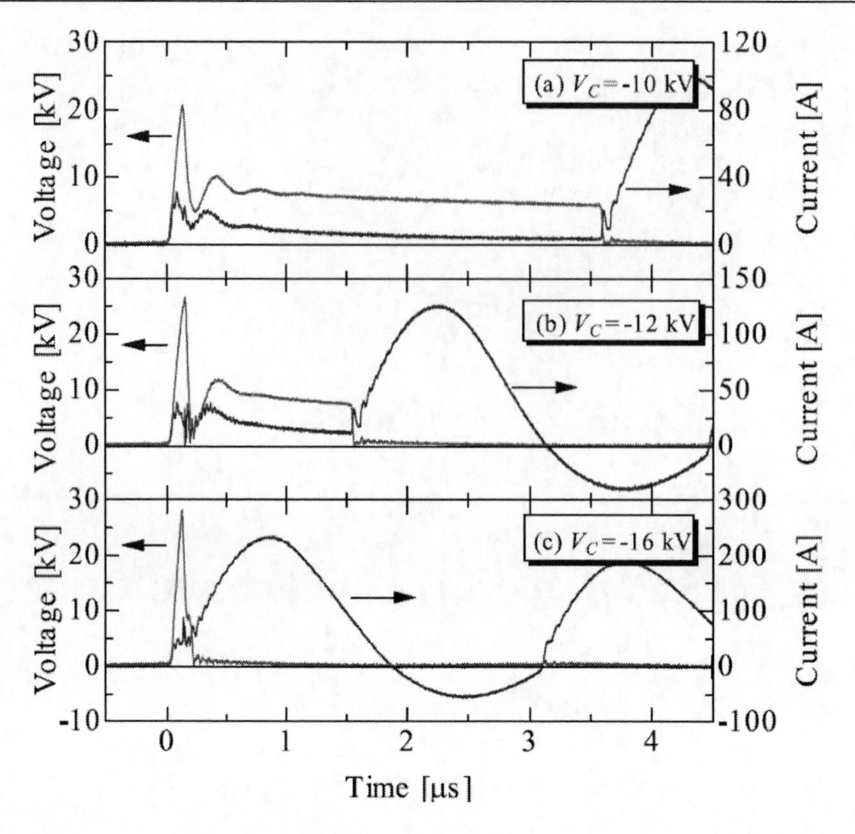

Figure 20. Voltage and current waveforms of a transient atmospheric pressure glow discharge at various charge-up voltages (gas: N_2, gap length: 1 cm) [39].

Figure 20 shows the waveforms of the voltage applied to the high-voltage electrode and the discharge current. The charge up voltages V_c are -10 (a), -12 (b) and -16 (c) kV, respectively. The capacitance and the inductance of the pulse power source are 8.0 nF and 20.0 μH, respectively. The impedance of the power source is 50 Ω after opening the SOS diode. The discharge duration time is ca. 3.5 μs, when V_c = -10 kV. When V_c = -12 kV, the discharge duration time becomes shorter and is 1.5 μs, while the maximum value of the current increases. On the other hand, when V_c = -16 kV, the glow phase almost disappears and the arc discharge appears immediately after the propagation of the streamers between the electrodes.

Figure 21 shows the relation between the charge-up voltage V_c and the energy consumed until the arc transition with various gap lengths between the electrodes at 760 Torr. The consumed energy was calculated by integrating the product of the applied voltage and the discharge current from the time of pulsed voltage application to the time just before the arc transition. The left hand side in Figure 21 where there are no plots indicates either no discharge ignition or no arc transition after generation of a glow discharge. The right hand side of Figure 21 corresponds to the cases without the glow phase as is the case of Figure 20 (c). It is seen that even with some scattering, the energy consumed until the arc transition is almost constant, independent of the charge-up voltage V_c. In addition, with the gap length increases, the energy consumption until the arc transition also increases. For instance, the energy consumptions are 60 mJ and 370 mJ at the gap lengths of 0.5 cm and 1.5 cm, respectively. The tendency of the increase in the energy consumption with increase in the gap length shows a good agreement with the tendency at low pressures [28].

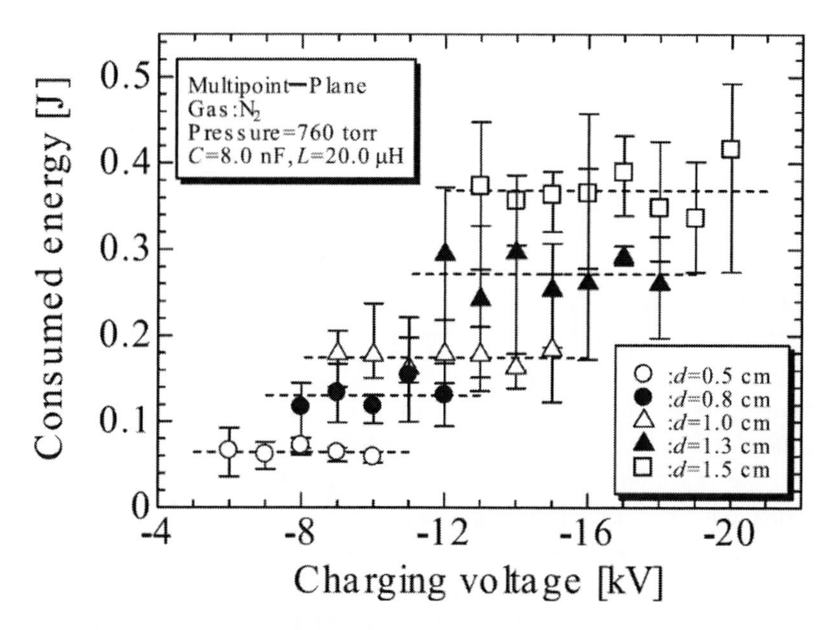

Figure 21. Consumed energy of the atmospheric pressure glow discharge in N_2 gas until arc transition [39].

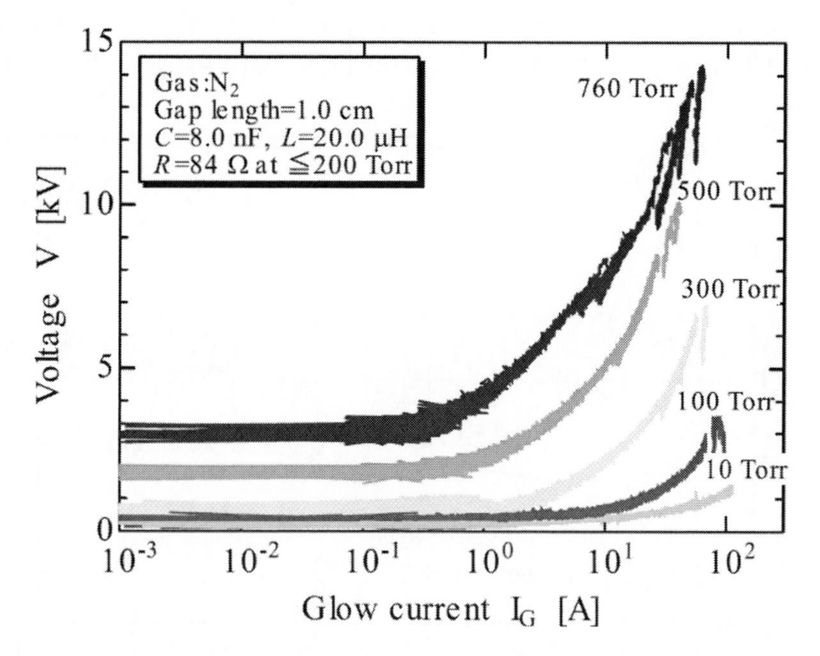

Figure 22. Voltage-current characteristics of the glow discharge in N_2 gas at various pressures (gas length: 1 cm).

Figure 22 shows the voltage–current characteristics of a transient glow discharge with various gas pressures. In each case there is a region showing the increasing voltage as the increasing current. It is suggested that the flat regions correspond to the normal glow discharge, while the regions of the increasing voltages with the increase of the currents correspond to the abnormal glow. For example, at 760 Torr the voltage increases once the current exceeds 0.3 A, and is 14 kV at the current of 60 A. The region of the current less than 0.3 A is assigned to the normal glow, while that over 0.3 A is to the abnormal glow. The threshold current which divides the regions of the normal and abnormal glow discharges tends to decrease as the gas pressure increases. Here the electron temperature and density of the APG discharge plasma in the region of the abnormal glow at the current of 60 A is 1.3 eV and 1.4 x 10^{12} cm^{-3}, respectively [39], estimated using equations (3) and (14).

When an APG glow discharge is generated using this technique described above, the SOS diode used for the opening switch plays an important role. With and without use of the SOS diode, the waveforms of the voltage and current are compared in Figure 23, and the peak value of the glow discharge current is shown in Figure 24 [41]. When the SOS diode is unused, the charge up voltage at the capacitor is applied between the electrodes directory after the closing switch works. On the other hand, when the SOS diode is used, the over-voltage for the static breakdown voltage is applied from the surge voltage similar manner to the spiker circuit at the beginning of the pulse as shown in Figure 20. In order to prevent the glow-to-arc transition, the charge up voltage is fixed at 13 kV in Fig. 23. The capacitance and the inductance of the pulse power source are 8.0 nF and 20.0 μH, respectively. As shown in Figure 23, when the SOS diode is used, the current is larger and the decay time of the voltage is shorter than those without the SOS diode. It indicates that the impedance of the glow discharge is lower and the plasma density is higher when the SOS diode is used. Figure 24 indicates that the glow current increases with the increase in the applied voltage. In addition, when the SOS diode is used, the current significantly increases, and thus the advantage for employing the spike pulse to generate the APG plasma with high density can be confirmed.

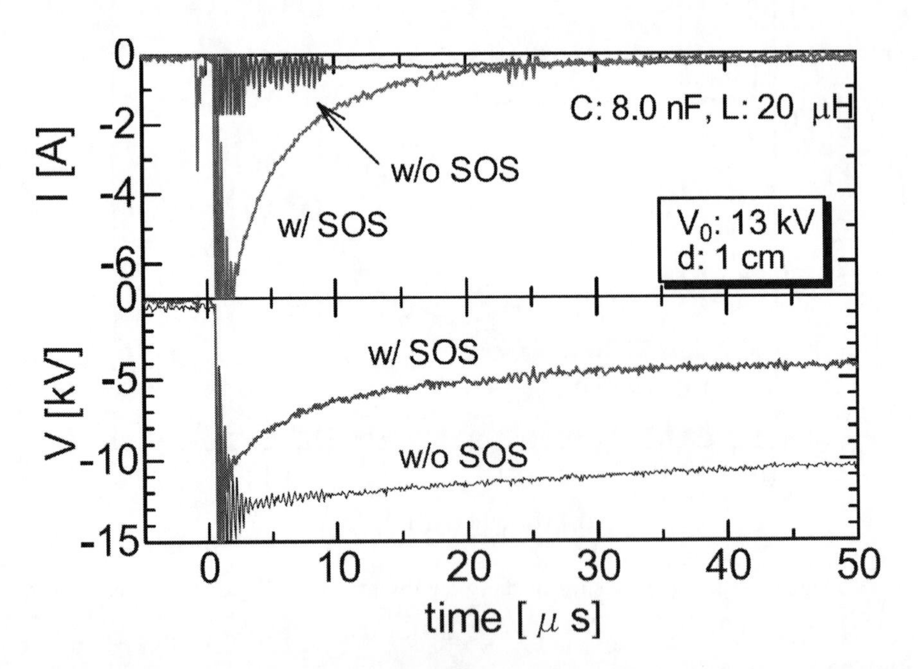

Figure 23. Voltage and current waveforms of the atmospheric pressure glow discharge in N_2 gas with (w/) and without (w/o) the SOS diode (gas length = 1 cm) [41].

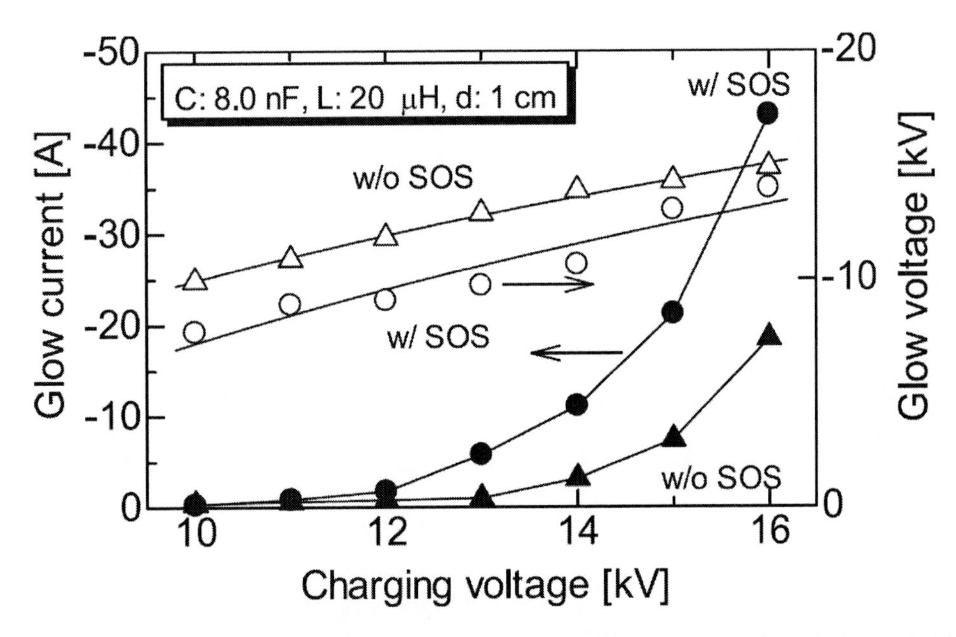

Figure 24. Sustaining voltage and peak current of the atmospheric pressure glow discharge in N_2 gas with (w/) and without (w/o) the SOS diode [41].

REFERENCES

[1] Kataoka,Y.; Kanoh, M.; Makino, N.; Suzuki, K.; Saito, S., Miyajima, M.; Mori, Y. *Jpn. J. Appl. Phys.* 2000, *39(1)*, 294-298.

[2] Prat, R.; Koh, Y. J.; Bubukutty, Y.; Kogoma, M.; Okazaki, S.; Kodama, M. *Polymer* 2000, *41(20)*, 7355-7360.

[3] Sawada, Y.; Souma, M. *J. Inst. Electrostatics Japan* 2002, *26(6)*, 258-259.

[4] Nozaki, T.; Miyazaki, Y.; Unno, Y.; Okazaki, K. *J. Phys. D Appl. Phys.* 2001, *34(23)*, 3833-3390.

[5] Gadri, R. B.; Roth, J. R.; Montie, T. C.; Kelly-Wintenberg, K.; Tsai, P. P.-Y.; Helfritch, D. J.;Feldman, P.; Sherman, D. M.; Karakaya, F.; Chen, Z.; UTK Plasma Sterilization Team. *Surf. Coat. Technol.* 2000, *131(1-3)*, 528-541.

[6] Laroussi, M. *IEEE Trans. Plasma Sci.* 1996, *24(3)*, 1188-1191.

[7] Kanazawa, S.; Kogoma, M.; Moriwaki, T.; Okazaki, S. *J. Phys. D Appl. Phys.* 1988, *21(5)*, 838-840.

[8] Okazaki, S.; Kogoma, M. ; Uehara, M.; Kimura, Y. *J. Phys. D Appl. Phys.* 1993, *26(5)*, 889-892.

[9] Sawada, Y.; Ogawa, S.; Kogoma, M. *J. Phys. D Appl. Phys.* 1995, *28(8)*, 1661-1669.

[10] Gherardi, N.; Gouda, G.; Gat, E.; Ricard, A.; Massines, F. *Plasma Sources Sci. Technol.* 2000, *9(3)*, 340-346.

[11] Tochikubo, F.; Chiba, T.; Watanabe, T. *J. Phys. D Appl. Phys.* 1999, *38(9A)*, 5224-5240.

[12] Golubovskii, Y.; Maiorov, A.; Behnke, J.; Behnke, J. *J. Phys. D Appl. Phys.* 2003, *36(1)*, 39-49.

[13] Gadri, R. *IEEE Trans. Plasma Sci.* 1999, *2(1)7*, 36-37.

[14] Korolev, Yu. D.; Mesyats, G. A. *Physics of Pulsed Breakdown of Gases;* URO-Press: Yekaterinburg, RU, 1998; pp 133.
[15] Schoenbach, K. H.; Habachi, A. E.; Shi, W.; Ciocca, M. *Plasma Sources Sci. Technol.* 1997, *6(4)*, 468-477.
[16] Shi, W.; Stark, R. H.; Schoenbach, K. H. *IEEE Trans. Plasma Sci.* 1999, *27(1)*, 16-17.
[17] Park, S. J.; Wagner, C. J.; Herring, C. M.; Eden, J. G. *Appl. Phys. Lett.* 2000, *77(2)*, 199-200.
[18] Endo, Y.; Yasuoka, O.; Ishii, S. *Trans. IEEJ* 2003, *123-A*, 364-469.
[19] Park, J.; Henins, I.; Herrmann, H. W.; Selwyn, G. S.; Jeong, J. Y.; Hicks, R. F.; Shim, D.; Chang, C. S. *Appl. Phys. Lett.* 2000, *76(3)*, 288-290.
[20] Kono, A.; Sugiyama, T.; Goto, T.; Furuhashi, H.; Uchida, Y. *Jpn. J. Appl. Phys.* 2001, *40(3B)*, L238-L241.
[21] Kunhardt, E. E.; Luessen, L. H. *Electrical Breakdown and Discharges in Gases, part B;* Plenum Press: New York, US, 1983; pp 5.
[22] Roth, J. R.; Rahel, J.; Dai, X.; Sherman, D. M. *J. Phys. D Appl. Phys.* 2005, *38*, 555-567.
[23] Chang, J.S.; Kelly, A. J.; Crowley J. M. *Handbook of Electrostatic Processes;* Marcel Dekker Inc.: New York, US, 1995.
[24] Braithwaite, N. S. J. *Plasma Sources Sci. Technol.* 2000, *9,* 517-527.
[25] Yushkov, G. Y.; Anders, A.; Oks, E. M.; Brown, I. G. *J. Appl. Phys.* 2000, *88(10)*, 5618-5622.
[26] Conrads, H.; Schmidt, M. *Plasma Sources Sci. Technol.* 2000, *9(4),* 441-454.
[27] Sugai, H. *Plasma Electronics;* Ohm-sha: Tokyo, JP, 2000; pp 70.
[28] Takaki, K.; Kitamura, D.; Fujiwara, T. *J. Phys. D Appl. Phys.* 2000, *33(11)*, 1369-1375.
[29] Gurevich, D.; Kanatenko, M.; Podmoshenskii, X. *Sov. J. Plasma Phys.* 1979, *5(6)*, 760-764.
[30] Babichev, V. N.; Golubev, S. A.; Kovalev, A. S.; Pismennyi, V. D.; Rakhimov, A. T.. *Sov. J. Plasma Phys.* 1980, *6(1)*, 111-112.
[31] Meyer, J.; Lee, C. *J. Phys. D Appl. Phys.* 1971, *4(1)*, 168-170.
[32] Takaki, K.; Taguchi, D.; Fujiwara, T. *Appl. Phys. Lett.* 2001, *78(18)*, 2446-2648.
[33] Takaki, K.; Kitamura, D.; Fujiwara, T. *Trans. IEEJ* 1999, *119-A*, 531-532.
[34] Hosokawa, M; Noda, C. ; Mukaigawa, S.; Takaki, K.; Fujiwara, T. *Trans. IEEJ* 2005, *125-A*, 993-1000.
[35] Takaki, K.; Hosokawa, M; Sasaki, T.; Mukaigawa, S. Fujiwara, T. *J. Adv. Oxid. Technol. Adv.* 2005. *8(1).* 11-17.
[36] Guerra, V.; Loureiro, J. *Plasma Sources Sci. Technol.* 1997, *6(3),* 361-372.
[37] Matsuda S.; Shimosato, K.; Yumoto, M.; Sakai, T. *Trans. IEEJ* 2003, *123-A*, 167-172.
[38] Pochner, K.; Neff, W.; Lebert, R. *Surf. Coat. Technol.* 1995, *74-75*, 394-398.
[39] Takaki, K.; Hosokawa, M; Sasaki,T.; Mukaigawa, S. Fujiwara, T. *Appl. Phys. Lett.* 2005, *86(15)*, P151501 1-3.
[40] Akiyama, H. *EE Text High pressure pulse power engineering;* Ohm-sha, Tokyo, JP, 2003; p80.
[41] Takaki, K.; Mukaigawa, S.; Fujiwara, T. *Tech. Meeting IEEJ* 2006, *1-S6-3,* pp 1S6(7).

Section 2
Applications of Atmospheric Pressure Plasmas

In: Generation and Applications of Atmospheric Pressure Plasmas ISBN: 978-1-61209-717-6
Editors: M. Kogoma, M. Kusano and Y. Kusano

Chapter 3

SURFACE CLEANING AND MODIFICATION OF ELECTRIC MATERIALS AND PARTS BY ATMOSPHERIC PRESSURE PLASMA IN A NOBLE GAS

Yasushi Sawada
Laboratory of Research and Development, Air Water Inc., Nagano, Japan

INTRODUCTION

As electrical equipments become smaller and have more functionality, the significantly high mechanical strength of electric parts and liquid crystal panels and the excellent adhesion at interfaces are demanded. In addition to the demands for improving the qualities, there are high demands for improving productivity and yield of products. Consequently, surface treatment techniques are highly demanded which are applicable to high-speed inline manufacturing process.

Current mainstreams of dry cleaning/surface modification techniques are ultraviolet (UV) radiation and low pressure plasma processing. In the UV radiation method, UV light at the wavelength of 254 nm is irradiated from a low pressure mercury lamp onto specimens. Organic contamination is dissociated by the UV, and ozone which is generated by the UV is used to remove the dissociated substances. However, the cleaning efficiency with ozone is not high enough. In order to solve this problem, Xe excimer lamp was recently developed which can emit photons of 172 nm wavelength. Owing to this development, ozone as well as reactive oxygen can be generated in air, and the cleaning efficiency was significantly improved. Nevertheless, the cleaning efficiency is not sufficient enough.

Low pressure plasma processing is widely used mainly for thin film deposition and surface modification in semiconductor industry. Here a plasma is a state in which atoms in a gas are excited to be reactive species such as ions, electrons and radicals by thermal or high frequency energy, and as a whole retained electrically neutral. In order to stabilize a discharge, a chamber is kept at high vacuum with a gas pressure less than 1 kPa using a vacuum pump. A plasma can be generated by applying a high frequency and high voltage between the electrodes, and specimens can be treated. This method shows high cleaning efficiency as gas is directly dissociated, being different from the UV radiation method. However, it has the disadvantage of its productivity because it is a low pressure processing. In addition up-scaling of the system is difficult.

On the other hand, atmospheric pressure plasma, which is the modification of a dielectric barrier discharge, does not require a vacuum pump or a large chamber, and its system structure can be very simple. Therefore its applications are rapidly spreading for the purpose of inline continuous treatment as the alternative of the UV radiation method or low pressure plasmas [1].

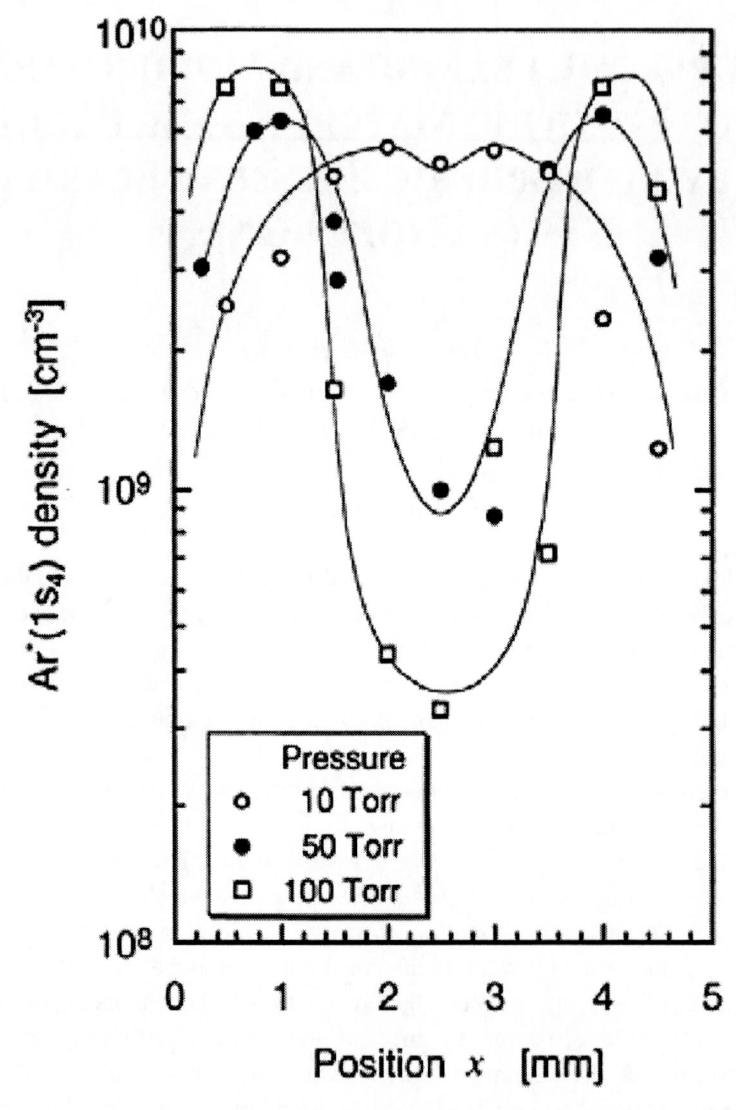

Figure 1. Ar*(1_{S4}) profiles in the x-position measured from the lower dielectric plate. Frequency: 13.56 MHz, Power density: 0.2 W/cm^2, A/He = 1/10. The position at x=0 is the ground side of the dielectric plate [2].

1. ATMOSPHERIC PRESSURE PLASMA

Even at atmospheric pressure, a uniform and stable glow discharge can be generated in a limited space by attaching dielectrics onto a pair of parallel plate electrodes, introducing a gas mixture containing a gas of low ignition voltage like a rare gas and a small content of a reactive gas, and applying high frequency electric field between the electrodes. Figure 1 shows the temporal and spatial distribution of long-lifetime excited Ar atoms in a dielectric barrier discharge with a gas mixture of Ar and He (Ar/He: 1/10), characterized using the absorption spectrometry of a GaAlAs wavelength variable semiconductor laser [2]. This Figure shows the density distribution of Ar($1s_4$) resonant state obtained from the absorption at 810 nm at different working pressures. The maximums of the density tend to localize near the dielectrics. In addition it is seen that temporally and spatially uniform plasma was generated. The radical density was approximately 8×10^{-9} cm^{-3}.

The atmospheric pressure plasma systems are categorized into the direct and remote types. Table 1 summarizes each structure and characteristics. In the direct method a substrate is treated with being exposed on a plasma. This method shows higher treatment speed and a less amount of gas used for processing than the remote method. However, the gap between the electrodes is between 5 and 10 mm, limiting the thickness and shape of the specimen to be treated. Furthermore, semiconductor devices such as integrated circuit (IC) and liquid crystal thin film transistor (TFT) are electronically damaged by the accumulation of charged particles.

Table 1. Characteristics of the direct and remote types of atmospheric pressure plasmas

	direct type	remote type
structure	electrode dielectric plasma substrate A substrate is directly exposed in a discharge plasma to treat the surface.	electrode dielectric plasma substrate Activated species generated in a plasma is blown to a substrate surface.
characteristics	· high treatment speed. · low gas consumption.	· low damage to a substrate · low restriction for the specimen shape.

In order to compensate such drawbacks, the remote method was developed [3-5]. In this method gas is fed between the electrodes and high voltage is applied for generation of a plasma. At the bottom of the electrodes the gas was subsequently blown out like a jet. The reactive radicals are fed to the substrate surface before inactivated, and the substrate surface can be cleaned and modified in an instant. The specimen can be continuously transported for the treatment, or can be treated only at the required parts by transporting it to a preprogrammed location. Consequently this method is not limited by a shape of a sample. In addition electrical damages to semiconductor specimens are very small as the process is radical dominated.

If O_2 is used as a reactive gas in an atmospheric pressure plasma, carbon atoms of organic materials can be oxidized, modified, or ashed by the reactive oxygen atoms. A small amount of organic contamination can be removed as CO_2. If H_2 is used as a reactive gas, metallic oxides can be deoxidized. The degree of the deoxidization depends on the stability of the metallic oxides. Physical effects like sputtering due to ionic impact which are seen in low pressure plasma treatment are not observed, while chemical effects dominate the processing. A variety of surface modification and addition of functionality is possible using not only O_2 or H_2, but also gases containing fluorine or special gas.

2. Application of Direct Type Atmospheric Pressure Plasma

Ensuring high adhesion strength between copper and resin is very important for manufacturing high density printed circuit boards and electric packages. Copper is oxidized in ambient air and naturally forms oxide film of Cu_2O. This natural oxide film can form a weak boundary layer due to the different lattice sizes of the oxide and metallic copper. In addition the poor affinity of the oxide to resin can be a reason of lowering adhesion strength [6]. Therefore, in the manufacturing of printed circuit boards, finely roughened copper oxide is formed on a copper foil using solution of sodium hypo-chloride. It is so-called black oxide and is generally applied. Due to the anchor effect of this rough surface, strong adhesion is achieved between a copper foil and the epoxy resin. However, copper oxides may be eluted from the substrate treated with black oxide during manufacturing process, and thus it is difficult to form fine circuits. Furthermore it is highly problematic for the treatment of waste water of the sodium hypo-chloride. On the other hand, a novel and simple adhesion method between copper and epoxy resin is proposed by using atmospheric pressure plasma [7].

2.1. Novel Adhesion Method for Copper and Epoxy Resin

This method consists of the following processes; dielectric plates (borosilicate glass, diameter 130 mm, thickness 1 mm) are placed inside a pair of parallel plate electrodes. A glass epoxy substrate with 35-μm thick copper foil is placed on the bottom dielectric and treated by using a plasma. The treatment condition was as follows; driving frequency 13.56 MHz, applied power 200W, gas mixture of He (5000 cm^3/min) and H_2 (50 cm^3/min). The treated substrate was immersed for 1 minute in a 2 % solvent of γ- aminopropyl triethoxysilane (γ-APS) which is one of silicon type coupling agent. After drying in air, γ-APS was cured at 120 °C. Prepreg containing epoxy was deposited on to a copper foil, and press-compacted at a pressure of 4 MPa at 170 °C for 90 minutes.

10 mm wide piece was cut from the specimen, and the 90° peel strength at the interface between copper/resin was measured. The result is shown in Figure 2. The horizontal axis indicates the preheat temperature before immersing the substrate into the γ-APS. The untreated substrate after immersing into γ-APS solvent shows lower peel strength than plasma

treated specimens. The peel strength of the untreated specimens significantly decreases as the preheat temperature increased before immerse. On the other hand, the plasma treated specimens show higher strength, which was almost unchanged at different preheat temperatures. The thickness of the oxide layer measured by x-ray photoelectron spectroscopy (XPS) is also shown in Figure 2. As the preheat temperature increased, the thickness of the oxide layer increased.

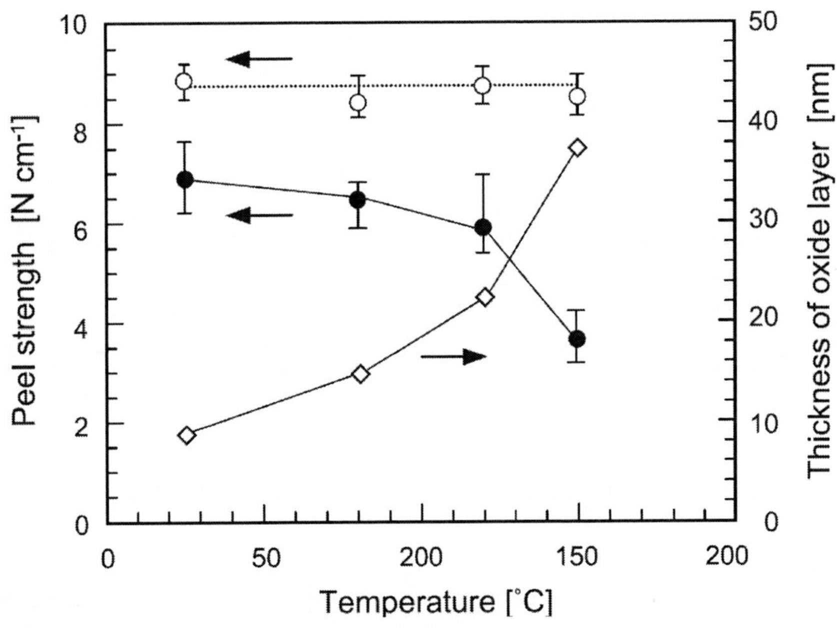

●: no plasma treatment.
○: plasma treatment for 5 minutes after preheated to each temperature.
◊: thickness of oxide layers formed on the copper foils.

Figure 2. Adhesion strength and thickness of oxide layers expressed as a function of preheat temperature.

Table 2. XPS compositional analysis for both copper and resin sides of the fractured surfaces after peel tests.

Pre-treatment of copper foils	Peel strength [N cm^{-1}]	Atomic concentration of elements [at%]											
		Copper side						Resin side					
		C	N	O	Cu	Br	Si	C	N	O	Cu	Br	Si
Preheating to 150°C for 3 hrs.	3.2	52.3	6.2	15.9	23.0	0.8	1.8	48.8	7.1	14.9	21.1	4.2	3.9
Plasma treatment for 5 min.	8.8	93.0	0.8	2.8	1.5	1.2	0.8	85.5	2.1	7.9	1.4	2.0	1.2

Table 2 shows the XPS results for copper and resin sides of the fracture surfaces after peel test for the specimen whose copper surface was heated at 150 °C for 3 h (peel strength: 3.2 N/cm), and the specimen which was subsequently treated by H_2 plasma for 5 minutes (peel strength: 8.8 N/cm). The copper content at the side of the copper foil of the former specimen is one order higher than that of the plasma treated one. In addition the contents of silicon and nitrogen composing the γ-APS coupling agent of the former specimen are higher. However, the contents of carbon and bromine in the resin substrate are low. It indicates that the specimen was fractured at near the interface between copper and the coupling agent. A significant amount of copper is detected at the side of the resin. This copper is assigned to Cu_2O, evaluated by the peak position in the XPS, indicating that the oxide layer formed by the heating was delaminated and left at the side of the resin. On the other hand, high carbon content and low copper content were observed at the side of the copper foil of the H_2 plasma treated specimen. It is indicated that the fracture is the cohesive failure at the resin. It is therefore concluded that the adhesive strength was increased using H_2 plasma by deoxidizing and removing oxide layer, which can form the weak boundary layer. In this method, the adhesive strength between copper and epoxy resin was approximately 9 N/cm. Although this value is slightly lower than that of the black oxide treated substrate (9 – 11 N/cm), it is practically sufficiently high.

2.2. The Deoxidizing Performance of Copper Oxide by H_2 Plasma

200 nm thick metallic copper was sputter deposited onto a silicon wafer, heated at 150 °C in dry air for 3 h, and then oxide film was deposited. This was used to study the deoxidizing performance of copper oxide [8.9]. The condition of the H_2 plasma was same as previously described one for the adhesion improvement.

Figure 3 shows the Cu (LMM) Auger electron spectra (AES) of the copper oxide films. The copper oxide observed was typical Cu_2O with the maximum peak at 336.8 eV. The peak at 336.8 eV disappeared after 2-minute treatment, while the peak at 334.9 eV corresponding to metallic copper became apparent as shown in Figure 3 (c).

The image of the transmission electron microscopy (TEM) at the copper oxide surface is shown in Figure 4 (a). 3 – 5 nm thick black layer is observed along the surface. This layer consists of Cu_2O (111) nano-crystalline (d = 0.247 nm) with different lattice orientation. This Cu_2O nano-crystalline layer disappeared after 2-minute H_2 plasma treatment as shown in Figure 4 (b). Furthermore, the Cu_2O nano-crystalline is rearranged to large Cu crystalline. Although the mechanism of this rearrangement is not clearly understood, it is suggested that the nano-crystalline surface became reactive by the removal of oxygen from the copper oxide due to de-oxidation, enhancing grain growth and large size copper grains were formed.

The de-oxidation would proceed in the following way; the hydrogen atoms dissociated in a plasma first deoxidize the copper oxide at the surface. As the de-oxidation proceeds, the hydrogen atoms diffuse the deoxidized copper layer, arrive at the interface between copper and the copper oxide, and then deoxidize the copper oxide. In such a way the interface gradually moves from the surface to inside, and finally all the copper oxide becomes metallic copper.

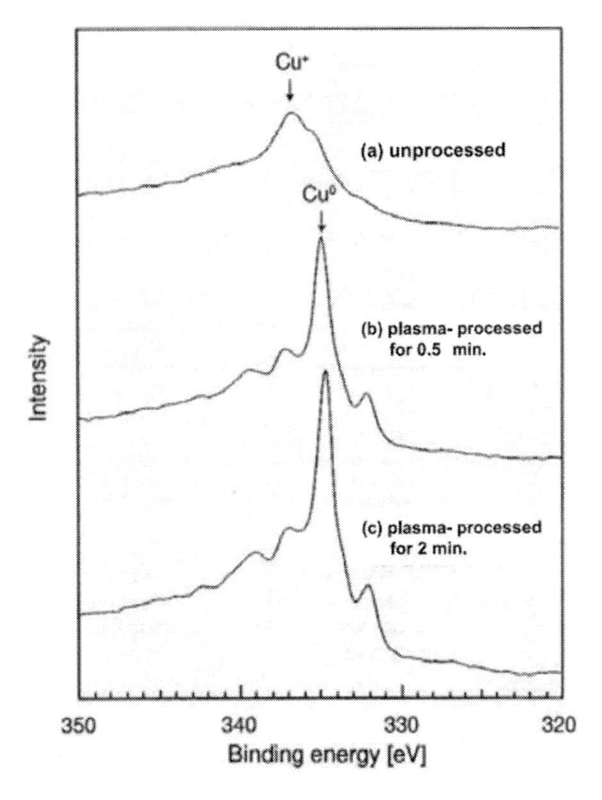

Figure 3. Cu(LMM) AES at the surface of the copper oxide layer. Power input: 200 W. He flow rate: 5L/min. H_2 concentration: 1 vol. %.

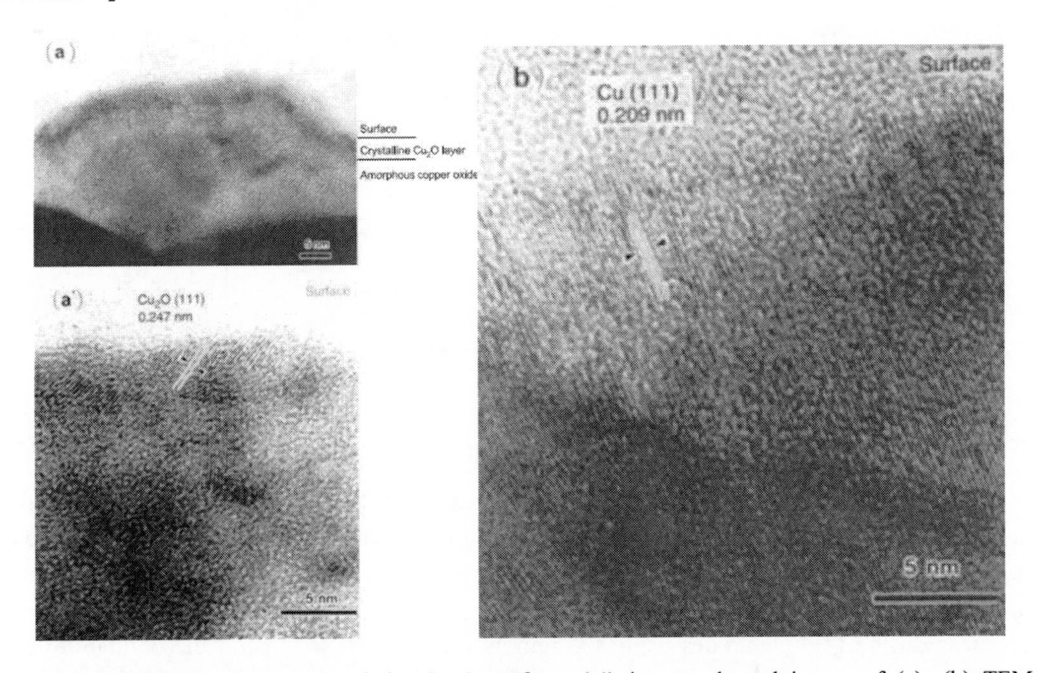

Figure 4. (a) TEM photograph of the Cu_2O surface. (a') is an enlarged image of (a). (b) TEM photograph of the copper oxide surface after H_2 atmospheric pressure plasma treatment for 2 minutes. Power input: 200 W. He flow rate: 5 L/min. H_2 concentration: 1 vol. %.

3. Application of the Remote Type Atmospheric Pressure Plasma

3.1. Cleaning Electrode Terminals of a Liquid Crystal Panel

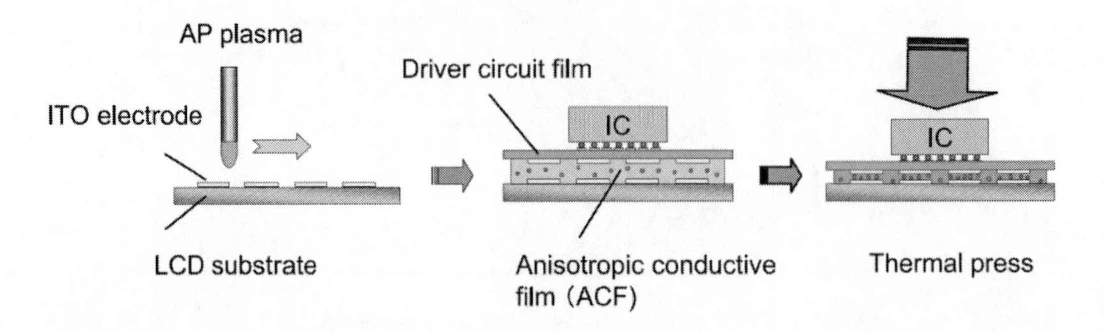

Figure 5. Interconnection of liquid crystal display (LCD) panel using anisotropic conductive film (ACF) between the electrode and the circuit film. The LCD drive circuits are anisotropically interconnected with the peripheral terminal electrodes using (ACF). At the interconnection section of these modules, the adhesion strength of ACF is substantially improved by the atmospheric pressure plasma treatment on the terminal electrodes in advance [10].

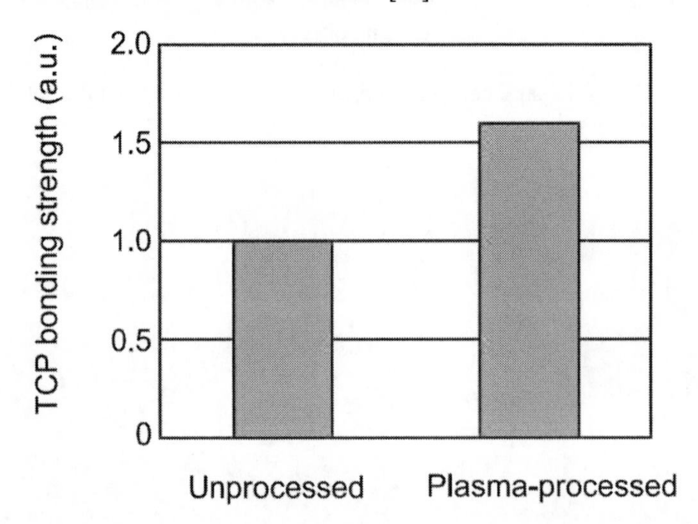

Figure 6. TCP bonding strength of liquid crystal display panel and polyimide film (unprocessed: 1). Power input: 100 W [10].

In liquid crystal panels and plasma display panels (PDPs) driving circuit is thermally pressed to the terminal indium tin oxide (ITO) transparent electrodes via the anisotropic conductive film (ACF) as shown in Figure 5. The ACF is manufactured by diffusing 3 – 15 μm diameter conductive micro-particles in a 15 – 35 μm thick insulating adhesive film. By pre-cleaning the ITO electrodes using the remote type plasma treatment, the bondability of the ACF can be drastically improved [10.11]. Figure 6 shows the bond strength of the ACF-bonded liquid crystal panel to the polyimide film. The bond strength can increase approximately 60 % by pre-cleaning the electrodes using the plasma at 100 mm/s before bonding. Here the bond strength for the untreated specimen is taken as 1. The reason of the adhesion improvement is the organic contamination attached at the surface is removed by the plasma. Furthermore the wraparound effect of the plasma is significant as the plasma is blown like a gas jet at atmospheric pressure. This enables the cleaning of the overlapping parts of the glasses. This method is applicable to the inline process of liquid crystalline panel mounting and is advantageous for high productivity.

3.2. Application for Parts of the Semiconductor Mounting

3.2.1. Improvement of the Bonding Reliability By Cleaning Of Electrodes at the Semiconductor Mounting Parts

In the semiconductor mounting process, the bond strength can be lowered due to the contamination of the electric parts deriving from various factors. The reliability of the bonding can be significantly improved by the plasma cleaning of the organic contaminants adsorbed at the bonding pads before IC bonding. Table 3 shows the wire bonding pull test results of Ball Grid Array (BGA) substrate golden pad before and after the plasma treatment [3,10]. The remote type treatment was performed at the driving frequency of 13.56 MHz, at the applied power of 700 W with the treatment speed of 30 mm/s. Argon gas mixed with approximately 1 % O_2 was used for the plasma gas. The value 30/1000 in the Table indicates that 30 in 1000 wires showed the failure-bonding mode and delaminated at the substrate surface during bonding. The result in Table 3 indicates that the plasma treatment reduced the fluctuation of the bond strength and the degree of the failure-bonding mode.

Table 3. Results of the wire-bonding test [10].

Substrate	Pull strength (g)	[a] C_{pk}	Number of bad detachment mode
Clean substrate	9.0	3.7	0/1000
Contaminated substrate	8.9	2.4	30/1000
Plasma-processed substrate	8.9	4.0	0/1000

[a] Process capability index calculated in lower standard limit of 5 g.

Figure 7 shows the depth profiling of C1s at the golden pad before and after the treatment. The substrate before the treatment shows the carbon content at the 60 nm region. After the treatment this organic substance was almost completely removed. It is considered to be the reason why the reliability of the wire bonding was improved.

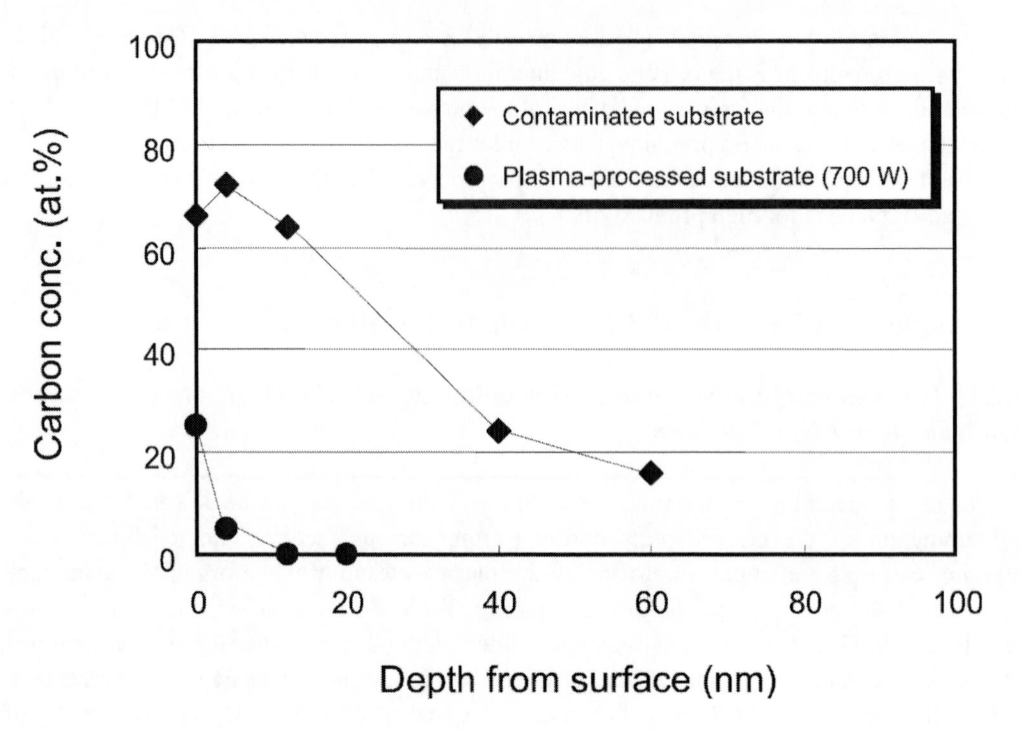

Figure 7. XPS C1s depth profile at the surface of an Au pad of the ball grid array (BGA). The depth from the surface is calculated by using the Au sputtering rate [10].

3.2.2. Evaluation of the Charge Up Damage of IC

The charge up damage is the degradation phenomena of the IC performances due to the damage of the oxide layer by the potential difference, accumulating charged particles in a plasma at the gate electrode of IC. The fluctuation of the performances and degradation of the reliability are concerned accompanied by reducing the size of semiconductor devices [12]. On the other hand, the test results of the change of the device properties after the remote type treatment are reported using the MOS-FET damage test device corresponding to 1G-DRAM [13]. More precisely, a device with 3 – 20 nm thick gate oxide film at the antenna ratio between 300 – 10000 was fabricated, and (1) threshold voltage (the gate voltage when the MOS-FET changes from off to on), (2) breakdown voltage of the gate oxide film, (3) the increase of the interface state density after the plasma treatment, and (4) the lifetime of the gate oxide film for the reliable performance (measurement of Q_{bd} lifetime) were evaluated. In each test, degradation of the oxide film was not observed even after 30 s irradiation of the remote plasma, which is 100 times longer treatment than that of practical plasma (approximately 0.3 s).

3.3. Application for the Pretreatment of Printed Circuit Board

The adhesive strength between a substrate and metallic coating for flexible printed substrates and buildup circuit boards used for mobile phones and IC packages can be improved by removing smear (residual of photo-resist and residual after laser processing) in a via-hole with using the remote type treatment as shown in Figure 8 [10,14]. As an example of removing laser smear Figure 9 shows the scanning electron microscopic (SEM) images before and after the treatment of the via-hole created by YAG and CO_2 gas laser at a polyimide film surface. It is seen that the smear can be almost completely removed by the treatment.

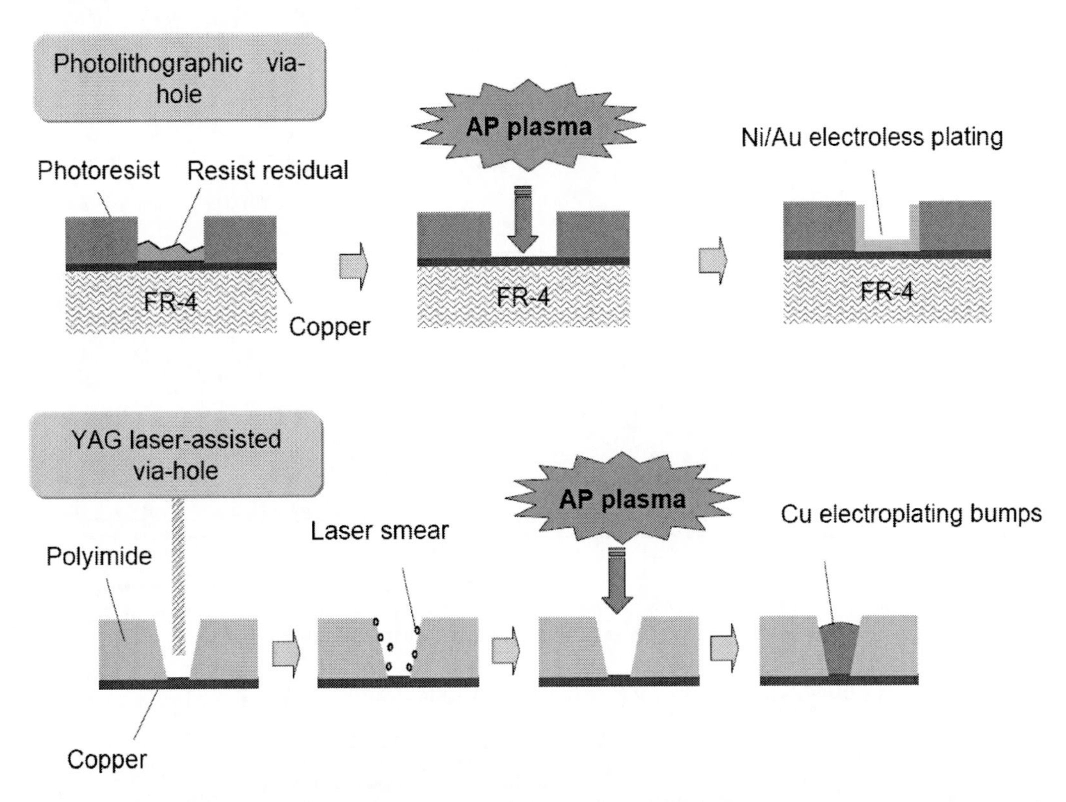

Figure 8. Application of the atmospheric pressure plasma to fabrication of printed circuit boards (PCBs). In the highly integrated PCBs, the combination of photolithographic or laser-assisted via-hole processing and the conductor build-up method has increasingly replaced the traditional drilling method [10].

Table 4 shows the deposit failure rate of electroless Ni/Au plating after atmospheric pressure plasma treatment of blind via-holes (BVH) created by photolithography at the photo-sensitive resist painted buildup print circuit board on a copper foil. As a comparison, the result for the wet desmear treated specimen using potassium manganate (VII) solvent is also shown. Reliability for coatings of the untreated substrates decreased as the via-hole diameter decreased. The coating failures of 0.8 % and 12.5 % were seen for ϕ100 μm and ϕ50 μm via-holes, respectively. The substrates after plasma treatment as well as wet desmear treatment showed 0 % failure. It is indicated that by applying the plasma treatment before coating, contamination can be removed which can not be completely removed by general water washing. Furthermore after the plasma treatment wettability was improved, the plating liquid can reach even small diameter BVH, and subsequently adhesive properties of plating would be improved. In addition the wet desmear treatment easily induces side etching, and the opening diameter of BVH increased by 15 %, while it was up to only 6 % by the plasma treatment.

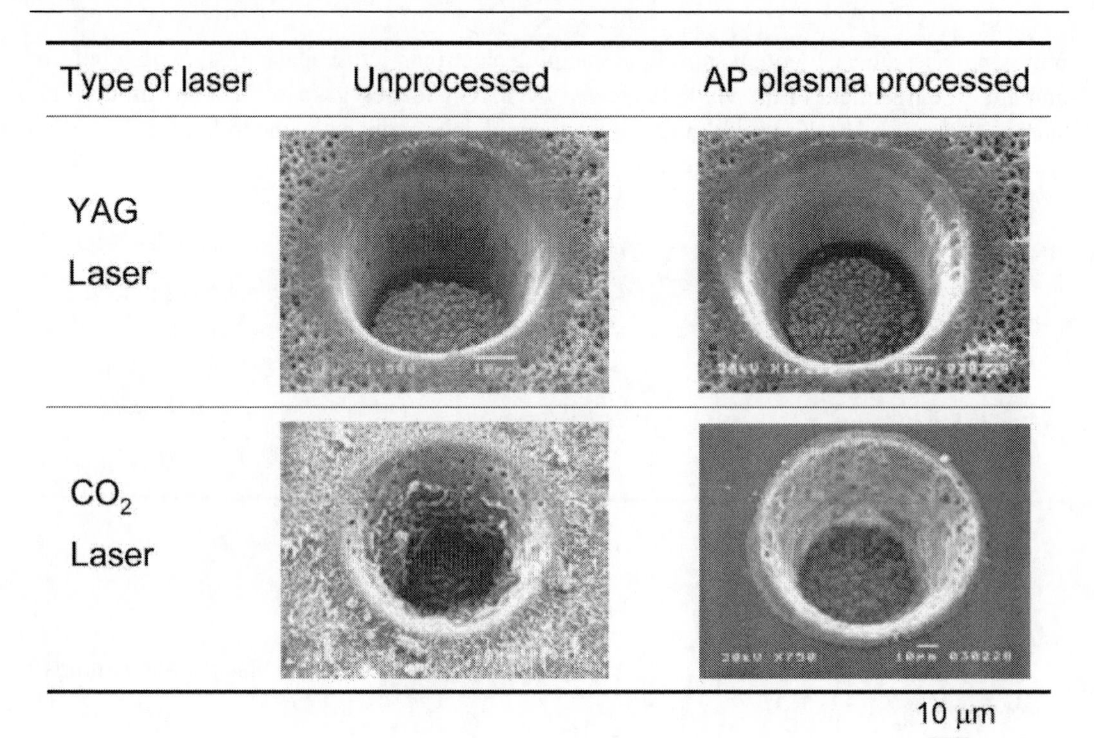

Figure 9. SEM photographs of laser-via holes before and after plasma treatment [14].

Table 4. Deposit failure rate of Ni/Au electroless plating [14]

Pre-treatment	Deposit failure rate (%)	
	50 μm	100 μm
Unprocessed	12.5	0.8
Wet desmear	0	0
AP plasma	0	0

3.4. Adhesion Improvement of Resin

The adhesive property and wettability of various resins can be improved by the remote type plasma treatment. It is reported that surfaces of poorly bondable poly-phenylene sulfide (PPS), which is one of general purpose engineering plastics, can be efficiently modified by the remote type plasma [15].

3.4.1. Evaluation of Adhesive Strength of PPS Resin

A couple of PPS strips were adhered using 1-liquid epoxy adhesive, and shear peeling strength suitable for JIS K 6850 were evaluated as shown in Figure 10. The untreated specimen was easily fractured at the interface, while the plasma treated specimens showed either cohesive failure or fracture at the bulk material. The atmospheric pressure plasma showed significant effect even with the power of 700 W at a speed of 30 mm/s. Comparing the 300-s treatment with low pressure Ar plasma which resulted in fracture of the main part, the adhesive strength can be improved with much shorter treatment.

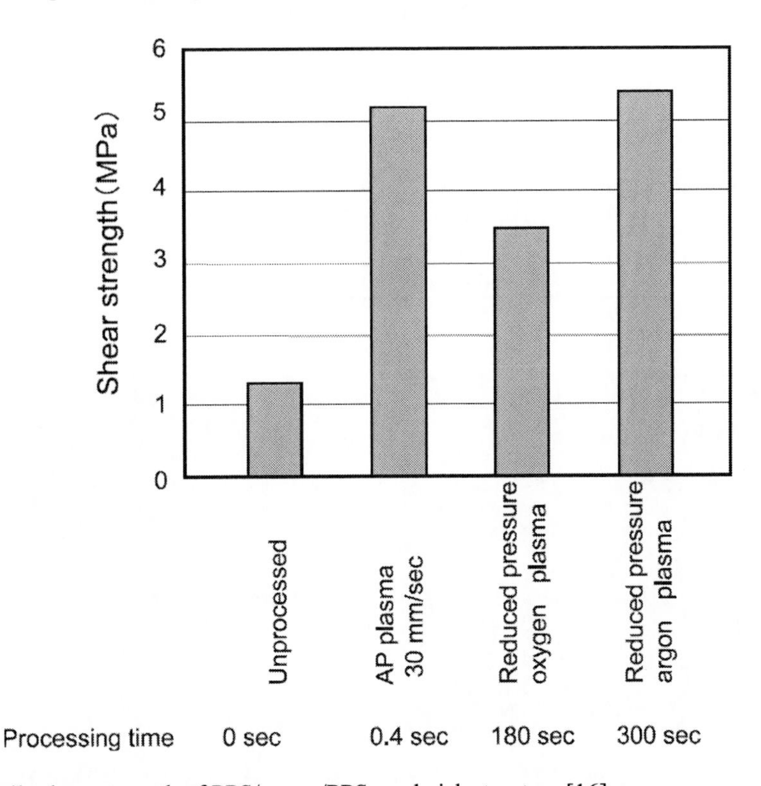

Figure 10. Tensile shear strength of PPS/epoxy/PPS sandwich structure [16].

3.4.2. Surface Analysis of PPS Resin

Table 5 shows the result of XPS analysis of PPS surface treated with atmospheric pressure plasma as well as low pressure plasma. It is seen that apart from the functional groups of –C-C-, -C-H, and –C-S- contained in the original PPS, hydrophilic functional groups of –COOH and –C=O were formed after atmospheric pressure plasma treatment. On the other hand, after low pressure plasma treatment, formation of –COOH was not observed, and the content of –C=O was less than that of atmospheric pressure plasma treated PPS. In order to form –COOH at the PPS surface, the bonding of benzene rings must be cut. As the benzene ring's binding energy is as high as 1514 kcal/mol (that of –C-S- is 71 kcal/mol), it turns out that the treatment efficiency of atmospheric pressure plasma is significantly high which can modify its bonding in an order of second.

XPS analysis indicates that content of hydrophilic functional groups of the PPS surface after atmospheric pressure plasma treatment is higher than that after low pressure plasma treatment. On the other hand, the adhesive strength of the one treated by atmospheric pressure plasma was comparable to that of ones treated for 300 seconds by low pressure plasma. Figure 11 shows the SEM images of the PPS surfaces. It is seen that micro-scale roughness increased after low pressure plasma treatment, resulting in increased surface area for the adhesion. It is suggested that the atmospheric pressure plasma mainly added hydrophilic functional groups while the low pressure plasma mainly modified surface morphology, and thus they would have different adhesion mechanisms.

Table 5. XPS C1s regional analysis of the PPS surfaces bfore and after the plasma treatment [15]

		Composition ratio [%]				
		-COO	-C=O	-C-O-	-C-S-	-C-C-, -C-H
Untreated		0.0	0.0	7.2	30.1	62.7
Atmospheric pressure plasma	100 mm/s	4.1	3.2	7.9	27.7	57.1
	30 mm/s	5.5	4.6	6.5	28.0	55.4
Low pressure plasma	O2 180 s	0.0	0.5	3.5	25.0	71.0
	Ar 300 s	0.0	1.0	2.9	30.0	66.1

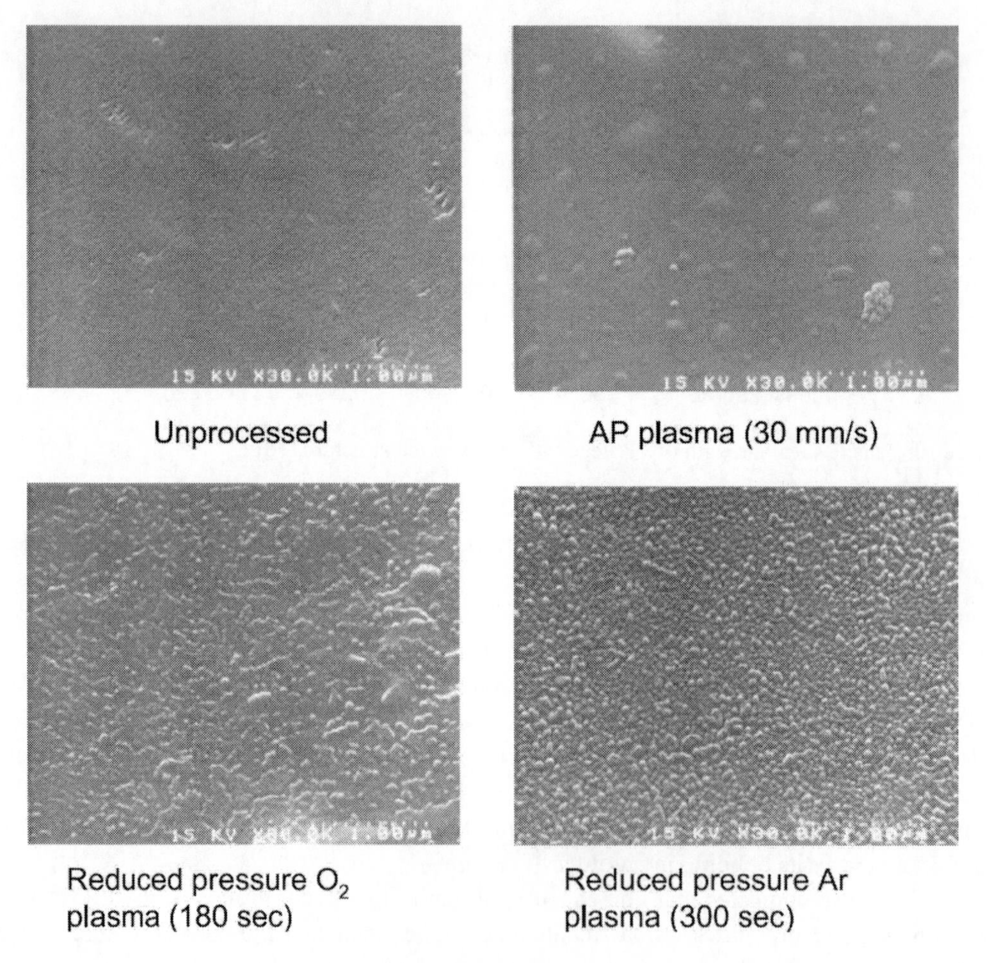

Figure 11. SEM photographs of the PPS surfaces before and after plasma treatment. Atmospheric pressure plasma, power input: 700 W, transport speed: 30 mm/s. Low (reduced) pressure O_2 plasma, treatment time: 180 s. Low pressure Ar plasma, treatment time: 300 s.

REMARKS

The atmospheric pressure plasma technique can add a variety of functionalities such as surface modification, cleaning, improvement of bonding and adhesion, and thus significantly wide range of applications. Due to the recent serious concerns about the environmental issues, the expectation for this technique is significant, and the research and development of this technique is spreading rapidly not only in Japan but also in the USA, Europe and Korea. The prototype system for the treatment of no less than 2-m wide specimens is already constructed. The future objectives of this technique are to reduce the gas consumption and to increase the treatment speed by increasing plasma density. For these purposes the basic research including plasma diagnostics is important. By achieving them, it is expected that the applications of this technique would further extend to a promising method for surface modification of various materials as well as electronic parts and liquid crystal panels.

REFERENCES

[1] In *Polymer surface modification and polymer coating by dry process technologies;* Iwamori, S.;Ed.; Research Signpost: Kerala, INDEA, 2005; Ch. 2.

[2] Tachibana, K.; Sugimoto; Sawada, Y. *Abstract 42nd Jpn. Soc. Appl. Phys. Annual Meeting*. 1995, 33.

[3] Sawada, Y.; Yamazaki, K.; Inoue, Y.; Kogoma, M. In *Proc. 8th Micro Electronics Symp*. 1998, 213-216.

[4] Catalogue of Matsushita Electronic Machine and Vision Ltd. *Atmospheric pressure plasma cleaning system.*

[5] Frazer, L. *Environ Health Perspect*. 1999, *107(8)*, A414-A415.

[6] Kinloch, A.J. *Adhesion and adhesives*; Chapman & Hall: London, UK, 1987; pp 144.

[7] Sawada, Y.; Ogawa, S.; Kogoma, M. *J. Adhesion* 1995, *53*, 173-182.

[8] Sawada, Y.; Tamaru, H.; Kogoma, M.; Kawase, M.; Hashimoto, K. *J. Phys. D* 1996, *29*, 2539-2544.

[9] Sawada, Y.; Taguchi, N.; Tachibana, K. *Jpn. J. Appl. Phys*. 1999, *38*, 6506-6511.

[10] Sawada, Y. *J. Plasma Fusion Res*. 2003, *79(10)*, 1022-1028.

[11] Sawada, Y. *Denshi-gijutsu in Japanese*. 2001, *7*, 56-57.

[12] Fukumoto, Y.; Sumie, S. *Kobelco Technol. Rev*. 2002, *52*, 2.

[13] Takakura, N.; Keno, T.; Yasuda, S.; Sawada, Y.; Inoue, Y.; Taniguchi, K. *Inst. Electr. Inform. Commun. Eng*. 2001, *ED2001-36*, 37-43.

[14] Sawada, Y.; Yamazaki, K.; Taguchi, N.; Shibata, T. *J. Microelectronics Packaging* 2005, *2*, 189-196.

[15] Sawada, Y.; Souma, M. *Inst. Electrostatics Jpn*. 2002, *26(6)*, 258-259.

In: Generation and Applications of Atmospheric Pressure Plasmas ISBN: 978-1-61209-717-6
Editors: M. Kogoma, M. Kusano and Y. Kusano

Chapter 4

SURFACE CLEANING OF A LARGE AREA SUBSTRATE FOR LIQUID CRYSTAL DISPLAY BY ATMOSPHERIC PRESSURE PLASMA

Motokazu Yuasa
Sekisui Chemical CO., LTD, Minami-ku, Kyoto-shi, Kyoto, Japan

INTRODUCTION

Miniaturization for electric devices of semiconductors and liquid crystal displays (LCDs) is demanded increasingly, and ultra-precise cleaning becomes necessary. In addition to conventional wet cleaning by acid and alkali, dry cleaning is recently industrialized using low pressure plasmas and short wave-length ultraviolet energy. In general, the dry cleaning is inferior to the wet cleaning in terms of a large amount of removal of contamination, but enables high-quality cleaning of organic contaminations up to atomic-level at less than μm fine structure, where infiltration of chemical liquid is difficult. Furthermore, the wet cleaning requires treatment of used liquid. It is one of the reasons why the dry cleaning attracts interests, which has small influence on environment.

Optical cleaning process, which is one of the dry cleaning processes, is advantageous in good infiltration to microscopic parts and no damage at the surface. This is therefore established as an indispensable technique in most of the cleaning process such as pre-cleaning before wet techniques and deposition for thin-film transistor (TFT) liquid crystal (LC) glass substrates. As the size of the LC substrate increases, improvement of cleaning performance and low temperature cleaning are strongly required, and subsequently lamps emitting the wavelength from 254-nm low pressure Hg lamps to 172-nm excimer lamps vacuum ultraviolet have been used [1-3].

However, in order to sustain the performance of these lamps, exchanging lamps within 2000 – 3000 h and overhaul of windows for the lamps are necessary. Their running costs are high. In addition the process is not environmentally friendly, since each exchange produces industrial wastes. Furthermore, as the size of the LC glass substrates increases, the optical cleaning process has hardly met the requirement of high speed processing by increasing the transportation speed of substrates while keeping the size of apparatus as small as possible.

Plasma cleaning is the process supplying ions, electrons and radicals generated in a discharge to a substrate, and cleaning the surface by physical and chemical reactions. This method is generally faster than the optical cleaning mentioned above, and is already applied for manufacturing process of multilayered print substrates [4]. However, generation of a plasma is generally in vacuum, in-line processing which is achieved for the optical cleaning is difficult, and thus the plasma cleaning is industrially applied only for small substrates practically.

LCD process whose substrate size is rapidly increasing requires high speed processing with simple setup. In order to fulfil such a need, a stable plasma at atmospheric pressure is successfully generated by controlling a pulsed electric field which is independent of the gas atmosphere, and subsequently developed manufacturing equipment for CVD, etching, and ashing [5]. In this chapter, cleaning technique and equipment of atmospheric pressure plasma without using vacuum is introduced.

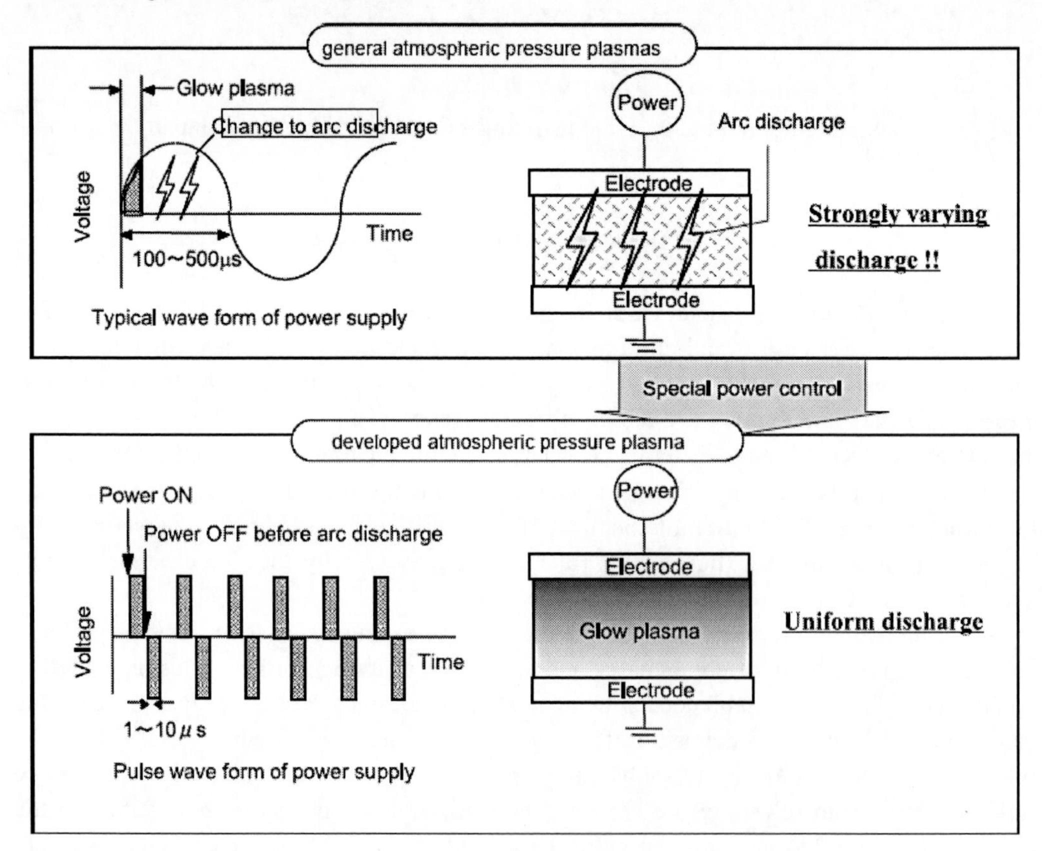

Figure 1. Principle of generating atmospheric pressure plasma by pulsed excitation.

1. Principle of Generation of Atmospheric Pressure Plasma

Generally plasmas immediately transfer from glow to arc discharges at atmospheric pressure. Therefore, it is considered that a stable glow discharge can be sustained by forcefully turning the electric field on and off repeatedly before transferring to an arc discharge as shown in Figure 1. However, lifetimes of meta-stable states of general gases excluding He are quite short. Therefore,

[1] pulsed generation of an electric field of high speed on/off depending on the lifetimes, and
[2] a power source which can supply high voltage (> several kV) and high capacitance for ionizing gas molecules at atmospheric pressure

are necessary. In fact they are substantially difficult. In this respect, we have noticed the powered electrode surface of solid dielectrics, and found an electrode material with which charge can be more easily stored and applied voltage can be suppressed. By arranging this material for facial electrodes, reducing the load for the pulsed power source and stabilizing the discharge plasma with the dielectrics which has higher charging efficiency, atmospheric pressure plasma was successfully generated without using He.

2. Cleaning Technique for LC Glass Substrate

The atmospheric pressure plasma processing for cleaning substrate which was developed here is the processing without using a vacuum or producing waste. In addition, different from conventional atmospheric pressure plasma processing [6-10] it does not need inert gases such as He and Ar. Furthermore it is the important characteristic that the process mainly uses inexpensive N_2 gas [11].

In order to apply atmospheric pressure plasma to the LCD manufacturing process,

[1] Avoiding damage of wiring and device by a plasma, and
[2] Uniform treatment by continuously transporting a large substrate

are necessary. In order to meet the requirement of low damage and continuous process which is free from the substrate thickness while shows excellent up-scalability in the direction of width, a remote plasma source is successfully developed which generates a plasma in a space far away from substrates, and then irradiating radicals on to substrate surface.

Figure 2 (a) shows an image of remote plasma source for G6 attached to transportation system with a belt conveyer. It can be easily attached to an existing substrate transportation system.

The characteristic of the developed system is removal of organic contamination and improving hydrophilicity at the substrate surface by extension of atmospheric pressure plasma outside of the electrodes along with high speed gas flow, which is generated between narrow gapped parallel plate electrodes. By introducing a gas mixture of N_2 and a trace of O_2, reactive species can contact the substrate surface. This enables cleaning of glass substrates, surface modification, and ashing as shown in Figure 2 (b).

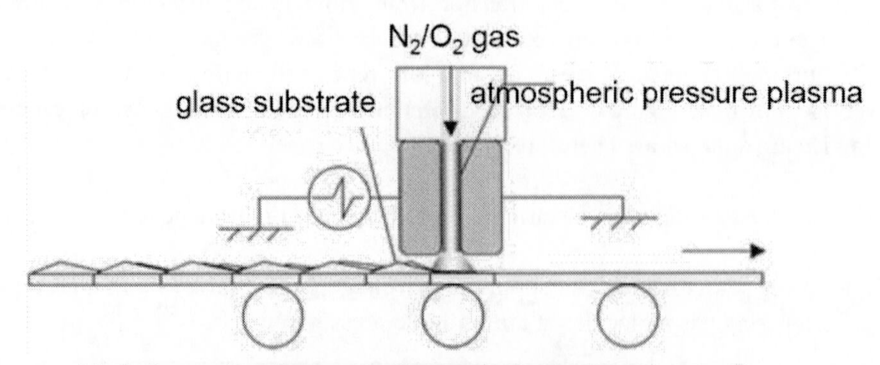

(b) Schematic diagram of remote type cleaning unit

(a) cleaning unit and transportation system for G6 substrate

Figure 2. Atmospheric pressure plasma cleaning unit.

A G6 size (1850 x 1500 mm) bare glass plate for LCD was treated using the system (0.02 % O_2/ 99.98 % N_2 atmospheric pressure plasma, 4 kW). The surface contact angle after the treatment was no more than 5 ° even when substrate transportation speed was no less than 5 m/min. This result indicates that cleaning is sufficient even at high speed treatment.

An ITO film is deposited on the glass surface for LCD as a transparent electrode. The purpose of our cleaning system in the LCD manufacturing process is to remove organic contamination at the surface and to improve hydrophilicity in order to enhance bondability at the next process. In order to evaluate the change of the chemical bonding states at the ITO glass surface after plasma treatment, x-ray photoelectron spectroscopic (XPS) analysis was performed.

Figure 3 shows C1s regional spectra of the ITO glass before and after the plasma treatment. It is seen that C-C, which is obvious in the spectrum before treatment, decreased significantly after the treatment, and is associated with the organic contamination. It is indicated that the organic contamination was removed after plasma treatment. At the same time, hydrophilic functional group such as C-O and O-C=O increased after the treatment, and thus oxidation of the organic contamination layer was found to be promoted. In conclusion, it is understood that cleaning treatment such as removal of organic contamination and improvement of hydrophilicity was achieved by the treatment. Since the change of the static charging was no more than 0.01 kV, it is clear that the glass substrate was not electrically charged up by the treatment.

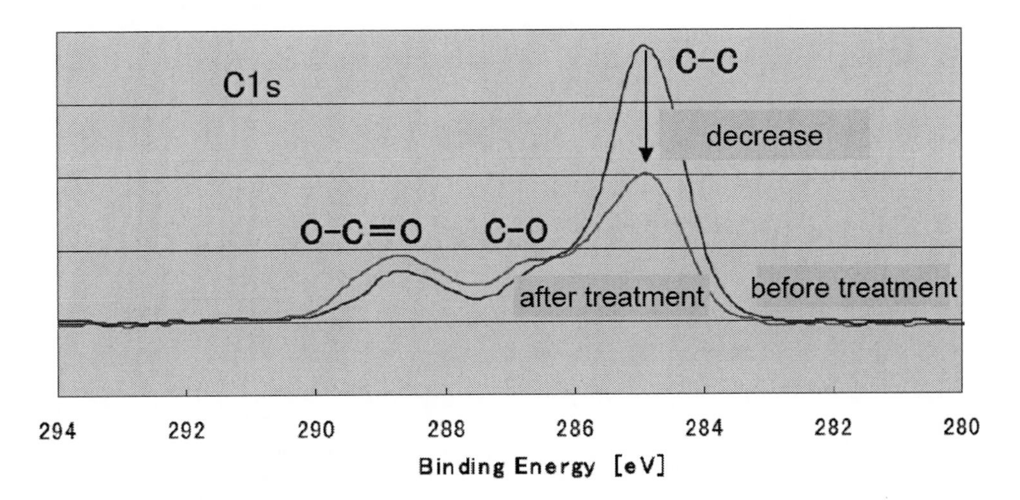

Figure 3. Regional C1s spectra of the ITO coated glass surfaces before and after the plasma treatment (G6 size substrate, transportation speed 5 m/min, gas content 0.02 % O_2/99.98 % N_2. 4 kW)

3. Investigation of Cleaning Mechanism

Afterglow, which is extended outside the electrodes, plays the major role in the processing. Optical emission of the afterglow was analyzed. Figure 4 shows a schematic diagram of the measurement system. A probe of the optical emission system was located 5 mm away from the outlet of the plasma. The optical emission spectrometry was carried out for the plasma in a gas mixture of O_2 0.02 % / N_2 99.98 % in a range between 200 and 900 nm. It is noted that the ratio of the gas mixture is the most effective for improving surface hydrophilicity of ITO glass used for LCD.

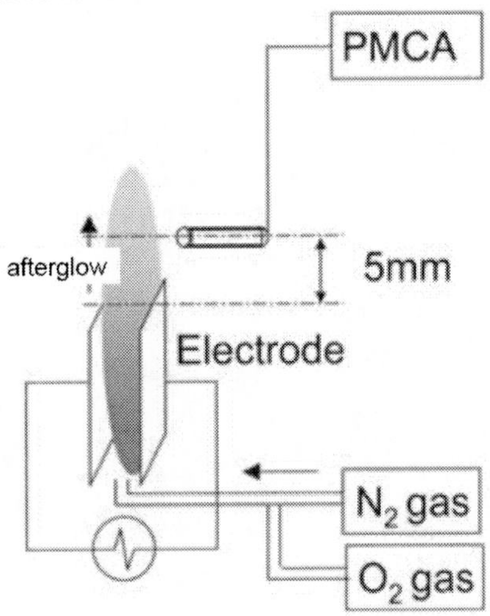

Figure 4. Schematic diagram of the optical emission analysis of atmospheric pressure plasma.

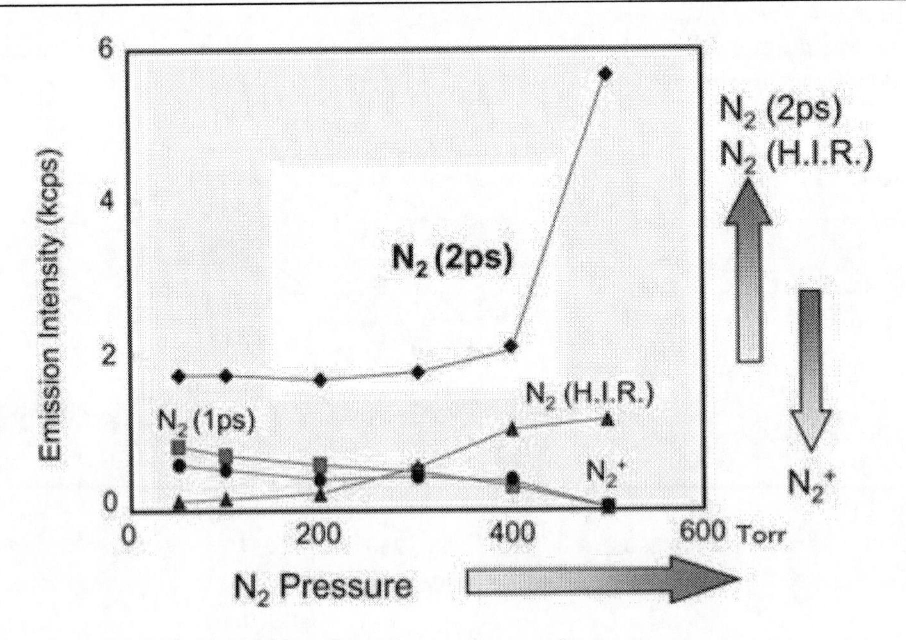

Figure 5. Optical emission intensities of reactive species at various gas pressures.

Figure 5 shows emission intensities of excited species in the afterglow at various gas pressures. As the pressure increases, the optical emission intensities of N_2 (1st P.S.) and N_2^+ decrease. They were too weak to be detected at gas pressures no less than 500 Torr. It is indicated that N_2 (1st P.S.) dominates in the emission spectrum as the gas pressure approaches to atmospheric pressure [12]. In addition, the results indicate that in the case of the afterglow treatment of atmospheric pressure plasma, surface treatment can be carried out at low temperature with no ion bombardment and no damage on to the substrate.

Optical emission from NO- γ, which is the characteristics when small amount of O_2 is added, was also measured. This emission appears in the UV range between 200 – 300 nm. It was found that the emission intensity depends on the amount of O_2 added in the gas mixture.

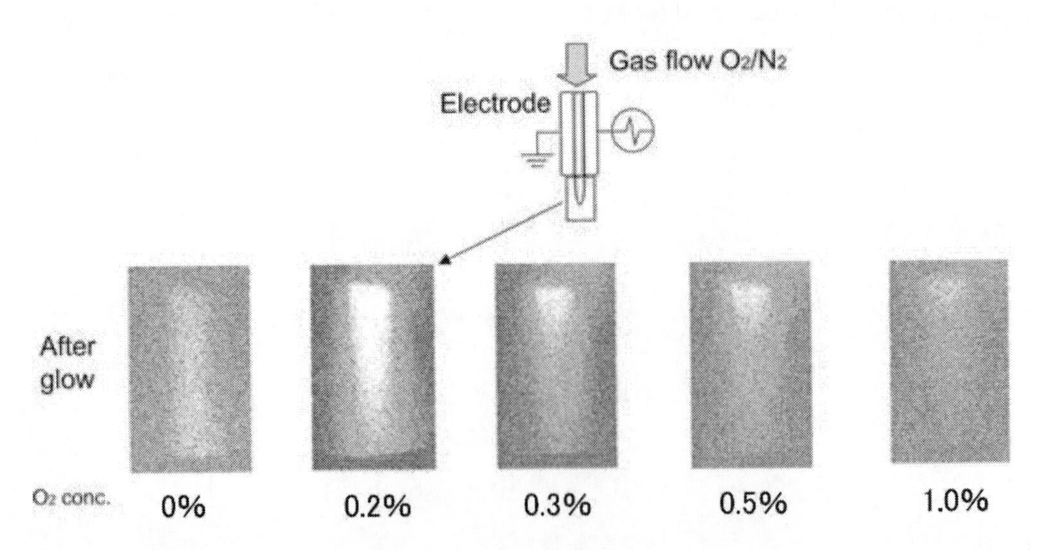

Figure 6. Images of optical emission of NO-γ at the afterglow as various O_2 content (Measurement performed at the Nagoya University).

Figure 6 shows the images of the optical emission of the afterglow through a filter whose wavelength is centred at 240.89 nm at various O_2 contents in the gas mixture. The UV emission intensity was found out to be the strongest when the O_2 content was 0.02 %. In this condition the extension length of the afterglow was the longest. As it was mentioned previously, the best surface cleaning performance can be demonstrated at this gas content. It is suggested that the NO-γ emission, which shows similar dependency on the O_2 content, would play an important role in the cleaning mechanism.

The following reaction model is considered as shown in Figure 7, based on the results mentioned above;

[1] The afterglow of the N_2 plasma mixed with 0.02 % volume fraction of O_2 extended down to just above the substrate surface. High energy neutral excited species accompanying N_2 (2nd P.S.) and NO-γ directly interact to the organic contamination at the substrate surface and cut its atomic bonding. The binding energies of C-C and C-H, associated with the organic components, are 83.1 and 98.9 kcal/mol, respectively. On the other hand, energies of the excited species of N_2 (2nd P.S.) and NO-γ, estimated from the wavelengths of the maximum optical emission intensity, are 335 nm (85.3 kcal/mol) and 254 nm (112.5 kcal/mol), respectively. These energies are thought to be enough for cutting the atomic bonding.

[2] The broad optical emission associated with NO-γ corresponds very well to the UV absorption band of ozone (Hartley band) generated in a discharge [13]. The reaction due to this absorption generates reactive atomic oxygen, which is known as;
$O_3 + h\nu$ (245 nm) → $O^* + O_2$.

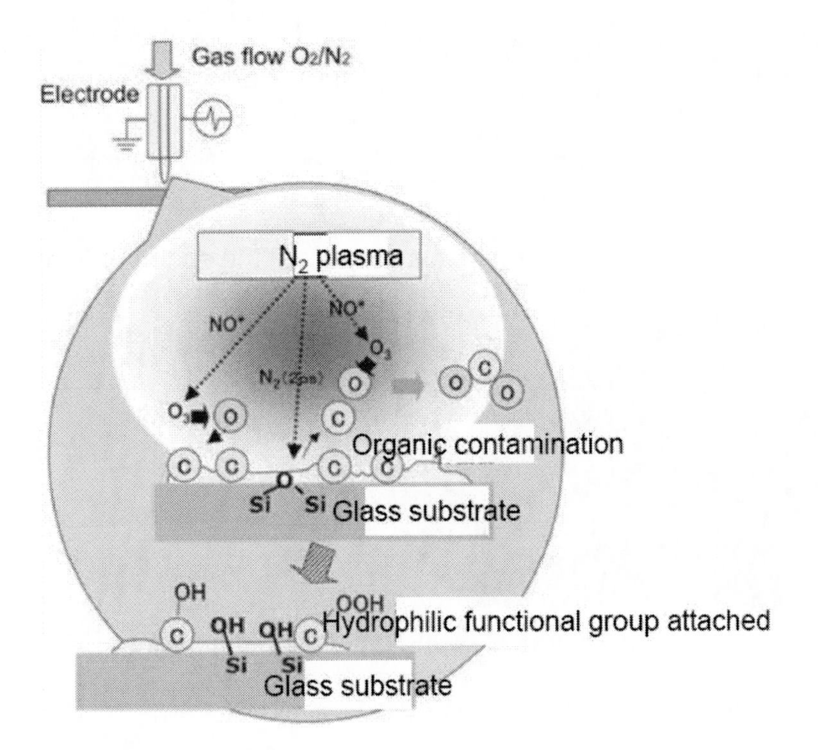

Figure 7. Reaction model.

Here, 245 nm is the central wavelength of the Hartley band. Since the Hartley band shows a broad spectrum in the range between 200 and 300 nm, the optical emission in this range induces this reaction.

The carbon atom generated by the bond secession described in the model (1) can react with O* in this reaction, generating CO_2, and thus carbon element can be removed. In addition, even if strong bonding such as C=C (145.1 kcal/mol) and C=O (173.0 kcal/mol) remains at the surface, it can become hydrophilic by its oxidation with O*.

4. Cleaning System

The developed atmospheric pressure plasma cleaning system was successfully able to generate a stable plasma by pulsed operation at atmospheric pressure without using vacuum and special gases, and enabled up-scaling up to the treatment of 2400 mm wide. Now the plasma cleaning units for the substrates between G2 and G8 are lined up. Figure 8 (a) shows an image of the plasma cleaning unit for the G8 glass substrates. The size of this unit is approximately one third of that of an excimer lamp system with the similar level of the capacity. It is seen that uniform treatment over 2200 mm wide of the LCD substrates such as bare glass and ITO glass is possible, as shown in Figure 8 (b).

The lifetime of the plasma electrode is more than several times longer than that of the excimer lamp, and thus approximately 50 % reduction of the annual running cost can be achieved.

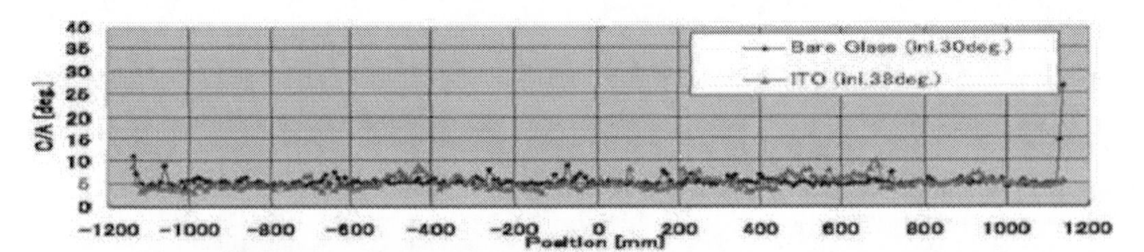

(b) Contact angle after plasma treatment at the speed of 5 m/min.

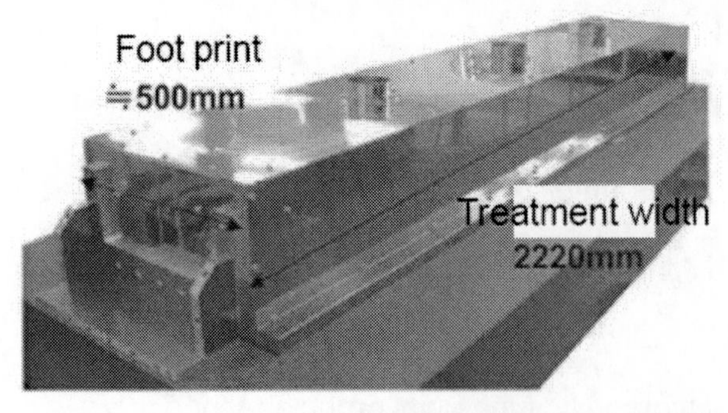

(a) Plasma cleaning unit for G8 substrate

Figure 8. Plasma cleaning unit for G8 substrates, and its treatment uniformity.

SUMMARY

The dry cleaning technique using atmospheric pressure N_2 plasma afterglow is being practically applied in various areas of the flat panel display (FPD) industry as the method of low substrate damage, high speed and low cost processing. The plasma diagnostics indicated that

1) The optical emission from the afterglow is dominated by the emission from N_2(2nd P.S.) and NO-γ. On the other hand, emissions from ions and excited atomic species were not observed.
2) When the gas mixture of N_2 plus 0.02 % O_2 was fed to the plasma, the plasma accompanying UV photons in the range between 200 and 300 nm derived from NO-γ extends outside the electrodes.
3) After the plasma treatment, C-C bond decreased and hydrophilic components increased.

The mechanism of the atmospheric pressure plasma cleaning can thus be explained in terms of secession of carbon containing components at the surface contamination layer, and the hydrophilic reaction by the reactive oxygen generated by the ozone/UV reaction.

Utilizing this cleaning technique, it is already applied for the ashing process of photo-resist residue. At the same time, development of chemical vapour deposition and etching using atmospheric pressure plasmas for the applications to the electronic material manufacturing process is also undergoing.

ACKNOWLEDGEMENT

Professor Fujimura and Dr Ryouma Hayakawa at the School of Engineering, Osaka Prefecture University, and Professor Hori at the School of Engineering, Nagoya University are highly acknowledged for their plasma diagnostics.

REFERENCES

[1] Honma, K., *Eco Industry* in Japanese 2000, *15(8)*, 45-54.
[2] Hishinuma, N.; Yoshioka, M. *Jpn. J. Optics (KOUGAKU)* in Japanese. 2001, *30(12)*, 790-794.
[3] Sugahara, H.; Takemoto, F. *J. Surf. Finishing Soc. Jpn. (Hyoumen-gijutsu)* in Japanese. 2002, *53(8)*, 497-501.
[4] Haji, H.; Arita, K. *Shinkuu* in Japanese. 2000, *43(6)*, 647-653.
[5] Yuasa, M. *Nikkei Micro-devices.* in Japanese. 2001, *4,* 139-146.
[6] Omi, T. *Ultra Clean Technol.* 1999, *11, Supplement 1,* 1-10.
[7] Okazaki, S.; Kogoma, M. *Proc. Inst. Electro. Statics Jpn.* in Japanese. 1991, *15(3),* 222-229.
[8] Uchiyama, H.; *et al. Converting.* in Japanese. 1994, *11,* 26-28.

[9] Sawada, Y.; *et al. Proc. Microelectronics Symp.* (*MES'98)* in Japanese. 1998, *12,* 213-216.

[10] Mori, Y.; Yoshii, K.; Yasutake, K.; Kakiuchi, H.; Kiyama, S.; Tarui, H.; Doumoto, Y. *J. Jpn. Soc. Precision Eng*. in Japanese. 1999, *65(11),* 1600-1604.

[11] Yuasa, M. *Monthly Display.* in Japanese. 2002, *8(11)* 29-35.

[12] Hayakawa, R.; *et al. Abstract 50th Jpn. Soc. Appl. Phys. Annual. Meeting.* 2003, *vol 2* 612.

[13] Griggs, M. *J. Chem Phys,* 1968, *49(2)* 857-859.

In: Generation and Applications of Atmospheric Pressure Plasmas ISBN: 978-1-61209-717-6
Editors: M. Kogoma, M. Kusano and Y. Kusano

Chapter 5

WETTABILITY CONTROL AND ADHESION IMPROVEMENT OF POLYMERS

Shigeki Ito
AirWater Inc., Matsumoto, Nagano, Japan

INTRODUCTION

Various products which we daily use are manufactured by processing materials, regardless of general-purpose or engineering. The materials are roughly categorized into four as shown in Figure 1 [1].

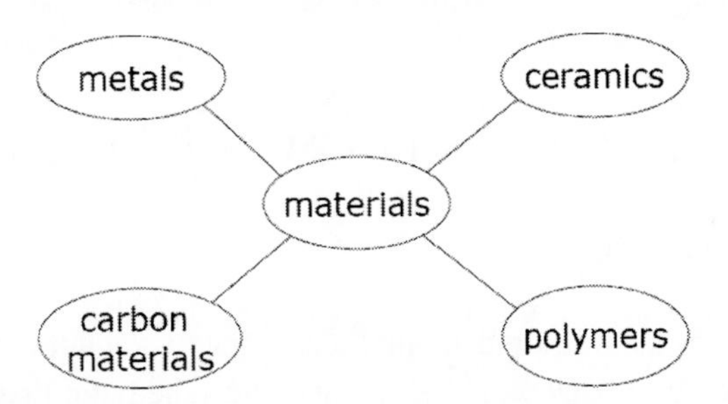

Figure 1. Classification of materials.

Among them, polymers have been long used as natural polymeric materials such as woods and textiles. On the other hand, synthetic polymers have been used for less than 100 years. However, polymers are now used for a plenty of products due to the rapid process in petro-chemistry.

As polymers are now used in many applications in such a way, surface treatment becomes more important than ever, which can add functionalities only at the surfaces without changing bulk properties of materials. Figure 2 summarizes surface treatment methods for polymeric materials [2].

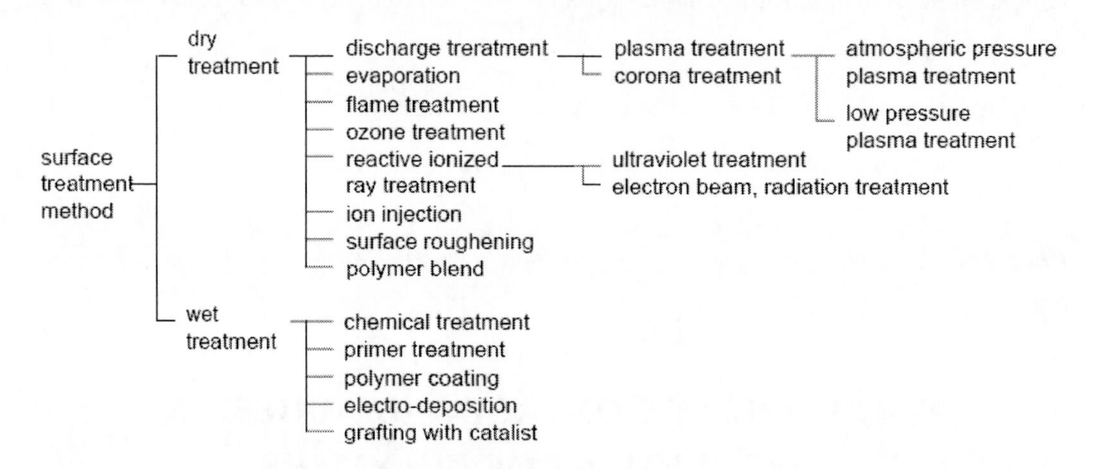

Figure 2. Surface treatment methods for polymers.

The surface treatment methods can be divided into dry and wet treatments. During wet treatment which is represented by chemical treatment, liquid chemical always contacts the polymer, and there are concerns of swelling to the polymer, dissolution, and elution of additives from the polymer. In some cases, the wet treatment might also influence bulk properties of the polymer. For this reason, the dry treatment is preferable to the wet treatment for polymers. Among dry treatment, plasma treatment which is one of the discharge treatments least affects the bulk properties of polymers. It was conventionally believed that plasma treatment could be carried out only at low pressure. However, Okazaki and Kogoma in Sophia University developed the first atmospheric pressure plasma treatment [3]. As atmospheric pressure plasma treatment can be easily applied for continuous treatment, it is industrially very advantageous treatment while keeping the advantages of plasma treatment.

1. Atmospheric Pressure Plasma Treatment Technique

1.1. Fundamental Technique

As was already explained, the fundamental technique of the atmospheric pressure plasma treatment was developed by Okazaki and Kogoma. The schematic diagram of the equipment used for this development is shown in Figure 3. Parallel plate electrodes are placed in the reactor, and a dielectric material is attached on the facing side of each electrode so that generation of an arc discharge is avoided. The gap between the electrodes must be filled with the gas for the generation of atmospheric pressure plasma. At the initial stage of the development, helium was used as the gas for the plasma generation. Subsequent development of the technique was carried out by various research institutes including the author's company as the engineering gas manufacturer, and it enabled the use of inexpensive gases such as Ar and N_2. After replacing the atmosphere gas between the electrodes with the gas for the generation of plasma, atmospheric pressure plasma is generated with time varying (for example alternating current) high voltage of over several kV applied between them, and a specimen placed between them can be treated by atmospheric pressure plasma.

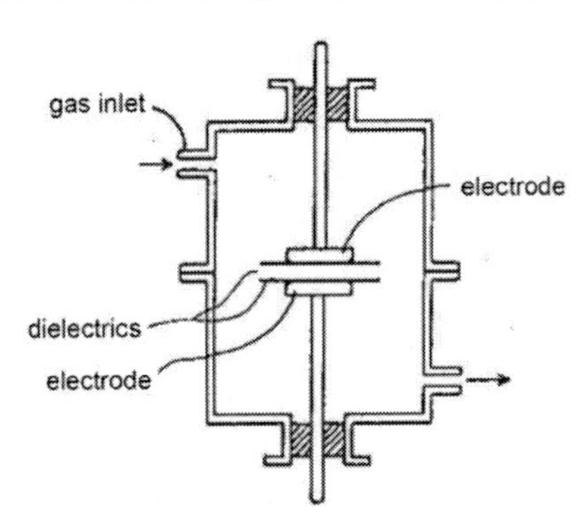

Figure 3. Schematic diagram of the atmospheric pressure plasma treatment setup.

It is possible to add various functionalities at the material surface using atmospheric pressure plasma. In order to add the functionalities, appropriate gas is mixed in the gas for generation of atmospheric pressure plasma. For example, if atmospheric pressure plasma treatment is performed in the process gas containing O_2, the surface often becomes hydrophilic. Furthermore, by simply replacing a component of gas (Instead of O_2, gases which will be discussed later can be used), hydrophobicity which is completely opposite of hydrophilicity, can be realized at the surface using the same equipment. The technique for the generation of atmospheric pressure plasma requires efforts for the choice of the material and shape of the electrodes, and the way of applying high voltage. However, in the case of surface treatment using atmospheric pressure plasma, the choice of the gas component is the most important.

1.2. Engineering Technique

E.C. CHEMICAL Co., LTD, which is one of the affiliated companies of ours, has worked on surface coatings on business and was much interested in the atmospheric pressure plasma treatment mentioned in section 1.1. E.C. CHEMICAL Co., LTD signed the mediation contract with the Research Development Corporation of Japan (whose current name is the Japan Science and Technology Agency) which owns the right of the license. E.C. CHEMICAL Co., LTD then started development of atmospheric pressure plasma treatment for industrial applications immediately.

In order to use atmospheric pressure plasma treatment industrially, treatment failure is the inevitable issue to tackle with. That is to say, it is necessary to treat the whole surface uniformly without fail. Considering this issue, neither three dimensionally shaped objects nor one-dimensional linear objects, but two dimensional objects such as sheets and films are most suitable as those to be treated by atmospheric pressure plasma treatment. Based on this idea, the development for industrial applications was carried out. Currently, the objects of atmospheric pressure plasma treatment are mostly two dimensionally shaped. Figure 4 shows the schematic diagram of the atmospheric pressure plasma setup for the continuous treatment of films. The film as the specimen is rolled out, passes through the slit into the reactor, and passes between the electrodes in the reactor. The atmospheric pressure plasma is generated between the electrodes where the gas for generating atmospheric pressure plasma is introduced and high voltage is applied as was also the case of the fundamental technique mentioned before. Therefore the film is treated when it passes the region between the electrodes. The treated film passes through another slit out of the reactor, and is rolled in.

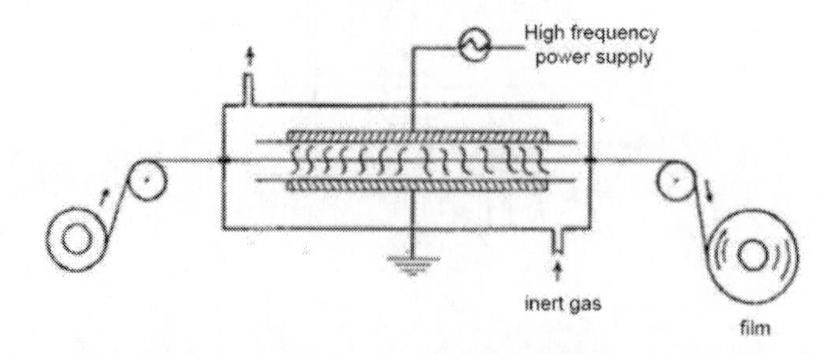

Figure 4. Schematic diagram of the atmospheric pressure continuous plasma treatment system for films.

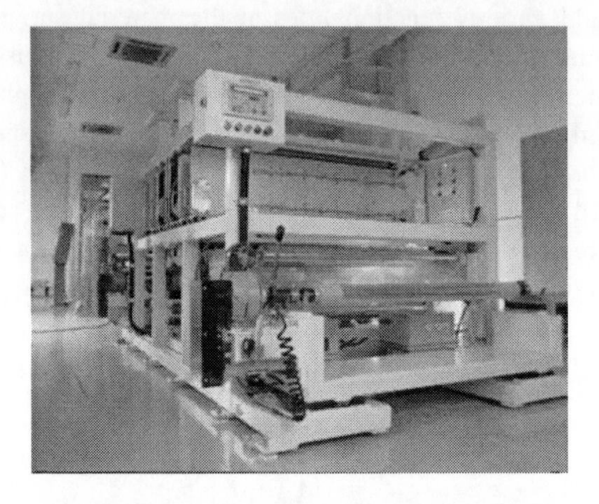

Figure 5. Atmospheric pressure continuous plasma treatment system for films at E.C. CHEMICAL Co., LTD.

Figure 6. The plant of the atmospheric pressure plasma treatment on consignment at E.C. CHEMICAL Co., LTD.

Figure 5 shows an example of the atmospheric pressure plasma setup for the continuous treatment of films. E.C. CHEMICAL Co., LTD has a plant (Figure 6) with class 1000 cleanliness environment for surface treatment business on consignment using atmospheric pressure plasma. This surface treatment on consignment is capable of treating maximum 2200 mm wide film as shown in Table 1.

Table 1. Specification of atmospheric pressure plasma treatment on consignment at E.C. CHEMICAL Co., LTD

Film width	300 – 2200 mm
Thickness	8 - 125 μm
Package diameter	> 500 mm
Mass of the film	> 500 kg
Core diameter	3 and 6 inch

2. ATMOSPHERIC PRESSURE PLASMA TREATMENT OF POLYMERS

2.1. Objectives of Atmospheric Pressure Plasma Treatment

Properties of polymer surfaces can be physically and chemically changed from the bulk properties by atmospheric pressure plasma treatment. However the material with treated surface itself seldom becomes the final product, utilizing the modified surface properties only. Instead, it is often the case that the material with treated surface becomes a practical product after using interaction with other materials such as adhesives and paints, in particular liquids. When the interaction with liquid is considered, wettability of a material surface is very important. As was explained before, atmospheric pressure plasma can control the wettability by tuning the gas content used for the plasma treatment. In the following sections, examples of atmospheric pressure plasma treatment are introduced; section 2.2 for hydrophilic treatment as the wettability control, and sections 2.3 and 2.4 for adhesion improvement which has a close relation to the wettability.

2.2. Hydrophobic Treatment

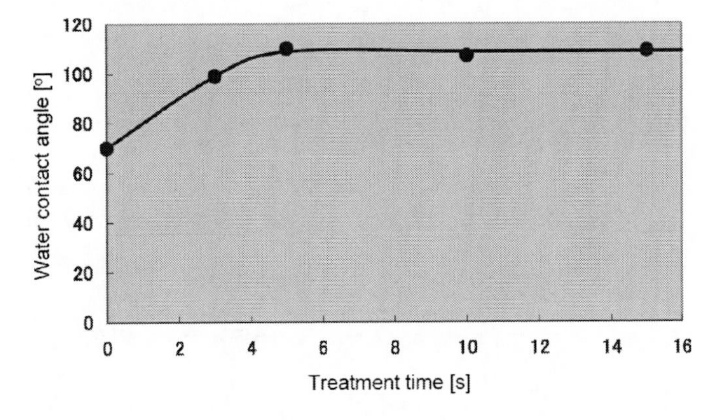

Figure 7. Water contact angle of polyimide surface after hydrophobic treatment in a gas mixture containing CF_4.

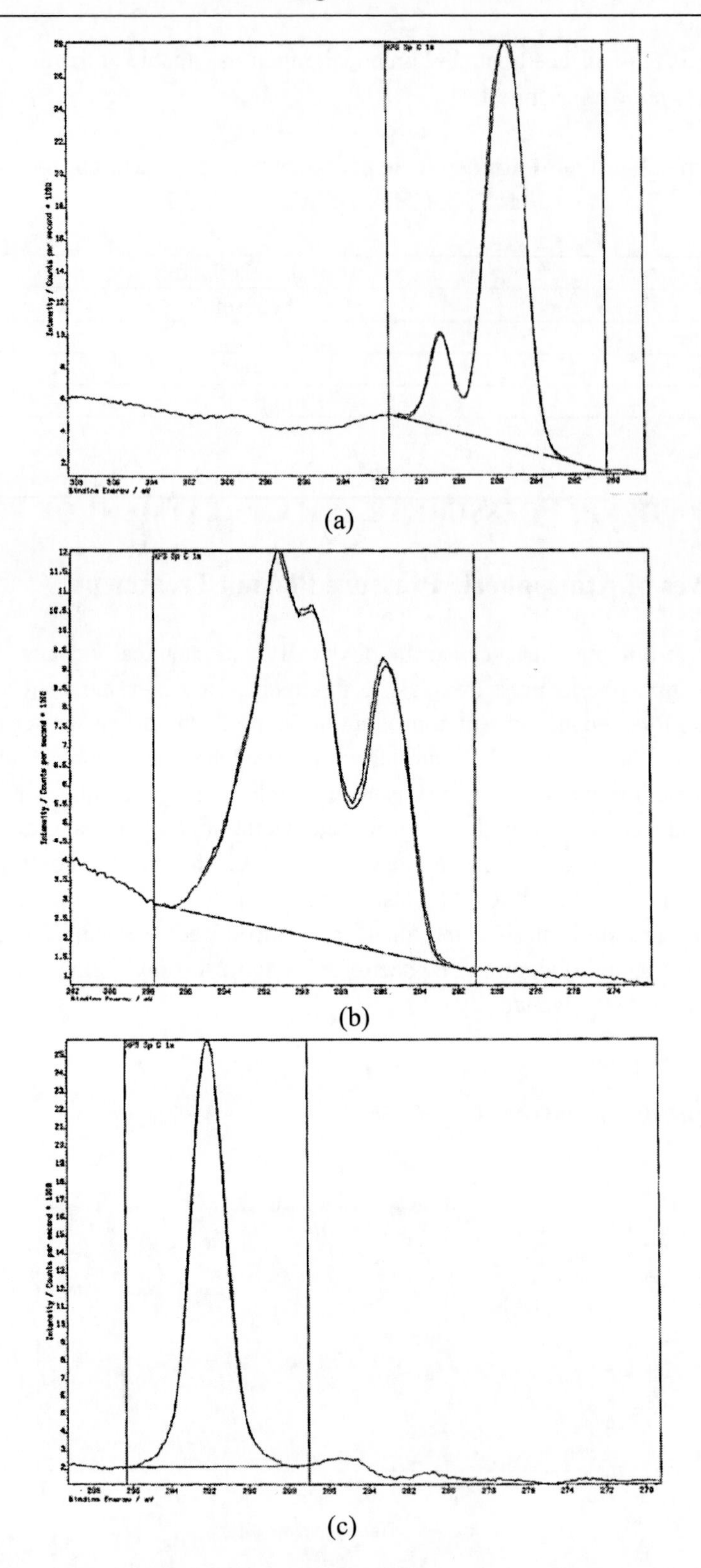

Figure 8. Regional C1s spectra in XPS measurement; untreated polyimide (a), polyimide after hydrophobic treatment (b), and PTFE (c).

When the fundamental technique of atmospheric pressure plasma treatment was developed, one of the aims was to achieve hydrophobicity by atmospheric pressure plasma which was already demonstrated by low pressure plasma treatment [3].

It is the most common for atmospheric pressure plasma hydrophobic treatment to use a gas mixture containing a small amount of tetra-fluoro-methane (CF_4) which is used in low pressure plasma treatment. In this case, reactive fluorine radicals are generated in a plasma by dissociating CF_4. The exposure of the polymer surface in this plasma results in immediate fluorination, showing hydrophobic surface.

Figure 7 shows an example of water contact angle of polyimide surface treated by this method for adding hydrophobicity. The result depends on the gas content and plasma energy. In this case, after 5-s treatment the water contact angle reaches 110° which is comparable to the contact angle of a poly-tetra-fluoro-ethylene (PTFE) surface. It is indicated that sufficient hydrophobicity can be added to a polymer surface in a short time of the treatment. The surfaces of the untreated polyimide, the plasma treated polyimide and PTFE were analyzed using x-ray photoelectron spectroscopy (XPS). Regional C1s spectra of these specimens are shown in Figure 8. In the C1s spectrum of the untreated polyimide surface, the strong peak is observed at around 285 eV. After the plasma treatment, the peak at 285 eV is weak, while the peak at around 292 eV is observed, which is also seen in the spectrum of the PTFE. This indicates that the polyimide surface was highly fluorinated by the plasma treatment.

Polymer surface treated by plasma, not only by atmospheric pressure plasma, sometimes change their properties significantly as time passes. This tendency is more significant for hydrophilic treatment. In some cases, the properties recover almost to the original ones one day after the plasma treatment. However, in the case of this hydrophobic treatment, even 3 days after the treatment, the hydrophobicity is sustained as shown in Figure 9.

The example above explained the atmospheric pressure plasma treatment in CF_4 containing gas. Similar results can be obtained when the gas mixture containing saturated aliphatic fluorocarbons is used.

Furthermore, hydrophobic polymerized coating can be synthesized using atmospheric pressure plasma for adding hydrophobicity. For example, it is reported that PTFE coating is formed on the material surface by atmospheric pressure plasma treatment using a gas mixture containing tetra-fluoro-ethylene (C_2F_4) [4].

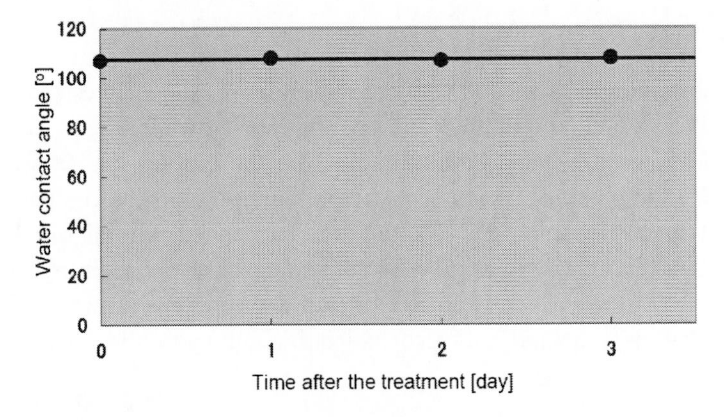

Figure 9. The sustainability of the water contact angle of polymer surfaces after hydrophobic treatment.

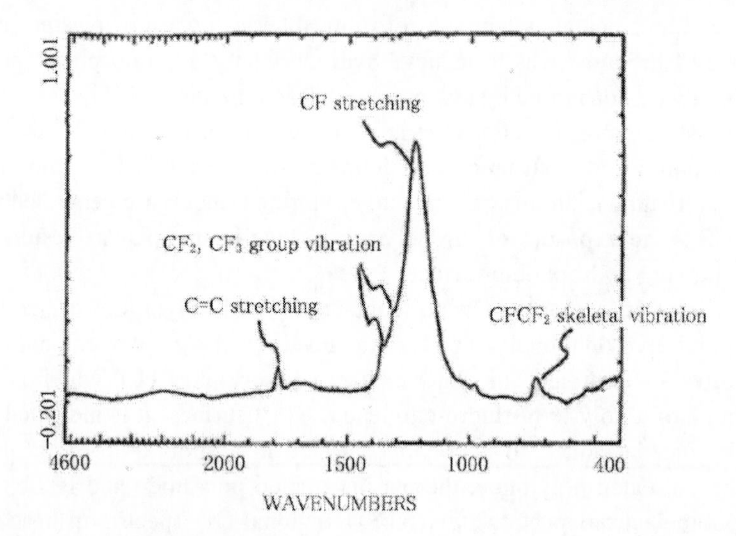

Figure 10. FTIR spectrum of PTFE coating.

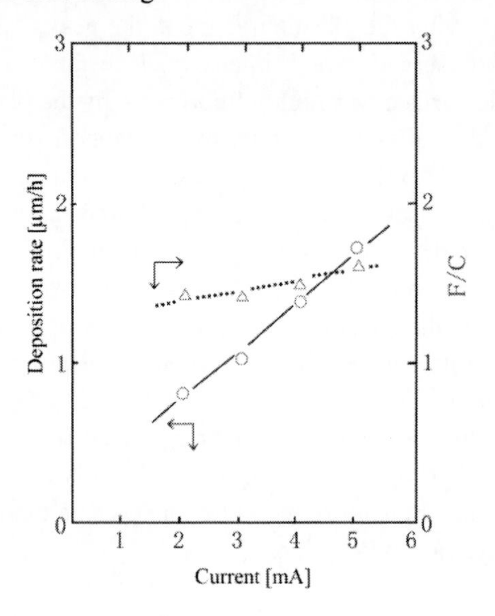

Figure 11. Deposition rate and F/C ratio of PTFE coating.

Figure 10 shows Fourier Transform Infrared spectrum of the synthesized PTFE coating. The deposition rate and F/C ratio of the PTFE coatings are shown in Figure 11. Generation of particles was not observed, which is often the case for the coating synthesis at atmospheric pressure. The F/C ratio obtained by atmospheric pressure plasma was comparable to that by low pressure plasmas. Hydrophobic coating was deposited on an acrylic plastic by atmospheric pressure plasma in a gas mixture containing hexa-fluoro-propylene (C_3F_6). The water contact angles of the surface at different treatment times are shown in Figure 12. This result indicates that this method can also add sufficient hydrophobicity at the polymer surfaces.

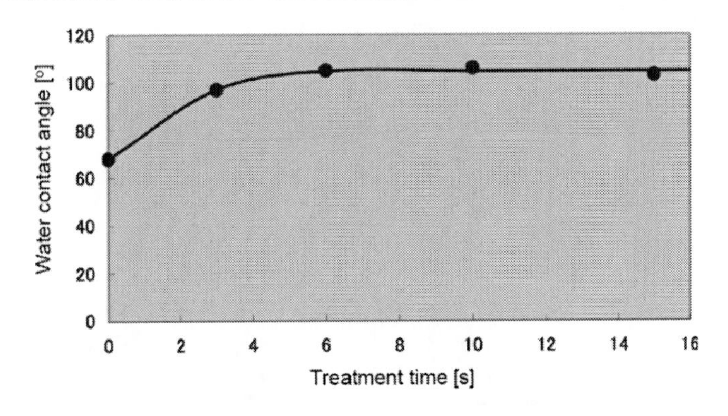

Figure 12. The water contact angle of hydrophobic polymerized coating.

Furthermore, atmospheric pressure plasma can make possible the method of synthesizing hydrocarbon-type polymerized coating and its subsequent fluorination [5], which is reported in the case of using low pressure plasmas. This method was applied to add hydrophobicity at polyimide surfaces. The water contact angle measured at each process is shown in Table 2, indicating that this method can also make the polyimide surfaces as hydrophobic as PTFE.

Table 2. Water contact angles of the polyimide surface at each step of the process.

Specimen	Water contact angle [°]
Untreated	70
After deposition of polymerized coating	86
After fluorination	109

2.3. Treatment for Copper Clad Laminate

In the area of electronic equipment, copper clad laminate used as the flexible print circuit consists of the laminate of copper with a polymer-base film, as the name indicates. There are two kinds of structures; the three layered structure with an adhesive layer between the copper and the base film, and the two layered structure without adhesive layer. The good adhesion of the base film with copper via the adhesive layer is needed in the three layer structure, and the direct adhesion with copper is required in the two layered structure. Here, it is introduced that atmospheric pressure plasma treatment can be a very efficient technique to satisfy these properties.

Figure 13 shows the adhesion property of the three layered structure of copper clad laminate. Polyimide was used as base film and thermosetting acrylic adhesive was used at the adhesive layer. The adhesive property was estimated by 90° peeling test. The specimen of the untreated polyimide fractured at the interface between the polyimide surface and the adhesive. On the other hand, as the atmospheric-pressure-plasma treated polyimide bonded tightly with the copper foil via the adhesive layer, it showed the fracture at the copper foil.

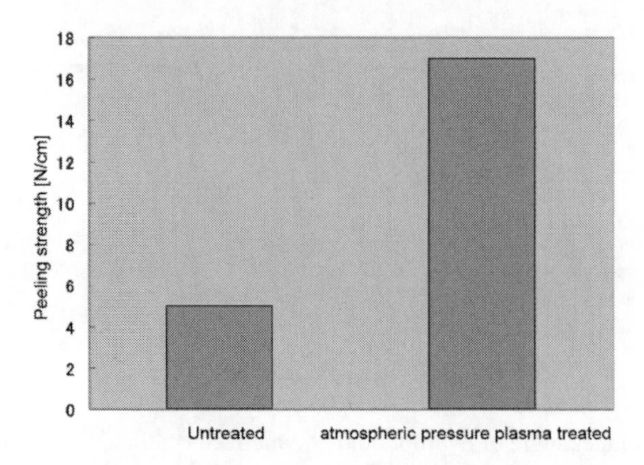

Figure 13. Adhesive property of the three layer copper clad laminate.

Figure 14 shows the surface properties of the treated and untreated polyimide surfaces. After the treatment, the polyimide surface became hydrophobic, and the water contact angle decreased. However, noticeable change was not seen the AFM images between the treated and untreated surfaces. It is therefore concluded that the demonstrated adhesion improvement by the atmospheric pressure plasma treatment is mainly due to the chemical modification linking to the atomic composition at the surface rather than the physical effect induced by the surface morphology linking to the anchor effect.

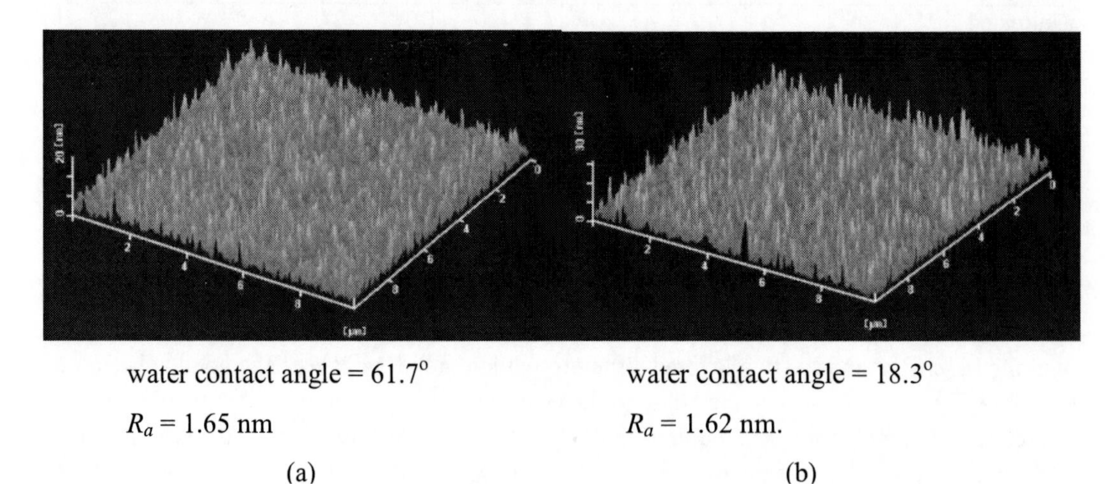

water contact angle = 61.7° water contact angle = 18.3°

R_a = 1.65 nm R_a = 1.62 nm.

(a) (b)

Figure 14. AFM images, water contact angles, and surface roughness (R_a) of treated (b) and untreated (a) polyimide surfaces for three layer copper clad laminates.

Electro-less plating of copper was applied on to the untreated and atmospheric-pressure-plasma treated polyimide specimens aiming for the two-layered-structure copper clad laminate. Table 3 compares the adhesive property of the Cu coating of these specimens. The adhesive property was evaluated by cross-cut test. The treated specimen showed no failure while the untreated specimen showed complete failure. Other metallic coatings were also tested and the results are listed in Table 3. It is confirmed that all the atmospheric-pressure-plasma treated specimens demonstrated excellent adhesive properties.

Table 3. Adhesion properties of the electro-less plated coatings

Coating material	Untreated	Atmospheric pressure plasma treated
copper	0/100	100/100
nikkel	0/100	100/100
silver	0/100	100/100

It is also reported that copper coating was evaporated onto the copper clad laminate, and that excellent adhesion was demonstrated between the polypropylene surface and the copper coating [6]. The adhesive property was evaluated by the critical load which is the minimum weight with which the polypropylene surface appeared at the scar during the load-changing-scratch test. The result is shown in Figure 15. It is found that the atmospheric pressure plasma treatment in N_2/He gas mixture is the most effective for the adhesion improvement. It is confirmed that XPS analysis detected the amino-group introduced at the polypropylene surface after this treatment. Figure 16 shows the ratios of N/C and NH_2/C at the polypropylene surfaces before and after the treatment. These ratios increased as the treatment time increased. Especially when the treatment time was short, it seems that the result in Figure 16 would be associated with that of Figure 15. As a result, it is considered that the reason of the tight adhesion between the polypropylene treated by atmospheric-pressure-plasma and the copper film is that the introduced amino group and copper would form the chelate bond. It is also reported that physical effect due to the change of the surface morphology would also affect the adhesion, as micro-scale needle like structure was formed on polypropylene surface after atmospheric pressure plasma treatment.

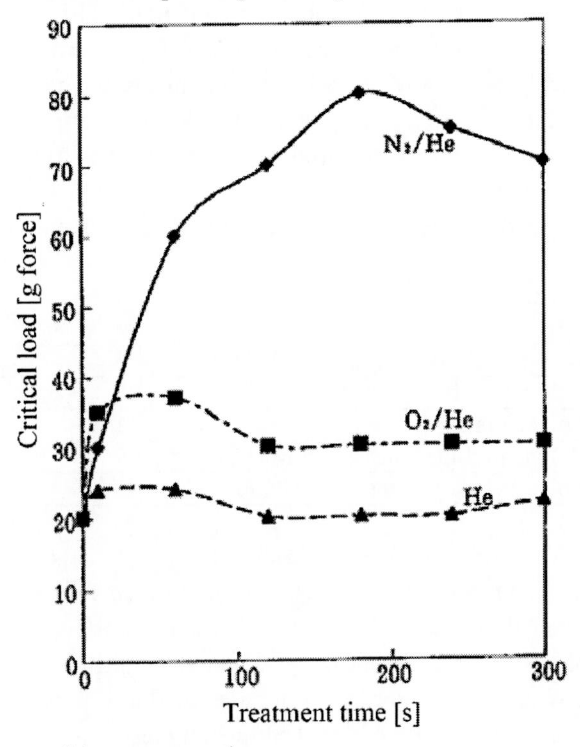

Figure 15. Scratch resistance of the copper coating.

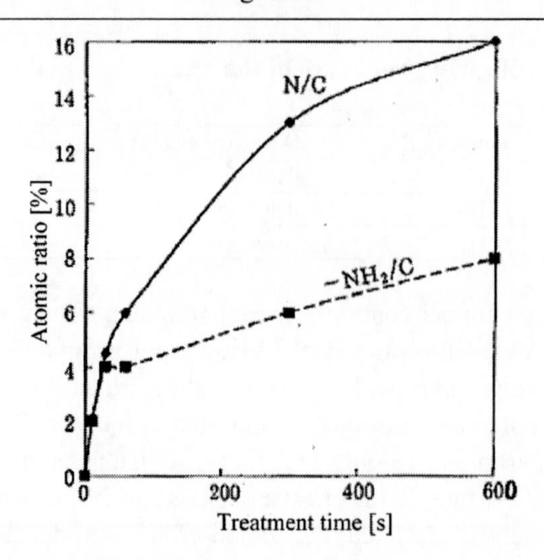

Figure 16. N/C and NH_2/C ratios of atmospheric-pressure-plasma treated polypropylene surfaces.

2.4. Atmospheric Pressure Plasma Treatment of Fluoro-Polymers

Fluoro-polymers represented by PTFE show excellent properties of thermal durability, chemical inertness and weather resistance. They have low surface tensions compared to other polymeric materials as shown in Table 4 [7]. They show high performances of non-adhesiveness, and stain resistance, and low coefficient of friction. Owing to these properties, fluoro-polymers are used as products in various applications. The use of the fluoro-polymers in laminated structures is very common, utilizing their surface properties efficiently. However, the property of the low surface tension makes the production of the laminate structures difficult. And thus atmospheric pressure plasma treatment is widely used for the surface treatment of fluoro-polymers for producing the laminate structures.

Table 4. Critical surface tensions of various polymers.

	Material	Critical surface tension [mN/m]
Fluoro-polymers	FTFE	18
	Poly-vinylidene- difluoride (PVDF)	25
	Poly-vinyl- fluoride (PVF)	28
Other polymers	Polyethylene (PE)	31
	Polymethyl-methacrylate (PMMA)	39
	Polyethylene-terephthalate (PET)	43

Table 5 shows adhesive properties of various fluoro-polymers to epoxy resin before and after the plasma treatment. 180° peeling test was performed for its evaluation. The peeling forces of the untreated fluoro-polymers were less than 0.2 N/cm, which are far from good adhesion. Atmospheric pressure plasma treatment improved the adhesion property significantly. Table 6 shows the adhesive properties of fluoro-polymers to various polymers using polyester type adhesive, evaluated by 180° peeling test.

Table 5. Adhesive properties of fluropolymer/epoxy/fluoropolymer sandwich structure (* the bulk fluoropolymer was fractured)

Material	Thickness [μm]	Peeling force [N/cm]	
		untreated	treated
PVDF	100	< 0.1	18*
Ethylene-tetra-fluoro-ethylene copolymer (ETFE)	60	0.2	9
Tetra-fluoro-ethylene-hexa-fluoro-propylene copolymer (FEP)	100	0.2	15*
Tetra-fluoro-ethylene-perfluoro-alkyl-vinyl-ether copolymer (PFA)	50	0.1	6
PTFE	100	< 0.1	4

Table 6. Adhesive properties of fluoro-polymer/polyester type adhesive/various polymers (*The bulk fluoro-polymer was fractured)

Fluoro-polymer	Thickness [μm]	Atmospheric pressure plasma treatment	Peeling force [N/cm]		
			Un-plasticized vinyl-chloride	Acryl	polycarbonate
PTFE	40	Untreated	0.2	0.2	0.2
		Treated	7.1	4.4	6.1
ETFE	40	Untreated	6.4	3.9	6.7
		Treated	8.3*	4.4	8.3*

The result also confirms that atmospheric pressure plasma treatment significantly improves adhesive properties of fluoro-polymers. Figure 17 shows the bonding states between PFA and silicone rubber. It is seen that atmospheric-pressure-plasma treated PFA adheres to silicone rubber at the whole area, and that the untreated PFA could not adhere uniformly and bubbles were generated between the PFA and the silicon rubber. The peeling test was performed for these specimens. The specimen with the untreated PFA fractured at the interface between PFA and the silicone rubber adhesive. On the other hand the specimen with the atmospheric pressure plasma treated PFA fractured the silicone rubber as shown in Figure 18. It is seen that silicone rubber attaches all over the colorless transparent PFA.

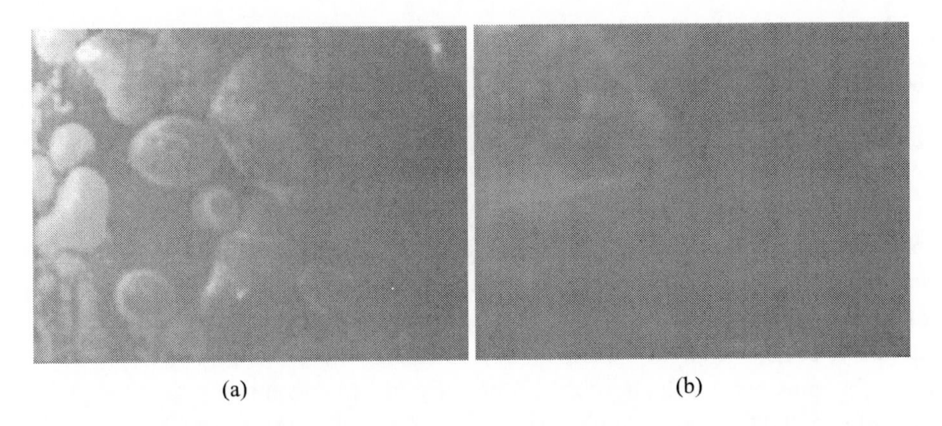

(a) (b)

Figure 17. Optical images of silicone rubber coated PFA. Untreated (a) and atmospheric-pressure-plasma treated (b).

Figure 18. Optical image of the silicone rubber/PFA structure after the peeling test.

Table 7 summarizes the wetting tension and atomic composition at the PFA surfaces by XPS. After the atmospheric pressure plasma treatment, the wet tension increased and the surface became hydrophilic. The analysis of the atomic composition at the surfaces indicates that after the atmospheric pressure plasma treatment, F atoms decreased compared with untreated PFA, while O atoms was detected which was not detected at the untreated PFA surface. However, it may not be concluded that the introduction of just approximately 5 % of O atom at the surface would improve the adhesive property drastically. Instead, it can be considered that this XPS result would contain substantial information of the bulk material. That is to say it is thought that atmospheric pressure plasma treatment would introduce high density O atoms just at the surface, and that subsequently the adhesive property would improve.

Table 7. Wettability and atomic composition of PFA surfaces

Specimen	Wetting tension [mN/m]	Atomic composition [%]		
		F	C	O
Untreated	< 22	69.4	30.6	0
Atmospheric pressure plasma treated	35 - 40	64.0	30.7	5.3

REMARKS

Atmospheric pressure plasma treatment was developed by Okazaki and Kogoma, and was for the first time industrially applied by E.C. CHEMICAL Co., LTD. It has mainly developed for the surface treatment of polymers. Recently, its applications are extended in other areas, such as cleaning of glass substrates for liquid crystal displays. However, development of polymers are progressed furthermore; optical films for high quality images in a display and base films for fabrication of flexible film in the display industry, and various polymers in other industry. Considering such a situation, it is expected that atmospheric pressure plasma treatment for polymers will be further progressed. My company together with E.C. CHEMICAL Co., LTD strongly thinks that we would contribute the progress.

REFERENCES

[1] Ogata, N. *Koubunshi Shinsozai Binran (Handbook of new polymeric materials)* in Japanese; Soc Polym Sci Jpn; Ed.; Maruzen: Tokyo, JP. 1989, pp 3.

[2] *Koubunshi Hyoumen No Kiso To Ouyou (Fundamentals and applications of polymer surfaces)* in Japanese; Matsuo, M.; Ikada, Y.; Ed.; Kagaku Doujin: Tokyo, JP. 1986, pp 31.

[3] Japanese patent No. 2138895.

[4] Okazaki, S.; Kogoma, M.; Yokoyama, T.; *Houden Kenkyuu (Discharge research)* in Japanese; 1990, *127*, 138-143.

[5] Takahashi, K.; Enishi, H.; Kogoma, M.; Moriwaki, T.; Okazaki, S.; *Nihon Kagaku Kaishi* in Japanese; 1985, *6(2)* 1916-1923.

[6] Kogoma, M.; *Ryuushisen Gijutsukaihatsu Teirei Kenkyuukai Shiryou Syuu* in Japanese, 1998, *5*, 1-5.

[7] *Nuregijutsu Handbook (Wetting Technology Handbook)* in Japanese; Ishii, T.; Koishi, S.; Kakuta, M.; Techno System, JP. 2001, pp 152.

In: Generation and Applications of Atmospheric Pressure Plasmas ISBN: 978-1-61209-717-6
Editors: M. Kogoma, M. Kusano and Y. Kusano

Chapter 6

IMPROVING WETTABILITY OF TEXTILE, PAPER AND TIMBER BY ATMOSPHERIC PRESSURE PLASMA TREATMENT

Tohru Uehara
Interdisciplinary Faculty of Science and Engineering, Shimane University,
Matsue-shi, Shimane, Japan

INTRODUCTION

Degradation of cellulosic materials such as paper, cotton, and timber is mainly governed by external factors such as water, heat, ultraviolet, bacteria, and termites. In order to avoid such degradation at the surfaces and extend lifetimes of the materials, these surfaces are protected. Surfaces of paper and cotton can usually be protected by coatings, while those of timbers by painting and overlays.

Plasma chemical vapour deposition (CVD) is recently used for synthesis of thin films for LSI etc. This technique can easily polymerize gas monomers in a plasma, and deposit coatings on substrates, enabling surface modification without changing the appearance of substrates and without using organic solvents. If the synthesized coatings are transparent without colouring, they can be hardly recognized with the naked eye.

It is known that polymerized ethylene films can be deposited by plasma treatment in ethylene gas. In fact ethylene is one of the general gas monomers [1].

Plasma CVD is usually employed at low pressure of approximately 100 Pa [2] in order to increase the efficiency of CVD due to long lifetimes of generated radicals and long mean free paths of activated species. However, low pressure CVD has a drawback requiring expensive vacuum system.

As atmospheric pressure plasma CVD easily leads to arc transition, and thus has been rarely studied. However, methods of avoiding arc transition are recently reported using developed novel electrodes, and treatment with stable glow discharges at atmospheric pressure have become possible [3].

Plasma CVDs are in most cases operated at the radio frequency ranging between several kHz and several hundreds kHz. As the frequency increases, the gas temperature tends to increase [4]. Therefore if organic materials are used lower frequency operations is preferable because they are degraded at high temperature.

In the present work, polymerized ethylene films are synthesized in ethylene gas by atmospheric pressure plasma treatment operated at the commercial frequency of 60 Hz, and are deposited on cellophane, paper, timbers and fibres. It is expected that hydrophobic surfaces are obtained because the material surface would be covered with synthesized polyethylene films.

1. POLYMERIZATION OF ETHYLENE AT CELLOPHANE SURFACE [5]

Since cellophane is a transparent film, optical spectroscopy and contact angle measurement are easily applicable. Therefore cellophane is chosen as a model specimen of paper, timbers, and textiles, and treatment conditions were investigated.

1.1. Specimens, Atmospheric Pressure Plasma Treatment, and Contact Angle Measurement

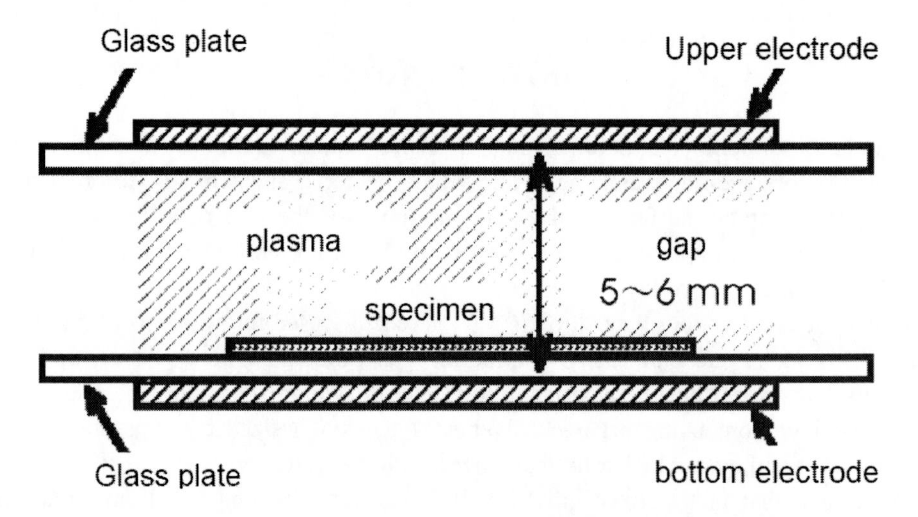

Figure 1. Structure of the electrodes and the position of the specimen.

Glycerol contained as the softener in cellophane (Nimurakagaku PF-3) was extracted away with distilled water. Figure 1 shows a parallel-plate-type discharge cell used for plasma treatment [5], which has a similar configuration to a conventional corona discharge. SUS304 plate electrodes (130 mm x 70 mm) were used with a gap ranging between 5.0 and 6.0 mm. In order to avoid the transition to an arc discharge each electrode was covered with a 1-mm-thick glass plate (170 mm x 110 mm). A specimen was placed on the bottom electrode and treated by a plasma. A commercial power source (100 V, 60 Hz) was used as a power supply. The Neon step-up-transformer was used to increase the voltage up to 10.5 ~ 15 kV. 99.9 % high purity ethylene (Nihonsanso) was used as a reactive gas with a flow rate of 4 L/min.

A contact angle measurement system (Elma Optics Ltd., Goniometer type contact angle measurement system G-III) was used to measure the contact angle of the specimen surfaces by sessile drop method at 20° 24 h after the plasma treatment. A contact angle of the surfaces was measured 10 seconds after putting a liquid-drop on the surface.

1.2. Surface Free Energy of Cellophane

Cellophane was chosen as a specimen for the plasma treatment since it is relatively stable against heat in plasma among organic polymer films. As cellophane swells water easily, distilled water is unsuitable for contact angle measurement. Instead glycerine, 1-bromonaphtalene and diiodomethane were used for the liquids of the contact angle measurement. The extended Fowkes equation [6] was used for the calculation of surface free energy.

In order to achieve high treatment efficiency in a short treatment time, plasma treatment should be performed at highest applicable voltage. However, the higher the voltage is, the easier spark transition the discharge shows as the treatment time passes. Therefore preliminary experiment was carried out to find a maximum sustainable voltage for 60-minute treatment. It was found that 12.75 kV is the maximum for the experimental setup used in this work.

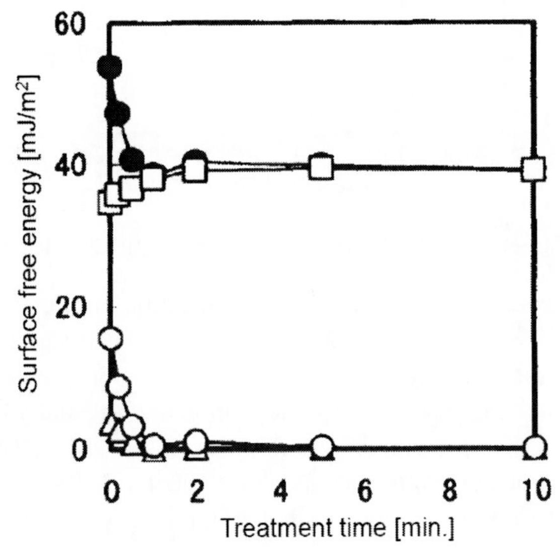

● : surface free energy (γ_{sv})
□ : dispersive component of surface free energy (γ_s^d)
○ : polar component of surface free energy (γ_s^p)
Δ : hydrogen bonding component of surface free energy (γ_s^h)

Figure 2. Surface free energy of cellophane.

Result of contact angle measurement indicated that wettability of glycerol decreased after longer treatment. Surface free energy calculated by the measured contact angles is shown in Figure 2. The surface free energy of cellophane (γ_{sv}) decreased as the treatment time increased. This change corresponds to the decrease in the hydrogen bond component (γ_s^h) and polar component (γ_s^h) of surface free energy. γ_s^h which contributes to hydrophilicity increased significantly, indicating that cellophane surfaces became hydrophilic after the plasma treatment.

1.3. Infrared Absorption Spectra

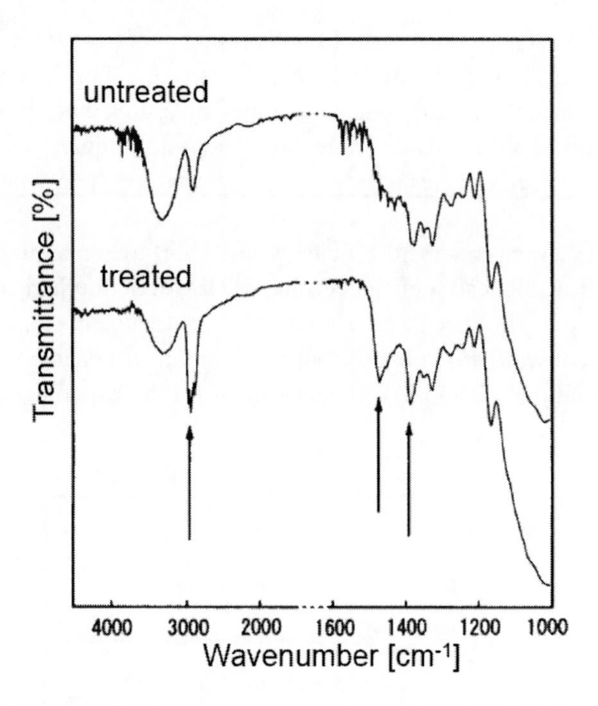

Figure 3. Absorption IR spectra of cellophane. Treatment condition: 12.75 kV, 60 min., 1.0 L/min.

Figure 3 shows the absorption attenuated-total reflection infrared (AIR-IR) spectra of the cellophane surfaces before and after plasma treatment. The spectrum of the untreated surface showed absorption associated with the chemical structure of cellulose, while that of the cellulose surface showed additional absorption at around 2940, and 1480 cm^{-1} treated at the voltage of 12.75 kV for 60 min. with the gas flow rate of 1.0 L/min. This absorption is often seen in an IR spectrum of polyolefin such as polyethylene [7]. It is therefore indicated from the IR spectra that materials containing a methyl group were formed at the cellophane surface after the plasma treatment.

1.4. X-Ray Photoelectron Spectroscopic (XPS) Analysis

The survey spectrum of x-ray photoelectron spectroscopy (XPS) of the untreated cellophane is shown in Figure 4. The peaks of carbon, oxygen, nitrogen and silicon are detected at the untreated cellophane surface. Among them oxygen's 1s peak was the highest, and carbon's 1s peak the second, while the peaks of nitrogen and silicon were low. As neither of nitrogen nor silicon is included in the chemical structure of cellulose, they can be attributed to the additives of cellulose. Figure 5 shows the survey spectrum of the cellulose surface after plasma treatment at the voltage of 12.75 kV for 5 min. with the gas flow rate of 1.0 L/min. As the treatment time increases, the intensity of C_{1s} increases, while that of O_{1s} tended to decrease. The peaks of nitrogen and silicon were not observed after the plasma treatment.

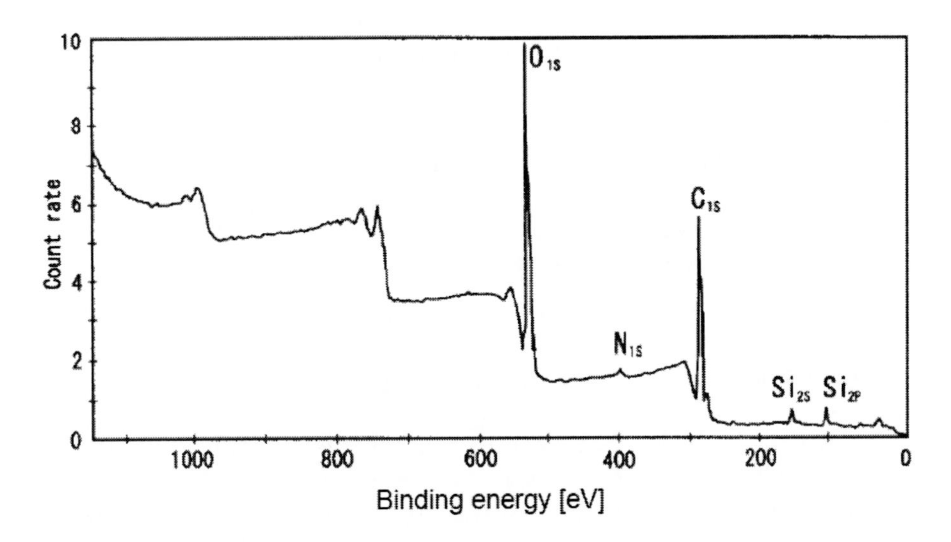

Figure 4. XPS survey spectrum of untreated cellophane.

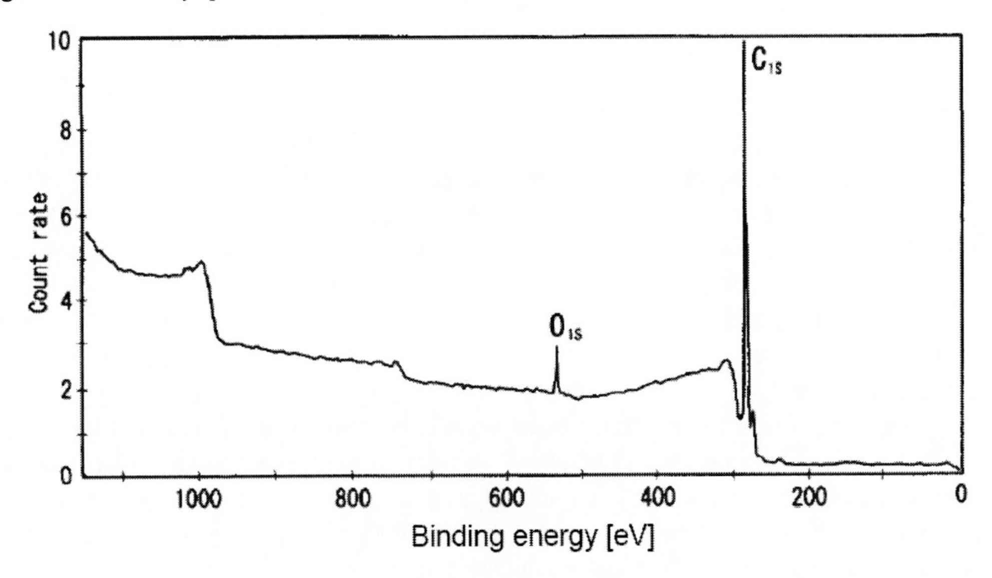

Figure 5. XPS survey spectrum of plasma treated cellophane. Treatment condition: 12.75 kV, 5 min., 1.0 L/min.

In summary, the results of surface free energy measurement, IR spectroscopy, and XPS analysis indicated that ethylene was polymerized by atmospheric pressure plasma, and that polymerized polyethylene film was deposited on to the cellophane surfaces. It is therefore concluded that plasma CVD of ethylene is possible even at atmospheric pressure.

2. Plasma Treatment of Paper

It was demonstrated that cellophane, which is recycled cellulose, can be treated by atmospheric pressure plasma. Therefore, plasma treatment of paper, which is a porous material of cellulose, is attempted below. In order to control penetrability of water-based ink for printing and drawing, paper is treated during its manufacturing process by inner sizing with which rosin is added in pulp slurry, or by surface sizing which applies rosin on the surfaces of pulp sheets. The sizing is commonly carried out by wet techniques, which have drawbacks of subsequent treatment of waste liquid, and also energy loss for drying is significant

In the present work, in order to perform surface sizing with a dry method, paper is treated by plasma polymerization in ethylene gas at atmospheric pressure.

2.1. Experimental

Filter papers (ADVANTEC 51B) were used as specimens, left at 20 °C, 76 %RH. Ethylene gas mixed with CO_2 or N_2 was used as a process gas. Stockigt sizing was compliance to JIS P 8112. Plasma treatment was applied to each side of the paper with the same condition because of the characteristic reason of the test in which liquid will penetrate from both sides.

2.2. Stockigt Sizing Test

Figure 6 shows the Stockigt sizing degree after each plasma treatment. Figure 7 shows the degrees of the filter papers after the plasma treatments, as well as an untreated commercially available paper (PPC) for comparing the degrees of the filters before and after the treatment. The Stockigt sizing degree shows the maximum at 21.8 s after plasma treatment in a CO_2/ethylene gas mixture, indicating that plasma treatment deposited polymerized ethylene coating which covered the paper surface, suppressing penetrability of liquid in to the paper.

In the present work, unsized filter paper which shows high penetrability of liquid was used. This can be the reason why the resulting Stockigt sizing degree was only approximately 22 s even after plasma treatment. It is likely that applying inner sizing would significantly improve the property against penetration, and that the Stockigt sizing degree would be similar to or better than that of the commercially available paper.

In addition, it was found that Penning effect of a gas mixture of ethylene and non-polymerizing gas promotes gas phase polymerization of ethylene, and consequently polymerized ethylene coatings can be deposited in a shorter time at lower concentration of ethylene. In a manufacturing process, this corresponds to diluting flammable gas with inert gas (satisfying the high pressure gas law), and thus it is expected that this technique can contribute to safe and economical operation in a manufacturing process.

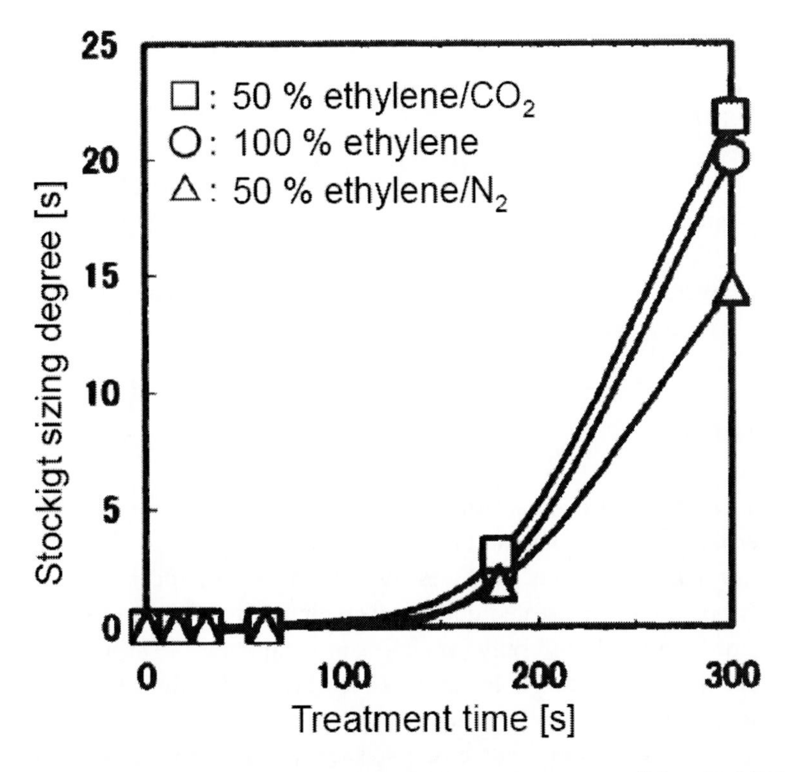

Figure 6. Stockigt sizing degree after various treatment time. Treatment condition: 13.5 kV, 0.1 L/min.

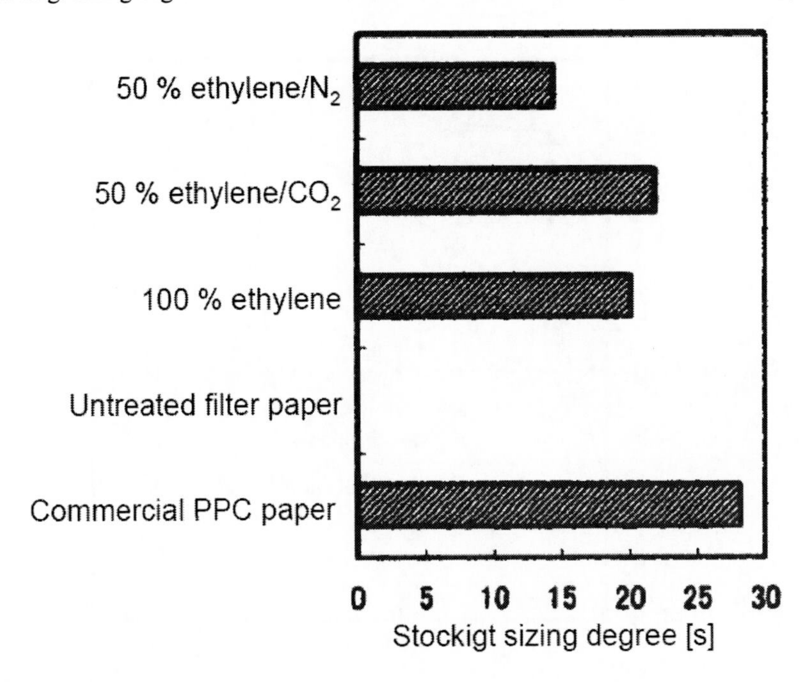

Figure 7. Stockigt sizing degree of PPC paper and treated/untreated filter papers. Treatment condition: 13.5 kV, 300 s, 0.1 L/min.

3. HYDROPHOBIC TREATMENT ON TIMBER SURFACES [9]

3.1. Experimental Methods

Hydrophobic treatment of 1 mm thick sliced plates of Konara (Quercus serrata) and Douglas fir was attempted by plasma treatment at the voltage of 12.75 kV and an ethylene flow rate of 0.1 L/min, whose conditions showed the highest deposition speed and best performance for hydrophobicity on cellophane. Other conditions, such as plasma treatment, and contact angle measurements were the same as those for the treatment of cellophane.

3.2. Hydrophobicity

Figure 8 shows measured contact angles of the sliced plates before and after the plasma treatment. After the treatment, both surfaces of Konara and Douglas fir showed hydrophilic rather than hydrophobic. Timber surfaces usually have significant roughness with in particular fine fibrils of fibres created during machining. Therefore it is anticipated that the surface could not be treated uniformly, or the resin components distributed at the surfaces might disturb deposition of coatings. In order to decrease the possibilities of these effects, the specimen surfaces were grinded with a sand paper, and subsequently immersed in acetone for 30 min. for leaching resin components at the surface. After these pre-treatments, they were plasma treated.

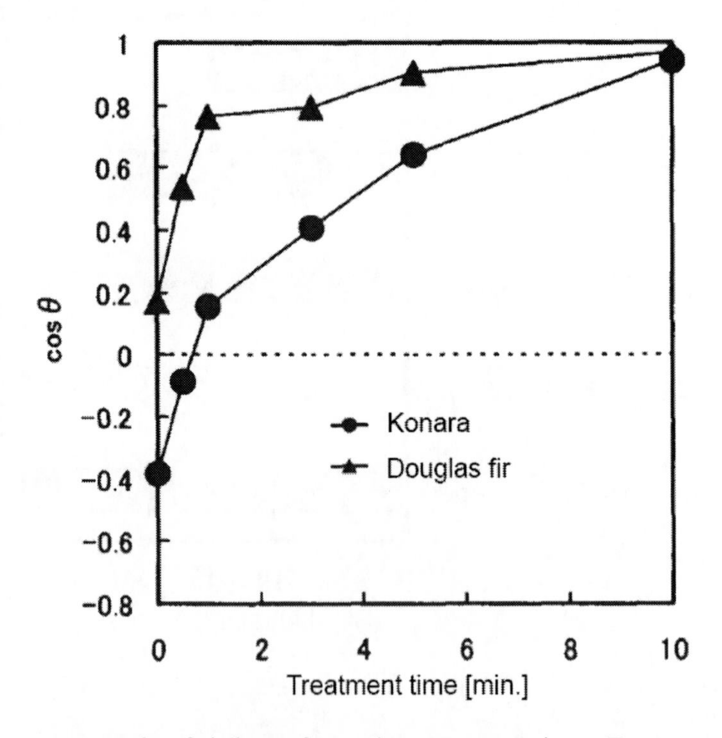

Figure 8. Water contact angle of timbers after various treatment times. Treatment: 12.75 kV, 0.1 L/min., without grinding.

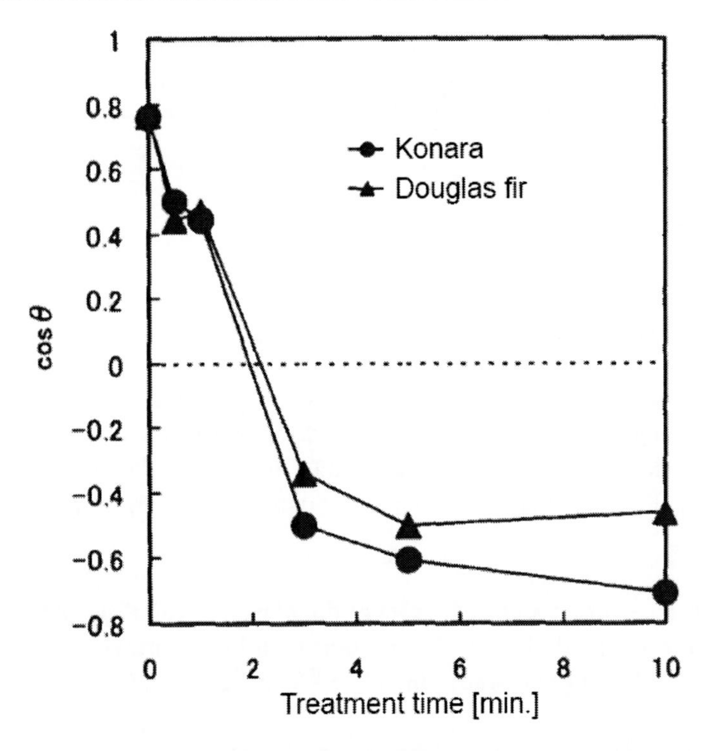

Figure 9. Water contact angle of timbers after various treatment times. Treatment: 12.75 kV, 0.1 L/min., Pretreatment: grinding and subsequent immersion in acetone.

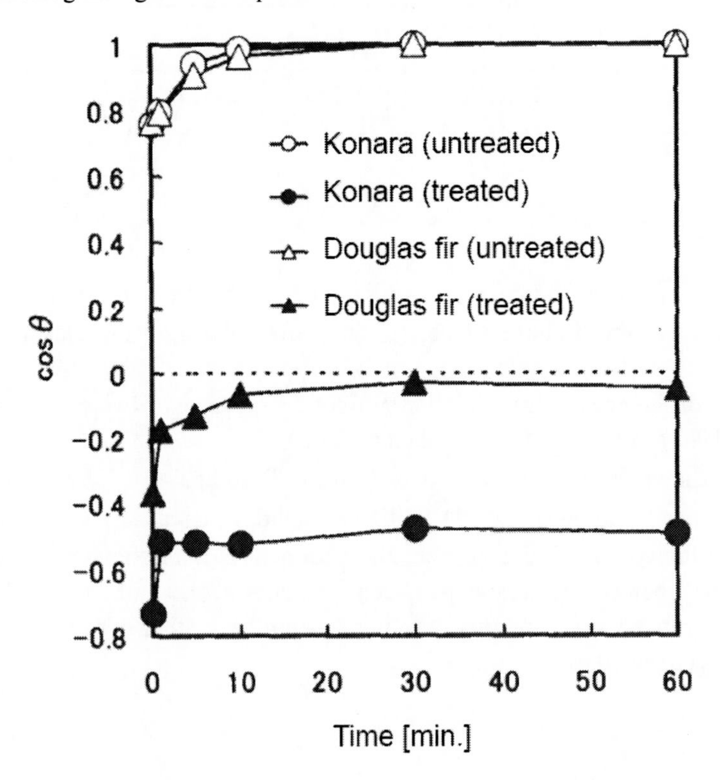

Figure 10. Change of water contact angle after contacting a water droplet. Treatment condition: 12.75 kV, 0.1 L/min., 5 min.

The timber surfaces became hydrophobic after plasma treatment as shown in Figure 9, as was also the case for cellophane. The contact angles of the plasma treated surfaces of timbers show significantly higher values than those of cellophane. It is reported that the contact angle of hydrophobic rough surface is more than 90o [10]. The present result seemingly has relevance to this report. As shown in Figure 10, water contact angle remained high values even 24 hours after the contact of the water droplet.

3.3. Water Resistance Test

The plasma treated timbers were dipped in water. After taking them out from water and dried, the contact angle of glycerine was measured. The measured results are summarized in Table 1. The contact angles of both Konara and Douglas fir after the plasma treated and dipped specimens are lower than those without dipping. However, they remained quite high even after 10 days.

Table 1. Glycerine contact angle (degree) of the plasma treated timbers. Some of them were dipped in water for 1, 5, and 10 days, dried, and the contact angles are measured. Treatment condition: 12.75 kV, 0.1 L/min., 5 min.

Specimens	Konara	Douglas fir
Untreated	40.5	40
Plasma treated (without dipping in water)	127	120
1 day	120	113
5 days	103	98
10 days	98	95

3.4. Colour Differences

When timber surfaces are treated, it is important to keep their appearance. In this respect, colour differences of the timber surfaces (⊿E*) before and after the plasma treatment was measured.

The colour differences quantitatively represent the distance of two colours in the uniform colour space (UCS). Here L*a*b* was taken as UCS.

Figure 11 shows the colour differences (⊿E*) of Korana at different plasma treatment times. The colour difference (⊿E*) initially slightly increased as the treatment time increased. And the colour difference after the treatment is approximately doubled from the colour difference before the treatment. Such a small increase of colour difference after treatment is interpreted as the temperature increase in the system during the treatment [11], and cannot be recognized by naked eyes.

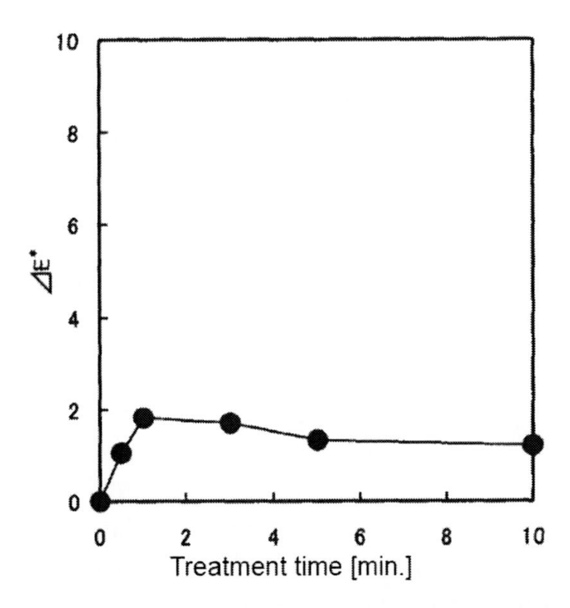

Figure 11. Change of colour differences of Konara after plasma treatment. Treatment condition: 12.75 kV, 0.1 L/min.

3.5. Specific Remarks for Timbers

Application of normal plasma polymerization for timbers is often difficult due to the emission of volatile components during processing. Furthermore it is impossible to introduce the reactive gas in to the inner surfaces (intracellular space) with the gas flow at low pressure. However at atmospheric pressure it is possible to introduce the reactive gas at least once into the inner surface when the gas is substituted.

As the polymerized ethylene coating is not durable against outdoor exposure test, the coating is applicable to the indoor use only. Considering the hardness of polymerized materials, polymethacrylic-acid etc. would be more desirable for the coating on timbers [12].

4. Moisture Permeable Waterproof Treatment to Cotton Cloth By Atmospheric Pressure Plasma [13]

Cellulosic textiles represented by cotton are widely used due to their water absorbance, comfortable texture, and good stainability. In addition, from the environmental point of view, attention is paid for the cellulosic materials as they are sustainably reproducible. If it is possible to add a moisture permeable waterproof property to cellulosic textiles, their applicability to clothing will be extended. In the present work hydrophobic treatment is attempted to cellulosic textile with using a plasma, which is environmentally compatible dry processing. Ethylene is chosen as the reactive gas which contains only carbon, hydrogen, oxygen, and nitrogen, is easily ionized.

4.1. Experimental

Tabby cotton broadcloth with single yarn 40, 130 x 70 inches, 122.5 g/m2, and diagonal cotton twill with ply yarn 30, 114 x 54 inches, 305.1 g/m2 were chosen as the specimens. They were washed with water, dried, and then left at 20 oC at 76 %RH. The plasma conditions are the same as those for the treatment of cellophane.

For the evaluation of hydrophilicity, the water drop method (JIS L1092, Appendix 3) was used, measuring the infiltration time to sink a water drop into the fabric. The moisture permeability was tested by the sodium chloride method (JIS L 1099A1).

4.2. Hydrophobicity of Cotton Textile [13]

Figure 12 shows the measured infiltration time for the cotton textiles after plasma treatment as a measure of hydrophobicity. The maximum infiltration time was as high as 360 min, which is comparable to the infiltration time (400 min.) for the commercially available treatments of Scotch guard® and Rain guard®. It is indicated that the textile surface is sufficiently covered with the polyethylene like coatings.

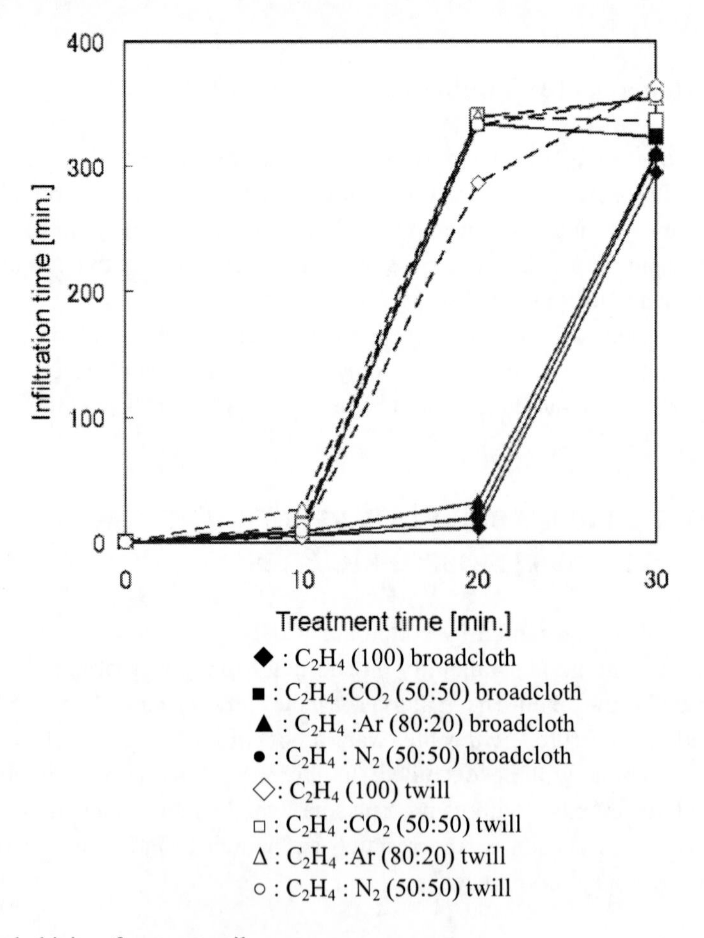

Figure 12. Hydrophobicity of cotton textile.

4.3. Moisture Permeability of Cotton Textile [14]

As shown in Figure 13, the moisture permeability was almost independent of the treatment, and the values are almost same as that before treatment and Scotch-guard treatment. Previous experiment indicated that the thickness of the plasma polymerized coating is approximately 10 nm or less [9]. It is suggested that the possible reason of the almost unchanged permeability after the treatment was that the coating hardly decreased the gap of the textile.

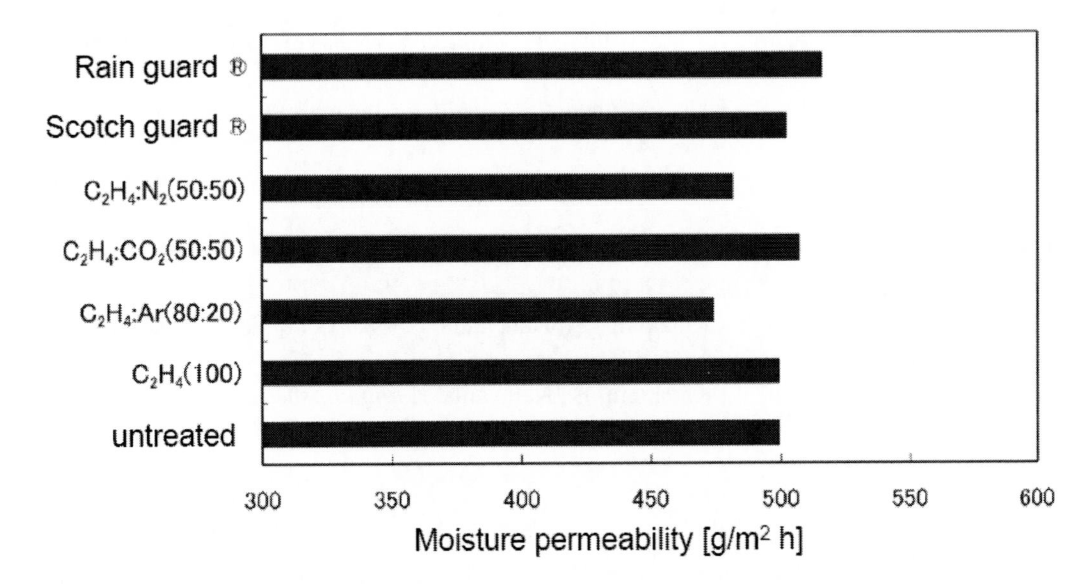

Figure 13. Moisture permeability of cotton textile. Treatment condition: 12.5 kV, 30 min.

REMARKS

The present work of atmospheric pressure plasma treatment was extended from the development of corona discharge treatment. In this method specimens are inserted between the electrodes, and thus are not free from the damage due to bombardment by electrons and reactive species. In an extreme case, the treatment can destroy the polymerized coatings. In the present work, the treatment conditions were carefully chosen in order to minimize the damage, while promoting the deposition of the coatings.

REFERENCES

[1] Millard, M. In *Techniques and Applications of Plasma Chemistry*; Hollahan, J. H.; Bell, A. T.; Ed.; John Wiley & Sons: New York, US, 1974; pp 177-213.

[2] Osada. Y. *et al.* In *Plasma Juugou (Plasma Polymerization)* in Japanese; Osada Y.; Ed.; Tokyo Kagaku Doujin: Tokyo, JP. 1986, pp 54.

[3] Yokoyama, T.; Kogoma, M.; Moriwaki, T.; Okazaki, S. *J. Phys. D Appl. Phys.* 1990, *123*, 1125-1128.

[4] Kogoma, M., In *Hyoumen-Shori-Gijutsu handbook)* in Japanese, Mizumachi, H.; Tobayama, K. Ed.; NTS Inc.: Tokyo, JP. 2000, pp 583-586.

[5] Uehara, T; Notsu, Y; Katayama, E; Katayama, H., *Adhesion (Nihon-Settyaku-Gakkaishi)* in Japanese, 2001, *37(2),* 52-56.

[6] Kitazaki, Y.; Hata, T., *J. Adhesion Soc. Jpn. Adhesion* in Japanese, 1972, *8(3),* 131-141.

[7] Nishkida, K.; Nishio, E, *Chart-De-Miru FT-IR* in Japanese; Kodansha Ltd.: Tokyo, JP. 1990.

[8] Uehara, T.; Teramoto, Y.: Katakami, E.; Katayama, H. *Trans. Mater. Research Soc. Jpn.* 2001, *26(3),* 841-844.

[9] Katakami, E.; Uehara, T; Katayama, H. *Adhesion (Nihon-Settyaku-Gakkaishi)* in Japanese, 2001, *37(10),* 380-384.

[10] *Shin-Jikken-Kagaku-Kouza 18, Kaimwn-To-Koroido* in Japanese; Chem. Soc. Jpn. Ed.; Maruzen: Tokyo, JP. 1977, pp 94.

[11] Uehara, T; Ito, T.; Goto, T. *J. Adhesion Soc. Jpn. Adhesion* in Japanese, 1984, *20(8),* 333-339.

[12] Uehara, T; Sugimoto, H.; Katakami, E.; Katayama, H. *Abstract of 37th Annual Meeting of Adhesion Soc. Jpn.*, 1999, 127-128.

[13] Uehara, T; Fukumoto, Y.; Katakami, E.; Katayama, H. *Abstract of 42nd Annual Meeting of Adhesion Soc. Jpn.*, 2004, 67-68.

[14] Uehara, T; Fukumoto, Y.; Katakami, E.; Katayama, H. *unpublished.*

In: Generation and Applications of Atmospheric Pressure Plasmas ISBN: 978-1-61209-717-6
Editors: M. Kogoma, M. Kusano and Y. Kusano

Chapter 7

TREATMENT OF POWDER BY LOW TEMPERATURE ATMOSPHERIC PRESSURE PLASMA

Masuhiro Kogoma
Department of Materials and Life-Sciences,
Sophia University, Tokyo, Japan

INTRODUCTION

A majority of the reports related to both plasmas and powders are devoted to production of powders using an inductively coupled plasma (ICP) at very high temperature (A.Takeda *et al.*), while plasmas are seldom used for surface treatment of powders. As the radius of each particle of the powder decreases, the net surface area of the powder increases inverse-square proportionally. In the cases of applications of ultra-fine powders dispersed into a liquid or solid, the properties of the powder surfaces which are exposed to the surroundings of the powder often play a more important role than its bulk properties. Therefore, surface treatment is indispensable for handling of nano-materials industrially and for adjusting surface wettability or affinity to the surrounding material. Since a plasma treats at and around surfaces only, significant efforts such as continuous stirring or pivoting the powder are needed to treat all particles in the powder uniformly. Ihara *et al.* reported that a low pressure plasma can be used for the surface treatments of powders with low melting points and of powders consisting of components whose bulk properties easily change at high temperatures. The examples include polymers and pigment powders. Such powders should be treated at low temperature. As low pressure plasmas are more diffusive than high pressure plasmas, they can easily extend the volume of the plasma treatment. However, at low pressures, a part of this powder can be transported to the downstream with the gas flow during the plasma treatment, and would contaminate the vacuum system, which may be problematic for industrial applications. In fact, plasma surface treatments of nano sized ultra-fine powders at low pressures are seldom reported. Even if a large scale vacuum system for continuous low pressure plasma processing is constructed for the treatment of powders, the maintenance of such a system is very difficult. In this respect, the difficulties that accompany the up-scaling of the system are rather reduced when atmospheric pressure plasma is applied for the treatment of powders, because the powder is handled at atmospheric pressure. In the case of powder treatment in air without plasmas, treatment efficiency can be improved by dispersing the powder using the gas flow and creating a fluidized bed. However, it is not easy to sustain the fluidized bed stably when there are certain types of unevenness in the shape, size and density of the powders. Atmospheric pressure plasma processing is not necessarily suitable for this fluidized bed method, as the method requires a large plasma volume, while atmospheric pressure plasma has difficulty in expanding treatment volume. Considering these issues, Kogoma *et al.* developed an atmospheric pressure plasma apparatus for the surface modification of powders by blowing-up and circulating powder particles [1]. This technique enables plasma treatment of powder surfaces by blowing up all the powder particles and subsequently transporting them to the discharge tube. This treatment is independent of the size or the shape of powder particles. Utilizing this technique, we carried out atmospheric pressure plasma enhanced chemical vapour deposition (CVD) at room temperature, and inorganic coatings such as silica (silicon oxide) and zirconia were deposited on to the powder surfaces [1]. It is desirable that the powder to be treated is non-aggregated in advance so that all surface area of the powder can be directly exposed in a plasma. Intense ultrasound irradiation by a horn-type generator, whose amplitude of vibration at the surface can be 50-60 μm, is useful for non-aggregation of powder particles. Surface treatment was accomplished efficiently by contacting the powder onto the ultrasound horn and subsequently sputtering and dispersing the powder particles into the plasma volume [2]. This method is particularly useful for plasma treatment of nano-powders which can easily aggregate.

1. Demonstration of Powder Surface Treatment and its Evaluation

Plasma surface treatment can be divided into two categories: (1) a chemical modification of the material surface and (2) a deposit of a different material than the bulk material. One example of the second case includes deposition of silica [3] and polymeric materials. When a coating is deposited on to a surface, a drastic change of chemical properties at the interface can occur, resulting in poor adhesion and cohesion and significantly affecting the chemical and physical properties of the treated surfaces. As described in a previous chapter, atmospheric pressure glow plasma is initiated by collision of accelerated free electrons colliding with helium atoms and subsequent electron avalanches. Emitted electrons and excited helium atoms bombard onto the substrate surface, creating radical sites. Additive gas mixed with helium can be excited by the collision with electrons, and excited helium atoms can react and bond to the reactive sites. As the excitation energy is generally sufficiently high in a plasma and is insensitive to the reactivity of the substrates, surface modification can be readily employed.

The following issues have to be considered when plasma treatment of powder surfaces is employed;

[1] as powder has a large value of surface/volume ratio, if the particles overlap or aggregate with each other, the treatment cannot be uniform,

[2] if plasma treatment is employed for powders such as polymers, they may be deformed or melted by the heat generated in a plasma, and
[3] depending on the properties and shapes of powders such as friction forces and dispersive properties, the treatment system can be choked. It is necessary to design a system appropriate for the powders to be treated.

1.1. Bulk Surface Treatment of Polymeric Powders

Polymeric powders can be potentially used for various applications. They could be applied for paintings, catalyst supports, and bio-compatible materials. For these practical applications, they have to be manufactured according to the actual usages without changing properties as the polymeric powders such as the densities, degrees of polymerization, diameters and shapes of the particles. For instance, in the case of production of water-based paints, if the surface of the powders chosen is hydrophobic, the powders are not usable, as they will be separated soon after mixing with water. Surfactants are usually used for dispersing them in water, but the application of surfactants often damages the properties of the powders. In this respect it is the best if the powder can be modified so as to suspend to water without changing the properties of the particles in the powder such as the densities and size. One should note that the desirable properties of powders such as hydrophilicity and liquidity are highly influenced by the properties of the surfaces. If it is possible to modify only the surface of powders, one can obtain the desirable properties without changing the densities and size of the particles in powders. In the last example, formation of hydrophilic functional groups by oxidizing the hydrophobic functional groups enables suspension in water. Such a surface treatment that changes only the surface properties becomes an important technique not only for the powder processing but also for general materials engineering.

Well-known surface treatment methods for polymers include oxidation using ozone that is produced by ultraviolet irradiation, and a method of dipping polymers into oxidizing and reducing agents. However, these methods generally need long times for reactions, and experience difficulties in uniform treatment due to low dispersiveness in water. There is a risk that inside the particles would also be treated by water permeation. In addition, effluent treatment after the use of the solvent has to be considered. Furthermore, surface reactions of inert powders such as polyethylene will not proceed with the normal chemical methods. Atmospheric pressure plasma can treat only the surface of the powders in a short time, and is thus useful as a surface treatment method. In this research, plasma surface treatment of high density polyethylene powder is attempted, and the relations of the properties of the treated surfaces to concentrations of functional groups at the surfaces are evaluated by the chemical modification method.

1.1.1. Experimental Methods

Figure 1 shows a schematic diagram of the atmospheric pressure plasma treatment apparatus used for the surface treatment of polyethylene powders. The main chamber is a 45 mm-outer and 41 mm-inner diameter quartz tube. A 36 mm outer-diameter stainless steel electrode connected to a radio frequency power source is inserted into the quartz tube. A stainless steel mesh electrode surrounds the quartz tube, and is connected to the ground. A plasma is generated between these electrodes. The length of the plasma is approximately 21 cm. Twenty μm diameter high density polyethylene powder is treated by the plasma. For prevention of possible melting of the polyethylene, both the electrodes and the quartz tube are water-cooled. The polyethylene powder is introduced into the quartz tube from a powder inlet using a vibrator. As the introduced powder is initially aggregated, it has to be non-aggregated using a powerful ultrasonic horn, and subsequently introduced into the plasma with a gas flow from the bottom of the quartz tube. O_2/He and NH_3/He gas mixtures are used for oxidation and nitridation, respectively. The polyethylene powder treated by the plasma is collected by a filter. The surface is characterized using x-ray photoelectron spectroscopy (XPS).

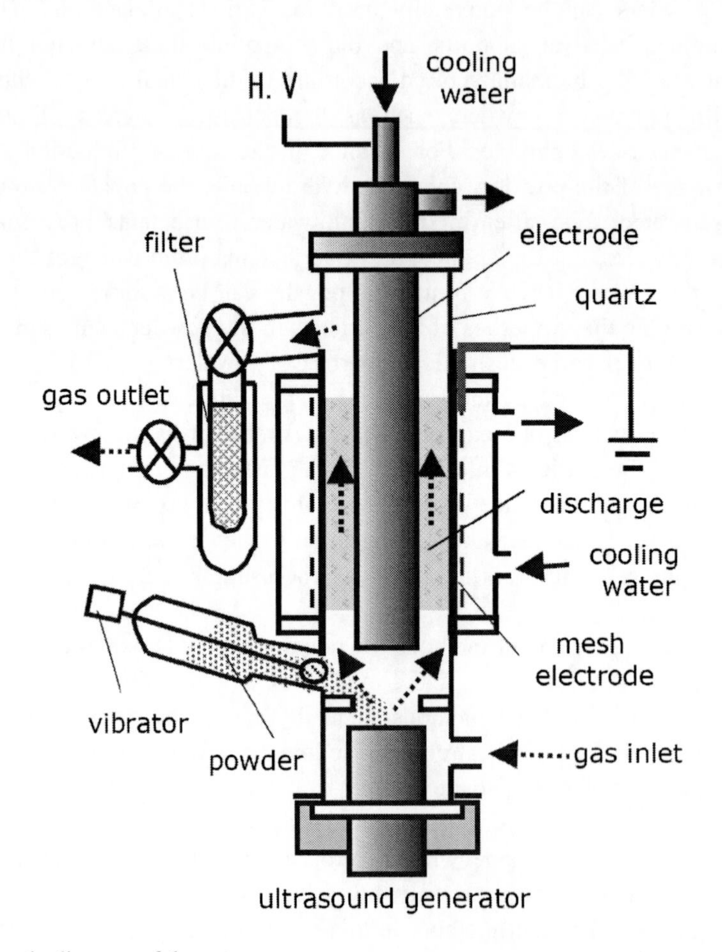

Figure 1. Schematic diagram of the setup.

1.1.2. Chemical Modification Method

The functional groups introduced by the plasma treatments have to be identified for this purpose, each functional group is labelled using a specific chemical which can selectively react with it, and then the amount is measured using the XPS. The reaction ratio between each functional group and its corresponding chemical is determined by the simultaneous reaction of a model specimen in which the amount of the functional group is known in advance. The powder sample is pressed for use. The chemicals used and the relevant reactions are listed in Table 1.

Table 1. Chemical agents and corresponding reactions

$$R-OH \xrightarrow{(CF_3CO)_2O\ (TFAA)} R-OCOCF_3 + CF_3COOH$$

$$R-C(R')=O \xrightarrow{NH_2NH_2\ (Hydrazine)} R-C(R')=N-NH_2 + H_2O$$

$$R-COOH \xrightarrow{CF_3CH_2OH\ (TFE)} R-COOCH_2CF_3 + H_2O$$

$$R-NH_2 \xrightarrow{C_6F_5CHO\ (PFBA)} R-N=CHC_6F_5 + H_2O$$

OH group: The oxidized polyethylene powder and polyvinyl alcohol (a model polymer) were introduced into individual test tubes, which were filled with vapour of tri-fluoro-acetic anhydride, and reacted at room temperature for 2 hours. The reaction rate χ and the concentration of the functional group $[R_{OH}]$ were evaluated using the following equations:

$$[C]_o + [O]_o + [N]_o = 1$$

$$[R_{OH}] + [R_{C=O}] + [R_{COOH}] + [R_{NH_2}] + [R_H] = 1.$$

$$[O] = \frac{[O]_o + \chi[R_{OH}][C]_o}{[C]_o + [O]_o + 6\chi[R_{OH}][C]_o}$$

$$[F] = \frac{3\chi[R_{OH}][C]_o}{[C]_o + [O]_o + 6\chi[R_{OH}][C]_o}$$

$$\chi = \frac{[O]_o[F]}{([F] + 3[O])[R_{OH}][C]_o}$$

Here [X] and $[X]_o$ represent the atomic contents [%] at the surfaces measured by the XPS before and after the plasma treatment, respectively. X is each atom such as C, O, N or F. $[R_{FG}]$ represents the content of carbon atoms bonded with a functional group FG which can be OH, C=O, COOH, or NH_2.

-C=O group: The oxidized specimen and poly-vinyl methyl-ketone (PVMK) as a model polymer were introduced into individual test tubes which were filled with hydrazine vapour. The contents were reacted at 50 °C for 4 hours.

$$\chi = \frac{[N]}{2[R_{C=O}][C]}$$

-COOH group: The oxidized specimen and of poly-acrylic acid (PAA) as a model polymer were placed in individual weighing bottles. First, 0.9 ml of tri-fluoro-ethanol and then, after 15 minutes, 0.4 ml of pyridine anhydride were introduced in each bottle. After further 15 minutes, 0.3 ml of di-tert-butylcarbodiimide was also introduced. Each mixture was reacted at 65 °C for 8 hours.

$$\chi = \frac{[F][O]_o}{3[R_{COOH}][C]_o[O]}$$

$-NH_2$ group: The nitride specimen and di-amino-dodecane were placed in weighing bottles which were filled with penta-fluoro-benz-aldehyde vapour; each mixture was reacted at 40 °C for 4 hours.

$$\chi = \frac{[F][O]_o}{5[R_{NH_2}][C]_o[N]}$$

1.1.3. Results And Discussion

1) Oxidizing Treatment

Polyethylene powder surfaces were treated with O_2/He plasma. Figure 2 shows the surface atomic ratio O/C with various discharge powers at the He flow rate of 8 SLM for O_2 gas composition of 0.38 %.

As the power increased, the oxygen content at the surface tended to increase. When the discharge power was less than 1000 W, the discharge was unstable. At discharge powers more than 2300 W, the powder was aggregated by the heat in the plasma.

Figure 3 shows the O/C ratios of the polyethylene powder surfaces with different O_2 contents in the gas mixture at the discharge power of 2100 W. The O/C ratio increases as the O_2 content in the gas mixture increases up to 0.38 %, while it gradually decreases when the O_2 content is larger than 0.38 %. This is attributed to the fact that the plasma becomes unstable at higher O_2 content in the gas mixture for this discharge power. The maximum O/C ratio was 0.38 after treating the powder in the O_2 gas content of 0.38 % at the discharge power of 2100 W. The powder treated with this condition was introduced into water, showing complete dispersion.

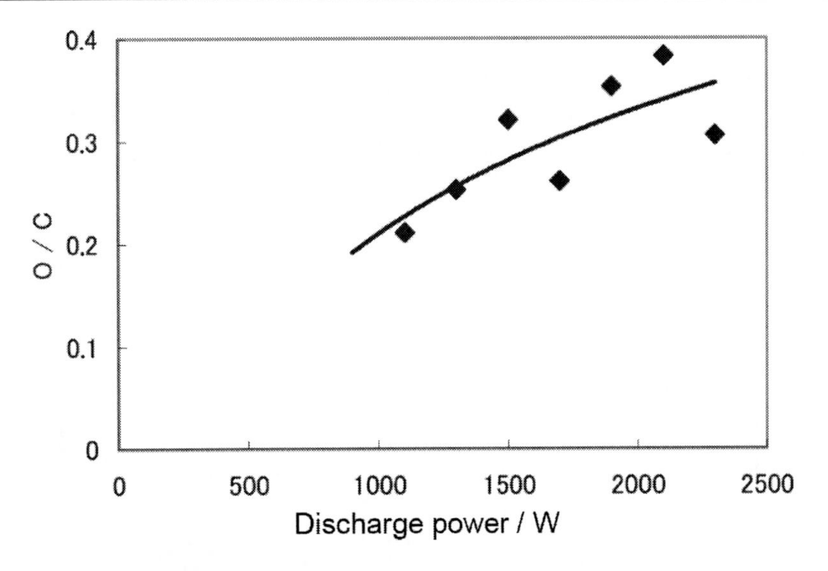

Figure 2. O/C ratios at different discharge powers.

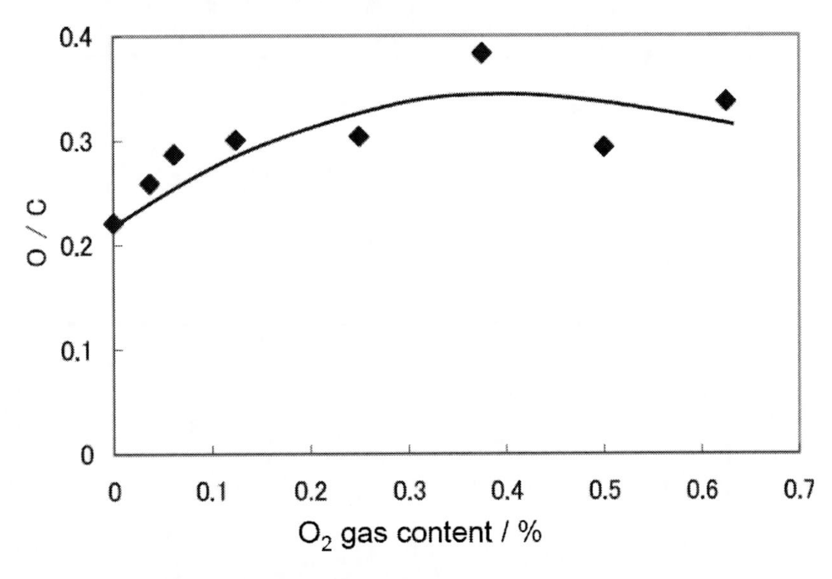

Figure 3. O/C ratios at different O_2 gas contents.

2) Identification of Oxygen-Containing Functional Groups By the Chemical Modification Method

The functional groups on the polyethylene powder surfaces were characterized by the chemical modification method after atmospheric pressure plasma treatment in the 0.38% O_2 gas with 8 SLM He gas flow rate at the discharge power of 1500 W. Figure 4 shows the contents of the oxygen-containing functional groups at the surfaces, estimated using the C1s regional spectrum and the chemical modification method. The C1s regional analysis shows the existence of three oxygen-containing functional groups; -C-O, -C=O, and –COO. According to the result obtained from the chemical modification method, the contents of –C=O and –COOH show good agreements with those of –C=O and –COO obtained by the regional C1s analysis, while the content of –OH obtained by the chemical modification is significantly lower than that obtained by the C1s regional analysis. The difference may be due to the fact that the –C-O structure contains not only OH group but also C-O-C. We therefore conclude that the plasma treatment in O_2/He gas mixture introduces oxygen-containing functional groups such as OH, C=O, COOH and C-O-C at the polyethylene surface.

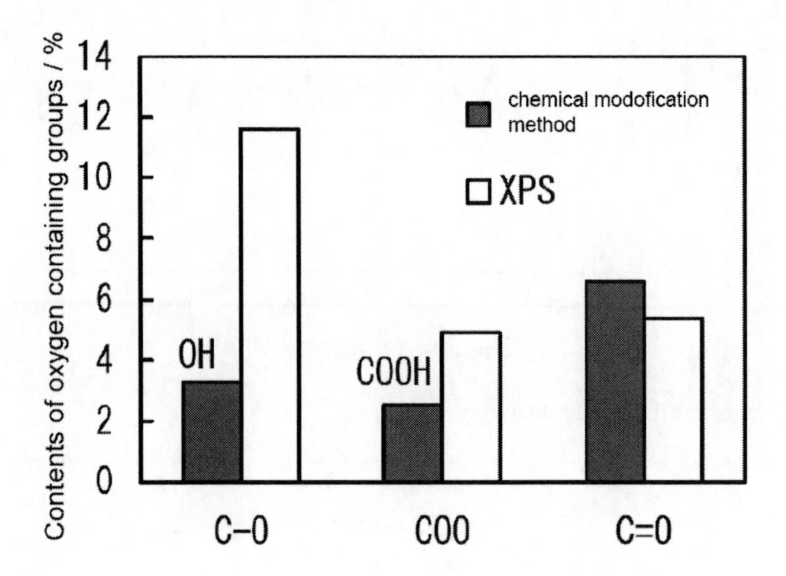

Figure 4. Contents of oxygen containing functional groups estimated by XPS and the chemical modification method.

3) Nitriding Treatment

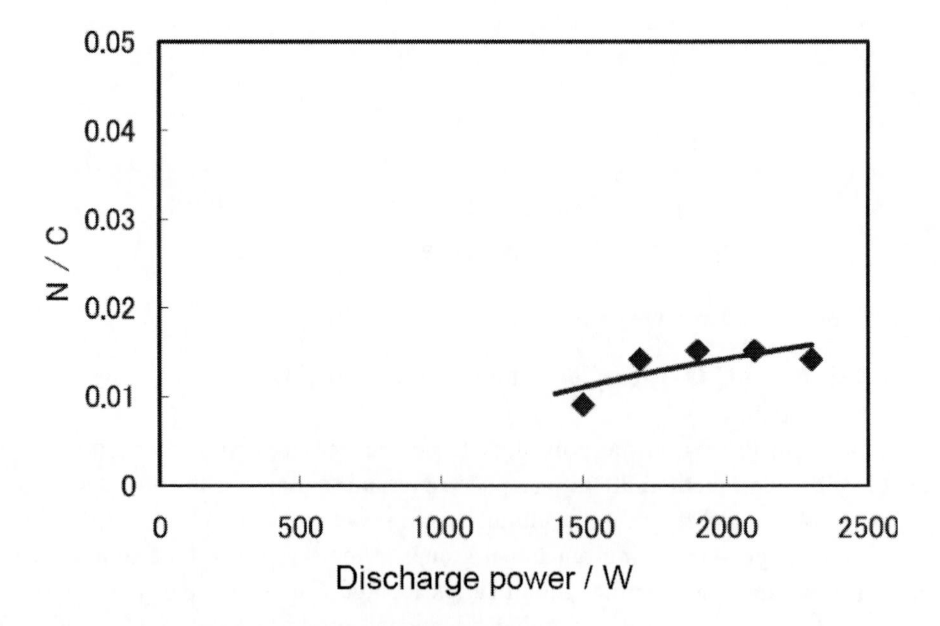

Figure 5. N/C ratos at different discharge powers.

Polyethylene powder was treated with NH_3/He plasma nitriding. Figure 5 shows the N/C ratios at the polyethylene powder surfaces after the plasma treatment in the NH_3 gas content of 0.75 % with the He gas flow rate of 8 SLM at different discharge power values. The N/C ratio did not significantly change as the power increased. This result is different from the oxidation described above. In addition, the N/C ratio did not significantly change as the NH_3 content changed. This may be due to the fact that the reaction rate of nitridization is smaller than that of oxidation. In order to increase the reaction time in the plasma, we repeated the plasma treatment with the same condition at the power of 2100 W several times. The N/C ratio increased as the number of the treatment increased, as shown in Figure 6. In addition, when the total flow rate is reduced to 4 SLM, further increase in the N/C ratio was observed. As a result, the maximum N/C ratio was seen when the powder was treated five times in the NH_3 gas content of 0.75 % with the He gas flow rate of 4 SLM at the power of 2100 W.

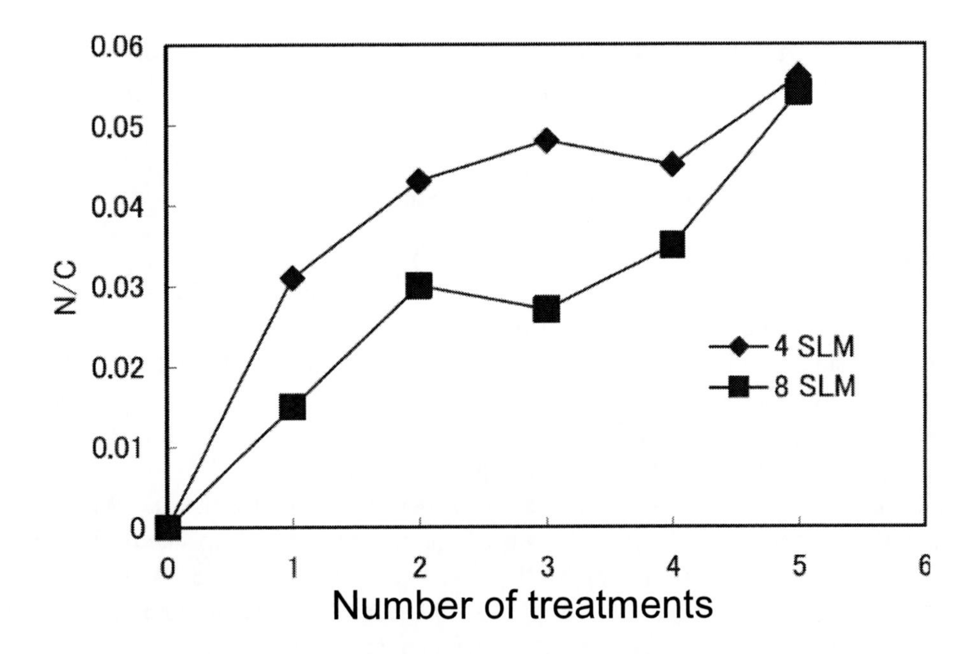

Figure 6. N/C ratios after various numbers of treatment.

4) Identification of Nitrogen-Containing Functional Groups by the Chemical Modification Method

The polyethylene powder surfaces were treated using a plasma having a NH_3 gas content of 0.75 % with the He gas flow rate of 8 SLM at the power of 2100 W. They were treated once, three or five times. The nitrogen containing functional groups at the plasma treated surfaces were characterized with the chemical modification method. Figure 7 shows the N/C and NH_2/C ratios as a function of the treatment time. As the treatment time increased, the content of amino (NH_2) group increased as the N/C ratio increased. It is shown that further treatment results in the formation of other nitrogen containing functional groups. About 80 % of the nitrogen-containing functional groups are assigned to the amino group after five-time treatment at the polyethylene powder surfaces.

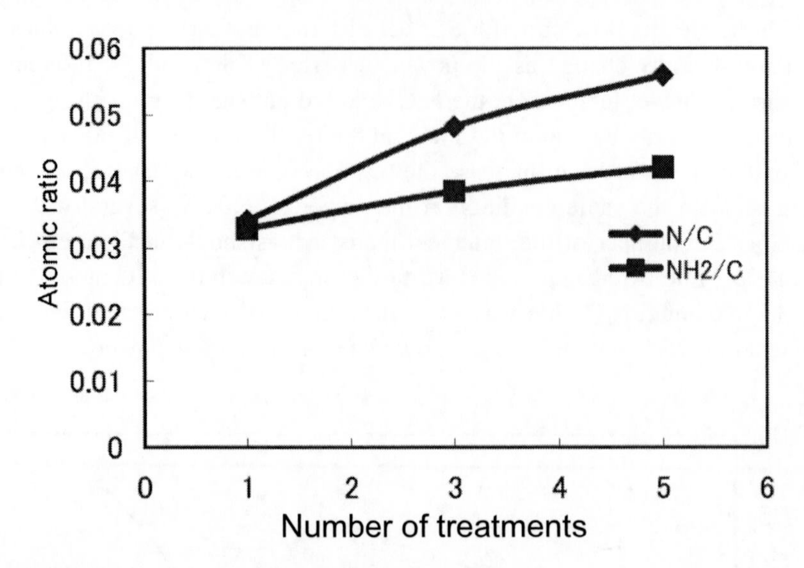

Figure 7. N/C and NH_2/C ratos after various numbers of treatment.

1.1.4. Conclusions

High density polyethylene powder surfaces were treated with atmospheric pressure glow plasma. Surfaces were oxidized using O_2/He plasma; we achieved the maximum O/C ratio of 0.38 by increasing the discharge power and the O_2 content in the gas mixture. The treated powders showed significant hydrophilicity. Oxygen-containing functional groups were identified using the chemical modification method, and the existence of the ether bonding was identified as well as OH, C=O and COOH groups.

The polyethylene powder surfaces were also treated by NH_3/He plasma. We achieved N/C ratio of the maximum of 0.056 by increasing the treatment time. The identification of the nitrogen-containing functional groups by the chemical modification method indicated that more than 80 % of the nitrogen introduced at the surfaces could be attributed to the amino groups. Additionally, other nitrogen-containing functional groups were also created as the treatment time increased.

1.2. Deposition of Silica Coatings onto Organic and Inorganic Super Fine Pigment Powders Used as Cosmetic Materials by Atmospheric Pressure Plasma [4]

Some components in cosmetic pigments can cause skin irritation problems such as cosmetic dermatitis or pigmentation, when they directly contact skin surfaces. When the responsibility of the product was claimed in 1978 in the US, the Product Liability Law for the United Model was created. It is the so-called PL Law. Based on the PL Law, the FDA in the US has made guidelines for the cosmetic raw materials stricter than before. For example, the FDA forbids manicures which include MMK monomers or more than 5 % of formaldehyde. In addition, it is demanded not to use any components which are reported to cause inflammation, nerve poisoning, inflammation due to photo-absorption, or other allergic reactions. The uses of dyeing additives for food, medicine, and cosmetics now require the safety tests and the approval by the FDA. Meanwhile, in Japan the Product Liability Law was established in 1995. Subsequently, the strict safety has been demanded in the cosmetic industry where only relatively loose regulations existed for the raw materials. The cosmetic raw materials are chemical substances, among which some pigment-tar dyes weie approved as additives for food. However, due to the recent increase in chemical sensitivity it is not very desirable that chemical raw materials including aromatics directly contact the skin. In order to avoid such a situation, it is the best for us to wait for the development of novel safe raw materials. However, waiting the new materials is practically unrealistic, as the regulation for new chemical substances tends to be far stricter. Consequently it has been preferred to coat the current raw materials with a substance whose safety is already confirmed, such as certain polymers or silica. In this section, silica treatment for super-fine powders of titanium oxides by an adsorption- plasma oxidation method (an adsorb- and-dry method) is presented.

A significant amount of titanium oxide powder has been used for cosmetics, since it shows high whiteness due to its high reflective index, and effective UV blocking capability. However the FDA in the US and other agencies have pointed out that its high capacity of catalytic photo-oxidation might cause serious problems at the skin. Although the powder surfaces are generally coated with resins, the capability of skin protection by such a coating itself is not perfect against long time photo-exposure, since it is also an organic material. In this work, silica coatings were deposited onto the titanium oxide powder surfaces using atmospheric pressure plasma, such coatings acted as perfect blocks against photo-oxidation. The experimental conditions are shown in Table 2. The discharge tube for the treatment of powder surfaces is similar to that in Figure 1.

Table 2. The experimental conditions for the deposition of silica coatings

Powder / g	10
Discharge Time / min	5
He Flow Rate / $cm^3 min^{-1}$	10000
O_2 Flow Rate / $cm^3 min^{-1}$	100
Discharge Power / W	2500
Discharge Frequency / MHz	13.56

The super fine titanium oxide powder with 200 nm diameter was obtained from Toho Titanium Ltd. (anatase 80 %, rutile 20 %, surface area 15 $m^2 g^{-1}$). Tetra-ethoxysilane (TEOS) dissolved in ethanol was adsorbed at the powder surfaces for a day, allowing the progress of hydrolysis there. Afterwards, the solvent was evaporated and the residual organic substance at the surfaces was oxidized in O_2/He plasma, forming a perfect SiO_2 coating (adsorb-and-dry method). This method enables one to synthesize thinner denser coatings than does the sol-gel method, by completely oxidizing the adsorbed organo-metalic compounds which are imperfectly decomposed during hydrolysis. The thickness of the coating can be controlled by the concentration of TEOS in ethanol. The discharge tube is made of quartz, surrounded by a copper mesh electrode, allowing the observation of the discharge from outside of the tube. The inner electrode is water-cooled stainless steel. When materials with low melting points are treated, the external electrode has to be water-cooled as well. The powder specimen introduced from the powder pool into the discharge tube by the vibrator is split into individual particles with intense ultrasound; these are sputtered upward and transported to the discharge zone with the carrier gas flow. The oxidized powder particles in the discharge were separated into gas and solid by the cyclone and the filter, and collected. The desirable O_2 content in the gas mixture is approximately 1 %. If the O_2 content is too high, the treatment speed decreases. The equipment used here does not recycle the carrier gas, but He gas is relatively easily recycled. Therefore the cost for the gas in the industrial scale system can be suppressed by recycling used He gas. The treatment speed of the present equipment is approximately 20 g/min. It can be treated faster at higher discharge power.

Compositional and morphological analyses were performed using XPS and transmission electron microscopy (TEM), respectively. In order to evaluate the photo-oxidation effect of the powders, we dispersed the powder (2,6,10,15,19,23-hexamethyl-2,6,10,14,18,22-tetracosahexaene) in squalene oil and exposed it to the UV irradiation from a Xe lamp for an hour. After that, the odour produced from the oil was characterized using GCMS. The photo-oxidation test procedure is shown in Figure 8.

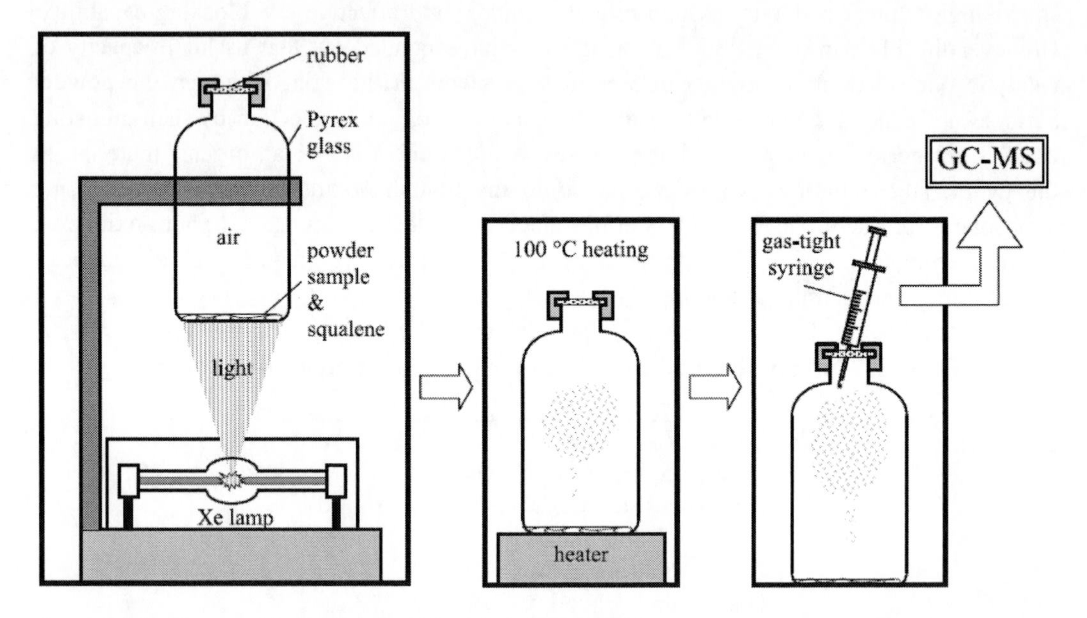

Figure 8. The photo-oxidation test procedure for powder specimen.

Figure 9 shows the Si/Ti atomic ratio at the surfaces of untreated (a), adsorbed (b), and adsorbed and subsequently plasma treated powders (c). 10 % TEOS adsorbed (conversion to SiO_2) powder particles are covered with hydrolyzed SiO_2. However, some carbon containing residue is left in the coating due to imperfect dissociation. Si content increases after the plasma treatment, because of the elimination of carbons by the plasma oxidation. Figure 10 shows the regional C1s spectrum of XPS for untreated (a), adsorbed (b), and adsorbed and subsequently plasma treated powders (c). The untreated surface is dominated by the contaminated carbon component that originates from the manufacturing process of the powder. The TEOS-adsorbed powder shows the peak shift at the higher binding energy corresponding to the TEOS component that is not dissociated yet. After the plasma treatment, this peak disappears, which indicates that the oxidation of TEOS has proceeded.

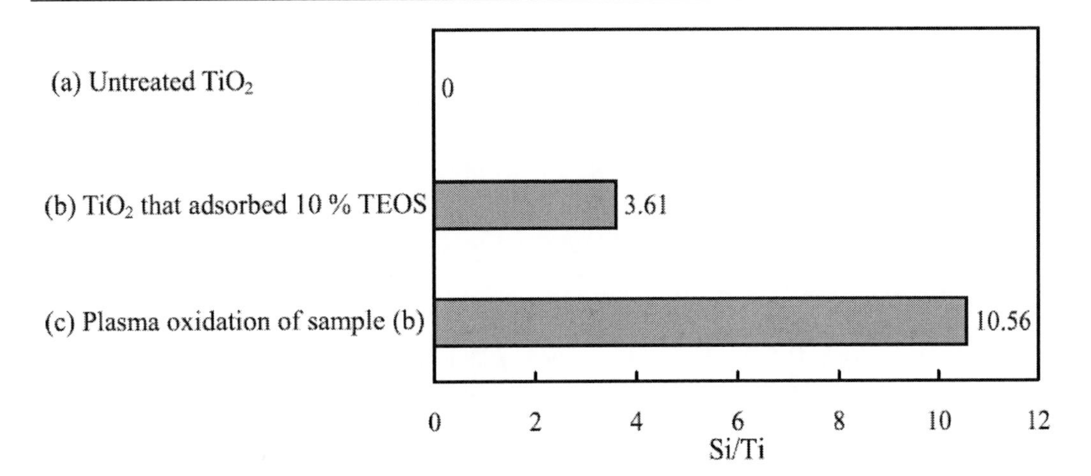

Figure 9. Si/Ti ratos at each procedure.

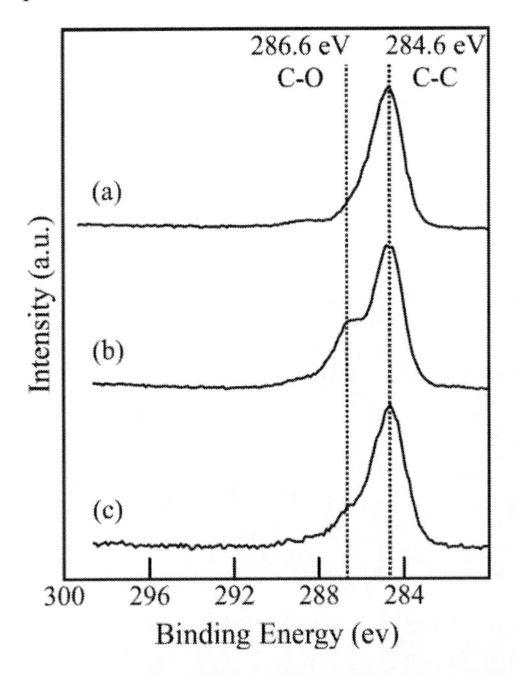

Figure 10. C1s regional specira of powder specimens. Untreated (a), TEOS-absorbed (b), and plasma-treated (c).

Figure 11 shows the corresponding regional O1s spectra of XPS. The O1s spectrum for the untreated powder consists of only one peak at 529.9 eV, corresponding to TiO_2. After the adsorption, the intensity of this peak decreases, and the peak assigned to SiO_2 dominates. After the plasma treatment, the peak at 529.9 eV disappears, indicated that the surfaces are completely covered with SiO_2.

Figure 12 shows the TEM image of the cross section of a plasma-treated powder particle. It is seen that an approximately 3 nm thick SiO_2 coating is deposited uniformly on the particle.

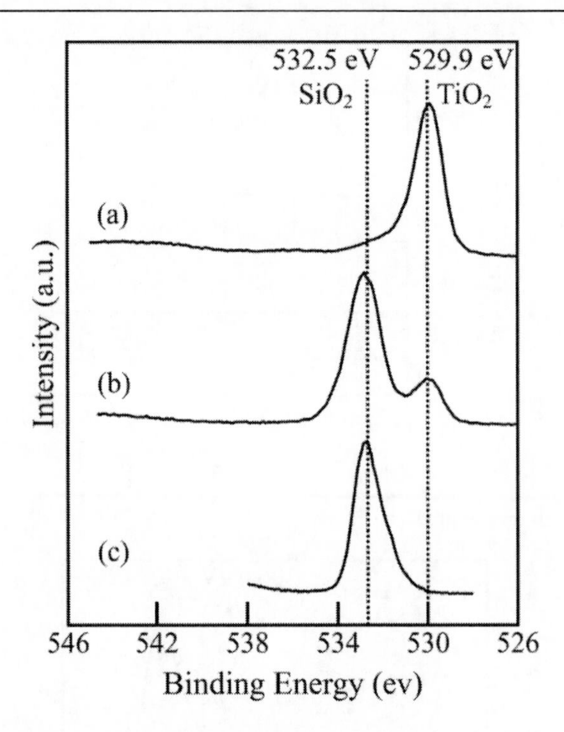

Figure 11. O1s regional spectra of powder specimens. (a), TEOS-absorbed (b), and plasma-treated (c).

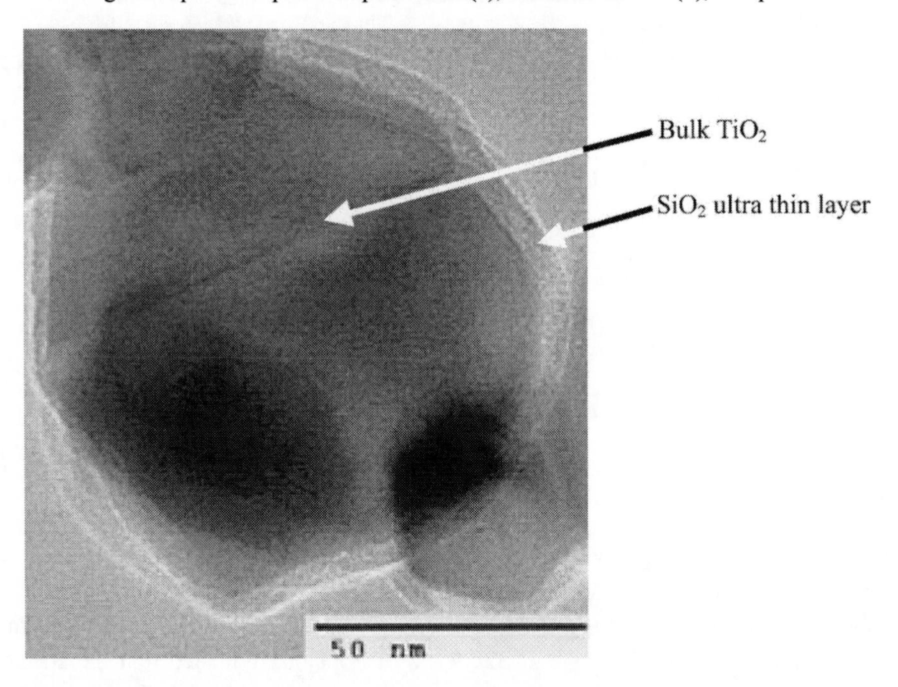

Figure 12. TEM image of plasma-treated powder.

For testing the usefulness of the silica coating as a protective layer, we dispersed the treated powder into the squalene oil and sealed it in a Pyrex bottle. The molecular structure of squalene is shown in Figure 13. UV light was irradiated from the bottom of the bottle for an hour, and emitted gas components were analyzed using GCMS, as shown in Figure 14. As can be seen in Figure 14, when the powder was not dispersed in the squalene oil, there was no emission of gases after the UV irradiation.

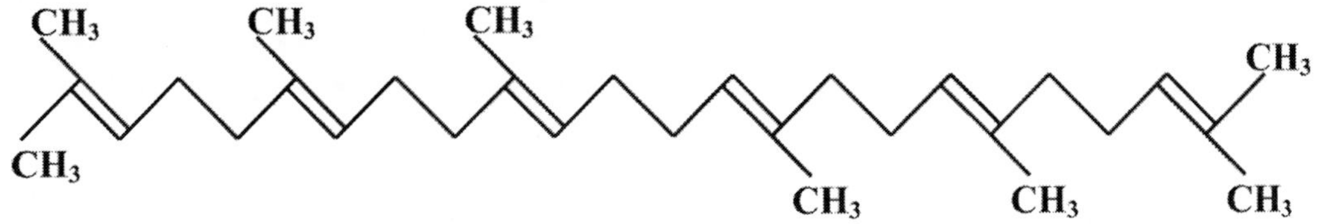

Figure 13. Molecular structure of squalene oil.

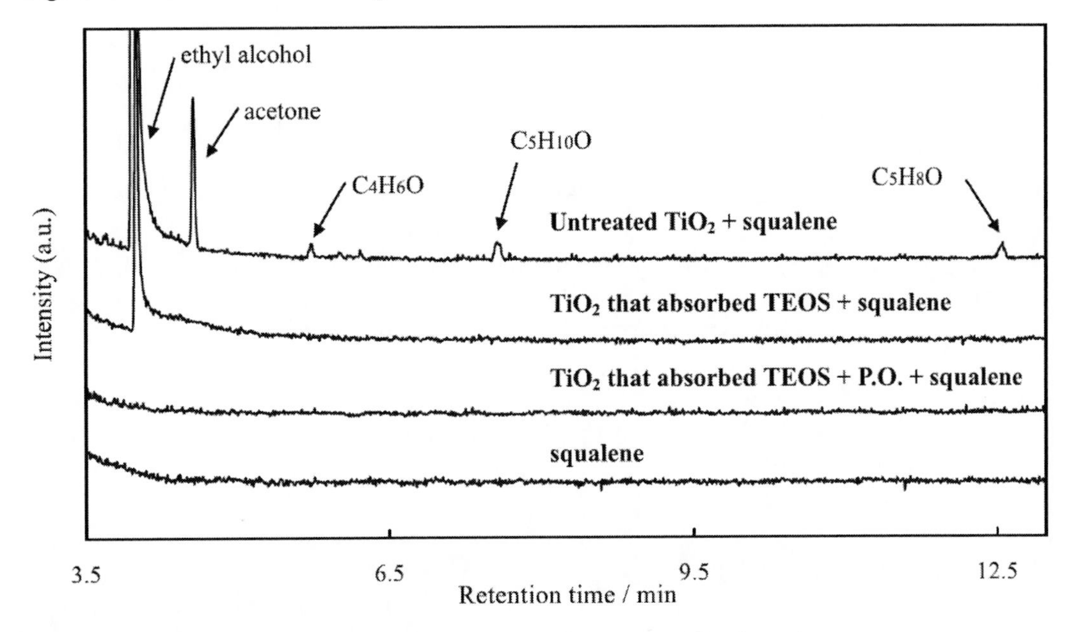

Figure 14. GCMS spectra of emitted gas components from UV-oxidized squalene.

On the other hand, after UV irradiation the squalene oil in which were dispersed the untreated titanium oxide powders emitted substantial amounts of various chemicals, such as alcohol, acetone and long-chain hydrocarbons, which are harmful to skin. The UV-irradiated squalene oil in which were dispersed the TEOS adsorbed titanium oxide powders did not emit oxidative compounds whose molecular weights are heavier than that of acetone. This may be due to the facts that the majority of the powder surfaces were covered with the hydrolyzed SiO_2 after the adsorption of TEOS at the surfaces, and that the edge of the squalene molecule migrated slightly into the remaining gaps, touched the titanium oxide surfaces, and subsequently decomposed partially. On the other hand, there was no emission of photo-oxidised substances from the UV- irradiated squalene in which were dispersed the plasma-treated powders. This therefore indicates a perfect protective effect of the plasma-oxidized silica coatings against the photo-oxidation effect by titanium oxide.

We conclude that the silica protective coatings by plasma oxidation can perfectly block the photo-oxidation effect of titanium oxide. The adsorb-and-dry method is able to synthesize various protective coatings involving multi-layer coatings onto surfaces of not only titanium oxide powders but also of most other powders, and thus will be an important powder treatment method.

REMARKS

About 17 years have passed since the development of atmospheric pressure plasma technique that started in Japan. Both basic and application oriented research activities are significantly spreading in the US, Europe and Asian countries recently. In Japan, this technique is highly appreciated as it is suitable to the application areas to which the low pressure plasmas cannot be applied due to the high cost. It is expected that industry will be further involved in the atmospheric pressure plasma techniques.

REFERENCES

[1] Mori, T.; Tanaka, K.; Inomata, T.; Takeda, A.; Kogoma, M. *Thin Solid Films* 1998, *316*, 89-92.

[2] Ogawa, S.; Kiuchi, K.; Tanaka, K.; Kogoma, M. In *Proc. 18th Symp. Plasma Proc.* 2001, 449-450.

[3] Prat, R.; Koh, Y. J.; Babukutty, Y.; Kogoma, M.; Okazaki, S.; Kodama, M. *Polym.* 2000, *41,* 7355-7360.

[4] Kogoma, M.; Suzuki, S.; Tanaka, K. In *Proc. ISPC16 Poc.* 2003.

In: Generation and Applications of Atmospheric Pressure Plasmas ISBN: 978-1-61209-717-6
Editors: M. Kogoma, M. Kusano and Y. Kusano

Chapter 8

SYNTHESIS OF VERTICALLY-ALIGNED SINGLE-WALLED CARBON NANOTUBES USING ATMOSPHERIC PRESSURE PLASMA ENHANCED CHEMICAL VAPOUR DEPOSITION

Tomohiro Nozaki and Ken Okazaki
Department of Mechanical and Control Engineering,
Tokyo Institute of Technology, Meguro, Tokyo, Japan

INTRODUCTION

Atmospheric pressure non-equilibrium plasma shows advantages that a highly reactive field can be generated at room temperature and atmospheric pressure using a simple setup. It is almost free from material restrictions of a reaction chamber, which are commonly the case at low pressure and/or at high temperature. At atmospheric pressure, gas breakdown is generally characterized by formation of streamers. In such a case, a large number of filamentary discharges, typically 10-100 μm diameter, are spatially and temporarily generated between the gap; consequently, large volume between electrodes can be ionized. Dielectric barrier discharge (DBD) is the well known among these discharge plasmas. DBDs are practically applied mainly for gas phase reactions such as ozone generation, air cleaning, excimer light sources, and chemical processes [1].

On the other hand, Okazaki, Kogoma and their group at Sophia University developed an atmospheric pressure glow (APG) discharge in the late 1980s [2]. The APG shows spatial uniformity similar to low pressure plasmas. Due to its simplicity for generating reactive plasma at atmospheric pressure and unique gas phase plasma chemistry compared to conventional low pressure non-thermal plasma the APG attracts keen attention in both academic research and industry applications. Both basic and application oriented researches are extensively progressing. Conventionally most of plasma processing has been performed at low pressure. It is not because low pressure plasma has an absolute advantage over atmospheric pressure plasmas, but because it was difficult to generate a stable glow-like discharge at atmospheric pressure. If a working pressure is extended from low pressure to atmospheric pressure, large productivity, mass treatment of materials, simplified processes, and drastic reduction of energy consumption is anticipated. In addition it is possible to establish unique plasma processing, which is enabled only at atmospheric pressure [3]. Furthermore, if the working pressure is extended between 10^{-3} and 10^{3} Torr (atmospheric pressure), it may enable a comprehensive understanding of plasma chemistry from low to atmospheric pressures. That is why APG attracts significant attention both from basic study and applications.

Based on these backgrounds, we have been studying controlled growth of single-walled carbon nanotubes (SWCNTs) [4-8] using Atmospheric Pressure Plasma Enhanced Chemical Vapour Deposition (AP-PECVD). At atmospheric pressure, molecules collide frequently in the plasma sheath, which exists between an electrode and a bulk plasma, and is important for AP-PECVD. Due to the highly collisional nature of such plasma sheath, acceleration of positive ions to a substrate is inefficient at atmospheric pressure. Therefore, ion-induced damage to substrate and/or CNTs, which is the main reason why synthesis of SWCNTs is difficult with low pressure plasmas, can be suppressed with atmospheric pressure plasma [9]. Such a unique reaction system at atmospheric pressure is also proposed for other material synthesis, and its usefulness has been investigated [10-12]. In the present work, basic properties of APG are introduced referring synthesis of CNTs. Attempt to establish successful synthesis of vertically-aligned SWCNTs, which was believed to be difficult with conventional low pressure plasma processing, is presented in this Chapter.

1. Basic Properties of APG

When the APG is driven by low frequency range between 10 kHz and 100 kHz, it is difficult to sustain a stable glow-like discharge without limiting discharge current by inserting dielectric barrier between metallic electrodes. The term "APG" is used because spatially uniform plasma structure resembles typical low pressure glow discharge. In addition, gas breakdown mechanism is clearly distinguished from streamer type DBD. However, in terms of electrode configuration, low frequency APG is categorized as a special case of DBD operated in diffuse mode. On the other hand, if the driving frequency increases up to radio frequency (13.56 MHz), the driving voltage significantly decreases due to charge-trapping, and the stability of APG is drastically improved. Consequently uniform glow discharge can be generated at atmospheric pressure without using dielectrics, and controllability of CVD process is markedly improved; RF-driven APG is no longer dielectric barrier discharge. In the next section, basic properties of APG at different driving frequencies are discussed.

1.1. Experimental Setup

Figure 1 shows the AP-PECVD setup used for the present work. A pair of 4 cm diameter parallel plate circular electrodes is installed in an approximately 10-liter metallic chamber. The upper electrode is used as the powered electrode. It is equipped with a water-cooled shower-head electrode so that the gas is uniformly fed onto the grounded bottom electrode where substrate is located. The bottom electrode is electrically heated up to 700 °C. The chamber was evacuated until 1 Pa using a rotary pump; the gas replacement is promptly carried out. Voltage and current waveforms were recorded using an oscilloscope at the location shown in Figure 1. The metallic chamber is connected to the ground in order to remove the floating capacitance of the system on the current measurement. A CCD camera (Andor iStar) was used for collecting the images of the plasma and optical emission spectroscopy. The emission distribution from excited species was measured using band pass filters and a spectrometer (ACTON SpectraPro 500i). Figure 1 also shows an image of the discharge during CNT synthesis. As plasma hardly diffuses at atmospheric pressure, emission from APG is limited between the electrodes, and thus deposition onto the chamber wall and the view port is almost negligible. Namely the process does not need frequent cleaning.

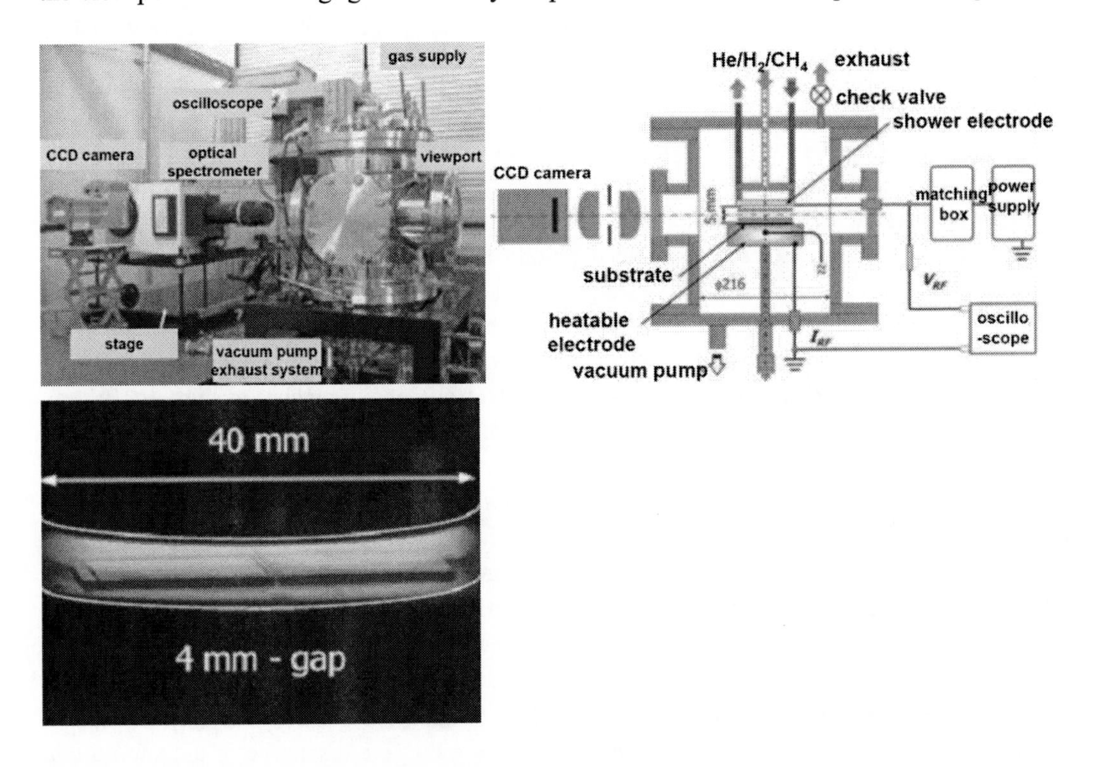

Figure 1. Atmospheric pressure glow discharge reactor setup. A digital image shows atmospheric pressure glow discharge in RF (13.56 MHz) operation during CNT synthesis.

1.2. Low Frequency APG and its Application to CNT Synthesis

Figure 2 shows the voltage and current waveforms and the emission profile of H_α at 656 nm between the electrodes, when an APG was generated in a mixture of He and CH_4 (He/CH_4 = 1000/10 cm^3/min). The APG was driven by 35 kHz sinusoidal voltage. The gas breaks down once every half cycle. However, as the dielectric plate of 1-mm thick alumina between the electrodes limits the discharge current and forms a reverse electric field, the discharge is abruptly extinguished, forming a single current pulse with nanosecond duration (half width ca. 500 ns). As the applied voltage increases, the net electrical potential between the gap increases repeatedly, then the second and the third current pulse is consecutively appeared [13].

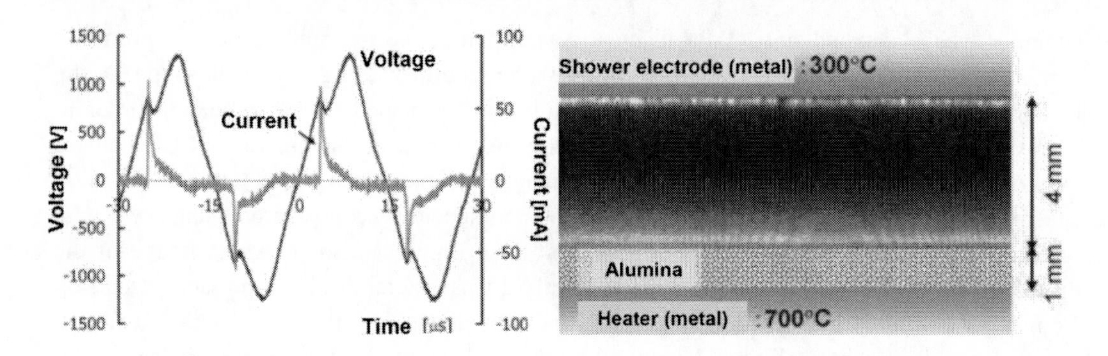

Figure 2. Voltage–current waveforms and emission distribution of Hα (656nm). 35 kHz, 2 W, 900 V (rms), 75 mA (peak value).

Helium metastable species (excitation energy of ca. 20 eV) have a long lifetime, and can ionize most gases by Penning ionization mechanism. Therefore breakdown voltage of helium containing gas mixture is substantially lower than those of other gases. In other words, when feed gas is diluted with a sufficient amount of helium, electron avalanche is not transferred to a filamentary discharge (streamer) while it readily forms glow-like diffused discharge. The major ionization mechanisms at the rising part of the pulsed current are due to electron collision and Penning ionization. When the current reaches at the pulse maximum, the negative glow, the Faraday dark space and the positive column are observed, which are similar to those in a low pressure glow discharge. Electron density reaches as high as 10^{11} cm^{-3} in the negative glow [14]. Although the current pulse is decayed, Penning ionization continues to produce secondary electrons and sustains electron density of greater than 10^6 – 10^7 cm^{-3}, which is necessary for the gas breakdown during the next half cycle. A next single current pulse will be formed at the next half cycle. The next discharge has to be ignited before the residual electrons are lost during no discharge period. It is shown by experiments and 2-dimensional numerical analysis that the driving frequency has to be more than 10 kHz in order to fulfil this requirement [14, 15]. The low frequency APG is the discharge in the sub-normal region, where the potential between electrodes decreases with the abrupt increase in the discharge current ($\partial V/\partial I < 0$). Therefore a low frequency APG is inherently not stable, and exists transiently. It is difficult to sustain a stable glow-like discharge in the sub-normal regime when high discharge current passes.

When the electrode becomes a temporal cathode, an intense luminous thin layer is formed near the cathode [16, 17]. This layer is considered to be the superposition of the cathode glow and the negative glow, commonly defined in a DC glow discharge at low pressure. However for simplicity it is called negative glow in this chapter. Because the negative glow is formed near substrate surface, the APG is suitable not for gas treatment but for surface treatment and CVD. On the other hand, when low frequency APG is applied for synthesis of CNTs, CNT growth is promoted only when the negative glow is formed on the substrate, namely the positive pulse current is passing. In order to investigate this phenomenon, unipolar pulse voltage (125 kHz) was applied and carbon nanofibres (CNFs) were synthesized onto a Ni coated (ca. 20 nm) quartz substrates [8]. A representative result is shown in Figure 3. The negative glow formed on the substrate enabled CNF growth. However, due to the small duty ratio of the pulse, the deposition rate was as low as 0.2 μm/min, and the duration of high electric field, which is necessary for vertical alignment of the growing CNFs, was similarly short. On the other hand, when the negative glow is formed near the counter electrode, CNFs are hardly grown. A beads-on-string structure is observed in Figure 3 (c). This indicates that CNFs are intermittently grown only when negative glow is generated near the substrate. In conclusion, in order to synthesize vertically-aligned CNFs, it is preferable to form the negative glow continuously near the substrate.

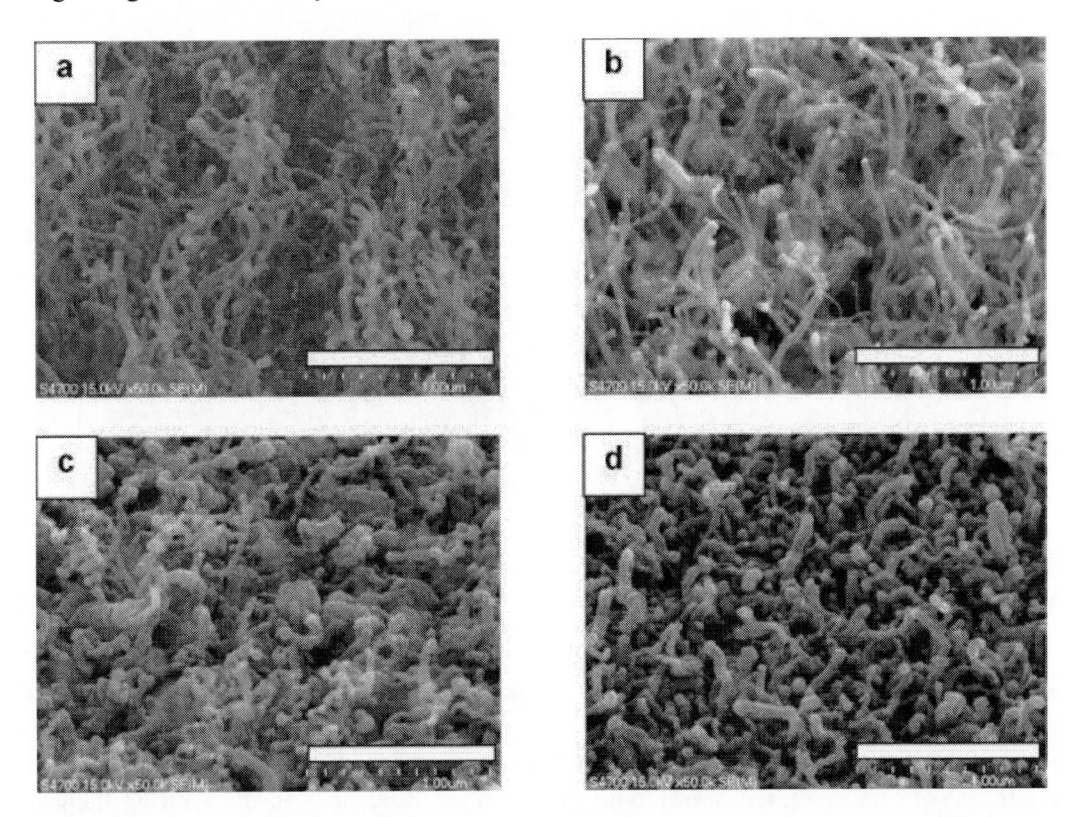

Figure 3. Synthesis of CNF by APG driven by unipolar pulsed voltage [8]: (a), (b) applying positive pulse voltage, (c), (d) applying negative pulse voltage. Frequency: 125 kHz, substrate temperature: 700 °C, $He:H_2:CH_4$ = 900:100:10 sccm, Ni catalysis (estimated thickness 20 nm), scale bar: 100 μm.

1.3. Development of Atmospheric Pressure Radio-Frequency Discharge (APRFD)

The APG was formed using radio frequency (RF) power source so that driving frequency is much faster than the characteristic frequency of current pulse observed in low frequency APG (see Figure 2), that is,

$$13.56 \text{ MHz} >> 1 / (2 \times 500 \text{ ns}). \quad (1)$$

Figure 4 shows the voltage and current waveforms and the emission profile between the metallic electrodes. The current waveform of the APRFD shows a sinusoidal wave with the forwarded phase shift rather than the single current pulse as shown in Figure 2. The plasma density in RF discharge is high due to accumulation of ions and radicals between the gap, and the electrical conductivity of gas media is steadily high. Consequently a plasma with lower voltage and higher current can be formed by RF than low frequency APG. In particular, if the concentration of metastable helium, which is necessary for Penning ionization, becomes higher, the glow discharge can be more stably sustained. In addition, since applied voltage can be drastically decreased owing to the residual ions, a stable glow discharge can be sustained even without inserting dielectric barrier between the metallic electrodes. Without dielectric material, controllability of PECVD is significantly improved. For example, low frequency operation likely induces filamentary discharge when an electrically conductive substrate is used. In addition, technical problems such as cracking of dielectrics due to substrate heating are inherently avoided.

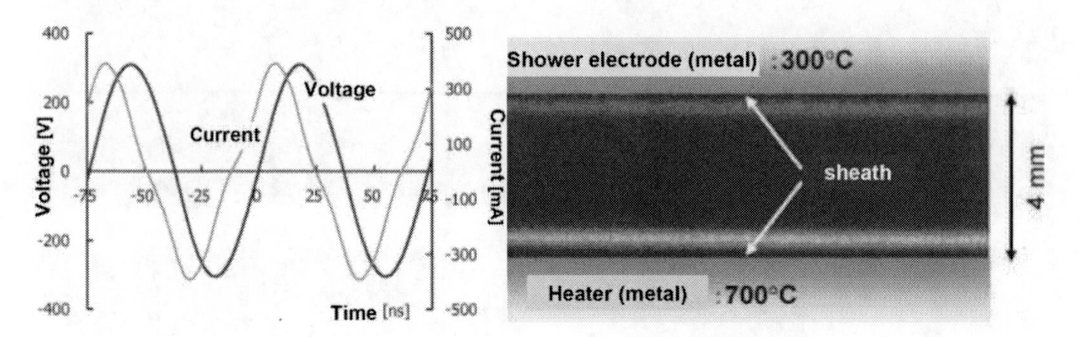

Figure 4. Voltage–current waveforms and emission distribution of Hα (656 nm): 13.56 kHz, 50 W, 200 V – 280 mA (rms value).

Although the emission profile of the APRFD and low frequency APG are alike; however, there is a distinct difference of the emission structure near the electrode. In the low frequency APG, ions are accelerated toward the momentary cathode; subsequently, ion-electron recombination takes place on the electrode. Therefore the negative glow looks attaching to the electrode with time average observation. On the other hand, ions in the APRFD trapped between the electrodes form a space charge. The plasma sheath is recognized as a dark space created between the negative glow and the electrode surface. In the plasma sheath a large electric field is created, which is used for vertical alignment of growing CNTs. The APRFD has been studied as a plasma source which is useful for the controlled growth of SWCNTs. In the next section, basic properties of APRFD are further discussed.

1.4. Mode Transition of APRFD and Sheath Parameters

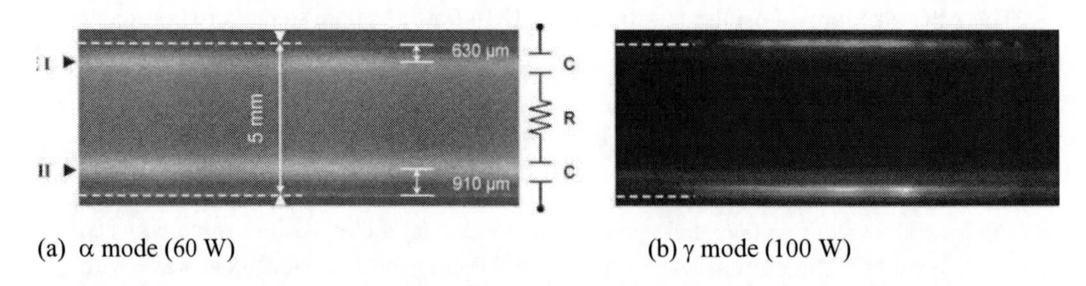

(a) α mode (60 W) (b) γ mode (100 W)

Figure 5. Emission distribution of CH (432 nm).

Figure 5 shows the emission profile of CH (432 nm) when APRFD is formed in different modes. The gas mixture of $He/H_2/CH_4$ = 1000/30/16 cm^3/min was fed between the electrodes. Since plasma species do not diffuse long distance at atmospheric pressure, a symmetric capacitively coupled discharge is formed between the gap as shown in Figure 5 (a). The sheath near the upper electrode is thinner than that near the bottom electrode. This asymmetric structure is due to different gas temperature near each electrode (bottom electrode: 700 °C, upper electrode: ca. 300 °C). Numerical simulation of APRFD indicates that 30 – 40 % of the applied voltage drops across the dark space (i.e. plasma sheath), and the electric field near the electrode surface reaches as high as 4 kV/cm [18]. However, as the electron density in the sheath is low, ionization does not occur in the sheath. Since the ionization is dominated in the gas phase due to electrons trapped between the electrodes, it is called the α–mode discharge (Figure 5 (a)) [19]. In the α–mode regime, the sheath is electrically equivalent to a capacitor and thus inserting dielectrics between the electrodes will not change the electrical properties of the discharge significantly. However, as input power is increased, the voltage drop in the sheath increases furthermore, and accelerates secondary electrons emitted from the momentary cathode, inducing significant ionization occurring in the sheath. As a result, an arc spot with strong emission is formed in the sheath. In such a case, as the secondary electron emission from the electrode dominates the ionization mechanism, it is called a γ-mode discharge (Figure 5 (b)) [19]. In the γ-mode regime, due to its high current and high thermal energy, substrates are seriously damaged. Therefore the α-mode discharge must be chosen for SWCNT growth. Note, emission intensity of Figure 5 (b) is approximately 10 times greater than that of Figure 5 (a).

For simplicity, the α-mode discharge shown in Figure 5 (a) is assumed to be a symmetric capacitively coupled discharge with the equivalent circuit of the sheath and the bulk plasma as a capacitance C and a resistance R, respectively (see inset of Figure 5 (a)). From the root mean square of voltage and current measured by oscilloscope, and their phase difference, the resultant capacitance C_{sum} is estimated using Equation 2. The thickness s of the collisional sheath is given by Equation 3 [20], where ε_o is permittivity of vacuum. The voltage through the sheath V_s is estimated by the capacitance C_{sum} and discharge current using Equation 4. Results are summarized in Table 1.

$$C_{sum} = \frac{I\sqrt{1+\tan^{-2}\phi}}{\omega V} \tag{2}$$

$$s = \frac{1.52\varepsilon_0 A}{2C_{sum}} \quad (C = 2C_{sum}) \tag{3}$$

$$V_s = \frac{I}{2\omega C_{sum}} \tag{4}$$

Table 1. Sheath parameters of atmospheric pressure RF discharge

Power [W]	Voltage (rms) [V]	Current (rms) [mA]	Sheath capacitance [pF]	Phase shift [°]	Sheath thickness [μm]	Sheath voltage [V]	Average electric field [V cm^{-1}]
40	356	147	49.84	6.35	1332	136	1021
60	377	156	45.17	6.85	1233	134	1084
80	398	180	40.58	8.16	1036	129	1249

The sheath thickness of APRFD, approximated by the equivalent circuit, shows a good agreement with the result of emission profile of APRFD: the dark space in Figure 5 (a) is equivalent to the plasma sheath and the negative glow corresponds to the boundary of the bulk plasma and the sheath. However, as the plasma is oscillating by radio frequency, it is difficult to distinguish boundary between the sheath and the bulk plasma. The results shown in Table 1 indicate that a high electric field of approximately 1 kV/cm is formed in the collisional sheath at atmospheric pressure. The order of this value agrees with the result of the numerical calculation for APRFD (4 kV/cm) [18]. At atmospheric pressure relatively high voltage gradient exists even in a bulk plasma, and secondary electrons are generated by different mechanism from low pressure plasmas. According to Equations 2, 3, and 4, about 30 % of the applied voltage drops across the plasma sheath. This estimation also shows a good agreement with a numerical analysis.

The mean free path of CH_4^+ in helium at 100 kPa and 700 °C is approximately 0.7 μm [21]. Transportation of CH_4^+ across plasma sheath with approximately 1 mm thickness results in more than 1000 times collisions with other molecules. Assuming that the drift velocity of ions in a collisional sheath is the product of the electric field and the mobility, the mean kinetic energy of CH_4^+ and He_2^+ in 1 kV/cm are estimated as 0.043 and 0.014 eV, respectively. The ion energy is lower than the kinetic energy of molecules (ca. 0.1 eV) at 700 °C. Non-energetic ions do not induce any damage to CNTs. The most important characteristic of the APRFD is that PECVD is carried out with negligible ion damage while applying high electric field to the substrate.

2. PECVD of Carbon Nanotubes

A carbon nanotube (CNT) is a nano material attracting a significant attention in a variety of areas including electric devises, sensors, biomedical applications and so on. CNTs have been synthesized using various methods. However, SWCNTs, which show unique electric, mechanical, thermal and chemical properties, are mostly synthesized using thermal CVD [9]. On the other hand, PECVD exclusively synthesizes multi-walled CNTs (MWCNTs) or CNFs which contain remarkable structural defects as shown in Figure 6 (b). Advantage of PECVD is synthesis of isolated and aligned MWCNTs and CNFs as shown in Figure 6 (a). However, high energy ions accelerated in the plasma sheath induce serious damage to nano-structured catalysts (Figure 6 (c)), which are indispensable for synthesis of SWCNTs. Therefore PECVD has been thought to be unsuitable for synthesis of SWCNTs until recently. In addition, if reactive species are excessively supplied, catalyst activity is readily degraded due to deposition of substantial amount of amorphous carbon. On the other hand, use of atmospheric pressure plasma avoids plasma-induced damage, enabling SWCNT synthesis at high yield because highly collisional plasma sheath acts as diffusion barrier of reactive species.

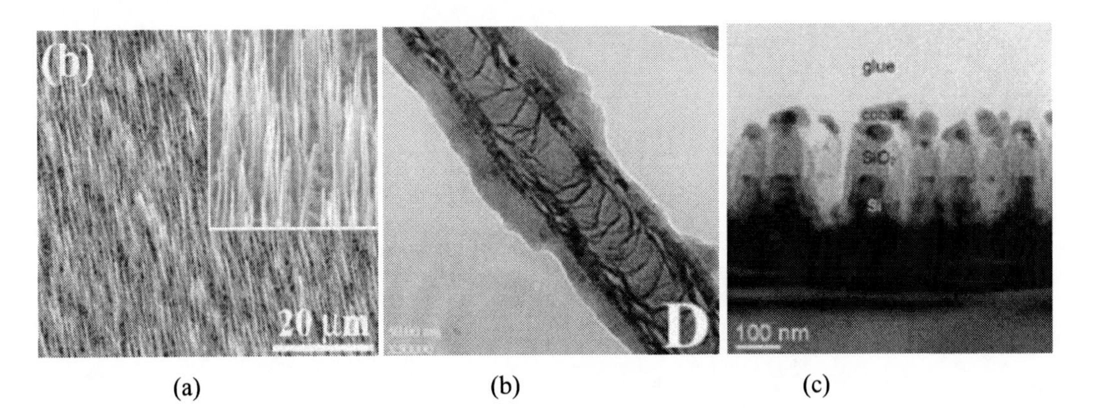

(a) (b) (c)

Figure 6. CNFs synthesized by low pressure PECVD. (a) Vertically-aligned MWCNTs [22], (b) high resolution TEM micrograph of CNFs [23], (c) substrate etched by ion bombardment before CNT synthesis [24].

2.1. SWCNT Synthesis by PECVD: A Review

Comprehensive reviews on synthesis, characterization and applications of CNTs including thermal CVD are available [25-28]. Here, a general review is presented focusing on PECVD. In details, please see Table 2. In 2003, Hatakeyama and his group in Tohoku University reported successful synthesis of isolated aligned SWCNTs, immersing Co-Fe catalysis supported at nano-sized pores of zeolite using magnetron-type remote plasma [29, 30]. This pioneering study is highly appreciated, showing the importance of controlling ion flux and energy onto the substrates [31, 32]. Point-Arc Microwave Plasma (PA-MP) CVD at low pressure recently developed by the group in Waseda University, enabling synthesis of highly aligned SWCNTs [33, 34]. Only reactive species are transported onto the substrate by generating microwave plasma around a needle electrode placed far above the substrate. Ions flow into the needle electrode, and thus the plasma and the substrate are not coupled electrically. They also developed the unique catalyst for SWCNT growth, which sandwiches an Fe catalyst with Al_2O_x. The catalysts maintain chemical activity more than 10 hours at 600 ^{o}C, demonstrating successful synthesis of 5 mm high SWCNT towers. It is also reported that double walled CNTs (DWCNTs) were synthesized using remote-type microwave plasma as a similar preceding technique [35-37]. It is important that a plasma ball generated by microwave irradiation and the substrate should not physically contact each other. Otherwise graphitic carbon nano-flakes were favourably synthesized [36, 37]. These groups used laser abrasion or plasma arc gun to deposit Co nanoparticles onto a substrate and used them as catalysts. Its difference with the PA-MP CVD is electric power consumption (heating substrate up by plasma) and the catalyst preparation methods, while they are essentially considered to be the same process. In summary there are two factors for the successful synthesis of SWCNTs using PECVD. The first is to mono-disperse a 1-2 nm transition-metal catalysts or its alloys onto the substrate. The second is to use a remote plasma so that generation of a plasma and synthesis of SWCNTs are spatially separated.

Table 2. Comparison of SWCNT synthesis method by PECVD

	Carbon structure	Catalyst	Buffer/ Substrate	Catalyst prep. method	Temp..	Pressure	Plasma source	Growth rate	Gas	Time
This work	*SWCNTs: crowding regime*	*Fe/Co 1-2 nm*	*Al_2O_3(~20nm) / Si-wafer*	*Dip-coating*	*700 ℃*	*810 Torr*	*APRFD 13.56 MHz 60 W*	*4 μm/min*	*$He/H_2/CH_4$ 1000/30/16*	*20 min*
[31]	SWCNTs: free-standing	Co/Fe pore size ~0.74nm	Zeolite/silver plate	Spin-coating	650°C	0.5 Torr	Magnetron 13.56 MHz 900 W	unknown	CH_4/H_2 30/70	0.5-30 min
[32]	SWCNTs free-standing	Fe < 1 nm	Al_2O_3(<20nm)/ Si-wafer	Vacuum evaporator	750°C		RF plasma 13.56 MHz 300 W			0.5-3 min
[34]	SWCNTs: crowding regime	Fe 0.5 nm	Al_2O_3(0.7nm) /Fe/Al_2O_3(5n m)/ Silicon-wafer	Sputtering	600°C	20 Torr	PC-MA 2.45 GHz 60 W	3.5-4.5 μm/min	CH_4/H_2 5/45	40-120 min
[35]	SWCNTs/ DWCNTs crowding regime	Fe 0.3-0.5nm	SiO_2(180nm)/ Si-wafer	Electron beam evaporator	850°C	20 Torr	MW plasma 2.45 GHa 600 W	7-12μm/mi n	C_2H_2/NH_3 1/4	30-40 s
[36]	DWCNTs,: crowding regime	Co 2-3 nm	TiN(20nm)/ Si-wafer	Arc plasma gun	700°C	70 Torr	MW plasma 2.45 GHz 900 W	36 μm/min	H_2/CH_4 70/50	20 min
[37]	MWCNTs: crowding regime	Co 5-10 nm		Laser ablation (KrF: 248 nm)	600-800°C			18 μm/min	H_2/C_2H_2 70/50	3 min
[42]	SWCNTs: crowding regime	Co/Mo 1-2 nm	Quarts	Dip-coating	800°C	10 Torr	Thermal CVD (w/o plasma)	~0.5 μm/min	Ar/H_2/EtOH 7/0.2/3	10 min

As for the synthesis of CNTs using atmospheric pressure plasmas, a remote microwave plasma [38, 39], atmospheric pressure plasma jet [40], and low frequency capacitively coupled discharge [41] are reported. The advantage of plasma jet is that they are applicable to treat 3-D shaped structures.

However, this method can not create plasma sheath which play a main role for the alignment of growing CNTs. In addition, species and heat flux are strongly coupled, thus individual control is not always easy. They are the reasons why the plasma jet has not synthesized SWCNTs yet. Kyung *et al.* proposes a DBD type reactor. However, due to inappropriate reaction system (C_2H_2 as a source gas combined with Ni catalysts) and low driving frequency (20-100 kHz), CNTs contain a substantial amount of defects. Even if APRFD is used, SWCNTs can not be synthesized if an unsuitable catalyst is chosen. Figure 7 shows the CNFs grown using Ni catalysts [6]. In this experiment, Ni thin film (ca. 20 nm) was initially deposited on Chromium which is used for Ni underlayer. After heating the substrate, Ni film breaks into nanoparticles and CNFs start growing (Figure 7 (a)). Although APRFD enables CNF growth, the deposition rate is low at 0.4 μm/min and the height did not exceed 2 μm. Transmission electron microscopy revealed that Ni catalysts attaching at the tip of CNF is completely covered by solid carbon; Ni catalysts lose their activity. However vertical alignment was drastically improved compared with CNFs synthesized by low frequency APG as shown in Figure 3. Raman spectrum in Figure 7 (d) shows the notable D-band peak indicating structural defects or synthesis of amorphous carbon. In addition the band width of the G-band peak is much wider than SWCNTs.

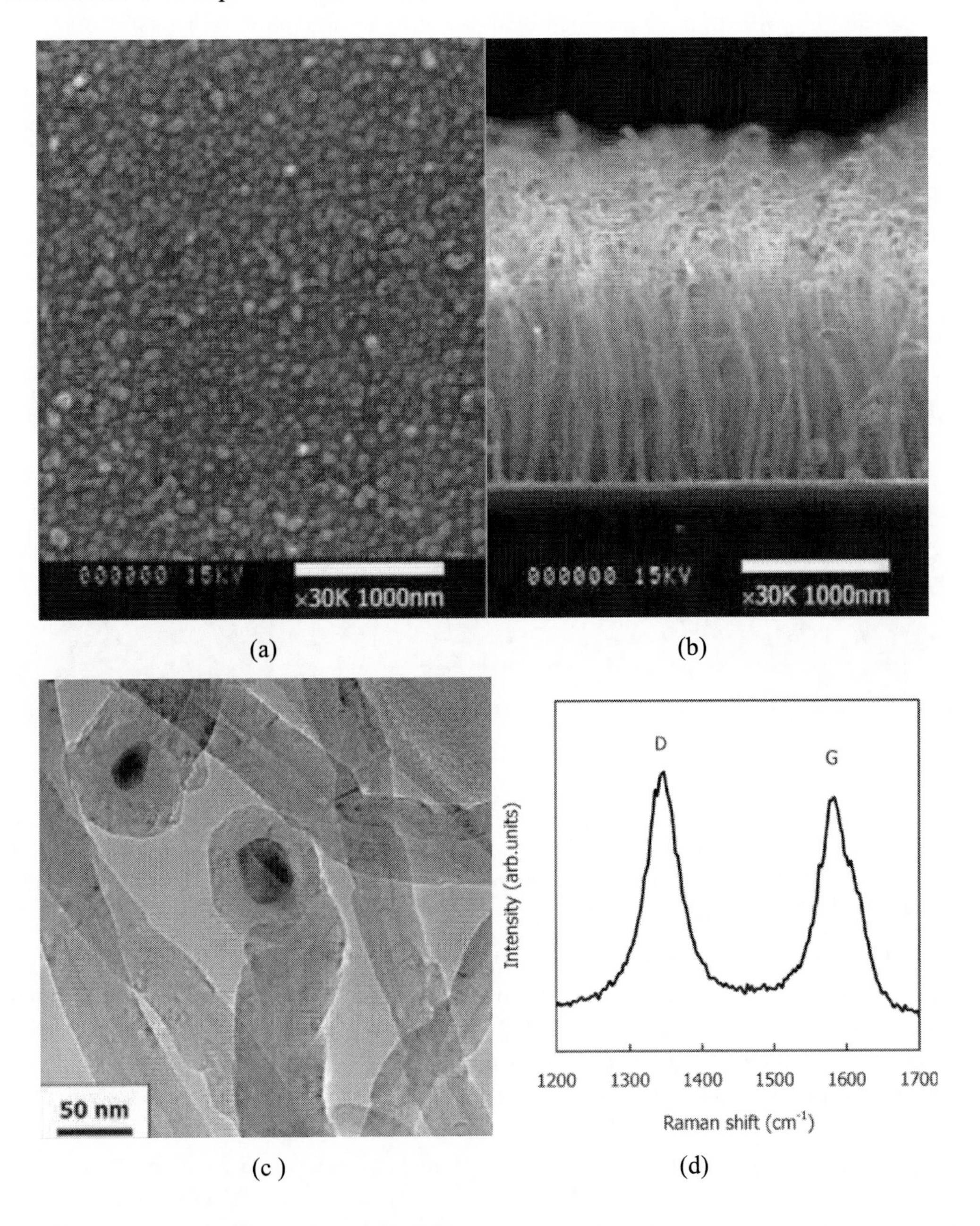

Figure 7. CNFs synthesized by APRFD with Ni catalysts originated from thick Ni film (ca. 20 nm) [6]. Substrate temperature: 700 °C, $He:H_2:CH_4$ = 1000:4:2 cm^3/min, power: 60 W, synthesis time: 10 minutes. (a) SEM micrographs of the substrate after 10-minute thermal CVD, (b) after 10-minute APRFD irradiation. (c) TEM micrograph and (d) Raman spectrum of sample shown in (b).

2.2. Experimental Methods

The AP-PECVD was performed using the experimental setup shown in Figure 1, and SWCNTs were synthesized in the α-mode of APRFD. Detailed information on the preparation of the catalyst is seen in [42]. Briefly, a 2-inch diameter silicon wafer on which aluminium is sputter-deposited is used as the substrate. It was first oxidized in air at 400 °C, and catalytic fine particles were prepared by dip-coating using 0.05 wt % Fe and Co acetate solution [43]. This method enables mono-dispersion of 1-2 nm metallic fine particles on the substrate. Before synthesis of CNTs, the chamber was evacuated down to 1 Pa, then He and H_2 mixture was fed until atmospheric pressure. Oxidized catalyst nanoparticles were reduced by APRFD at 400 °C for 5 minutes. Subsequently substrate temperature was elevated up to the desired level, CH_4 was fed into the chamber and AP-PECVD was performed for 5 - 60 minutes.

2.3. SWCNT Synthesis

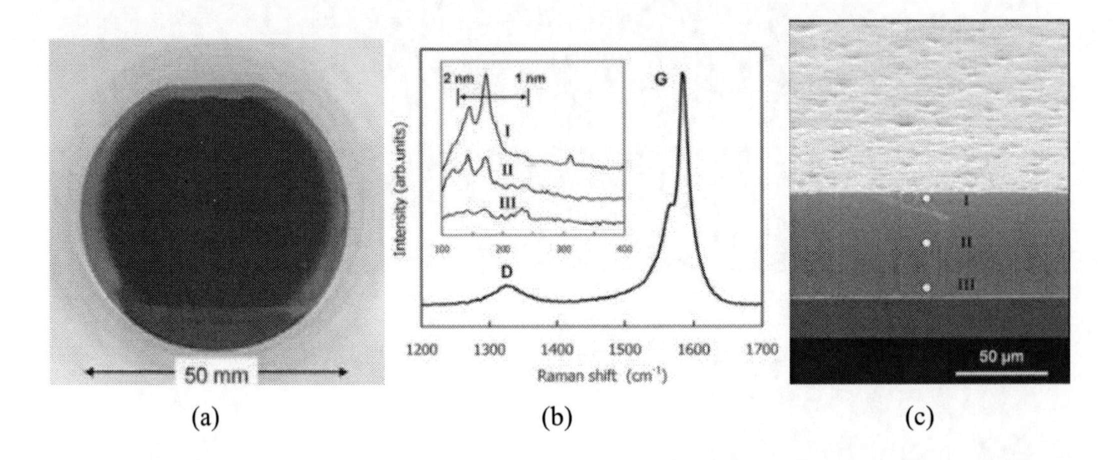

Figure 8. SWCNTs synthesized by APRFD with Fe/Co catalyst nanoparticles. Substrate temperature: 700 °C, He:H_2:CH_4 = 1000:30:16 cm^3/min, power: 60 W, synthesis time: 20 minutes. (a) 2-inch Si substrate after AP-PECVD. (b) Cross-section Raman spectra. (c) cross-section SEM micrograph after 20-minute synthesis.

CNTs synthesized with input power of 60 W, at the substrate temperature of 700 °C, gas flow rate of He/H_2/CH_4 = 1000/30/16 cm^3/min for 20 minutes was characterized by scanning electron microscope (SEM: HITACHI S-800) and Raman spectroscopy (JASCO NRS-1000: 532 nm excimer laser). Figure 8 (a) shows the image of the substrate after the synthesis. CNTs are uniformly synthesized on the 2 inch diameter substrate. The cross-section SEM micrograph in Figure 8 (c) indicates that 60 μm thick vertically-aligned CNTs were synthesized after 20-minute treatment. The low frequency region in Raman spectrum shows the radial breathing mode peaks (RBM: ω_{RBM} = 100 – 300 cm^{-1}) which reflect radial stretching vibration of SWCNTs. The diameter of the majority of SWCNTs is distributed between 1 and 2 nm, evaluated from the following equation [44]:

$$d_{SWCNT} = \frac{248}{\omega_{RBM}} \quad (5)$$

G-band peak located around 1580 cm^{-1} is observed, but D-band associated with amorphous carbon or structural defects of graphite is only slightly observed; therefore, SWCNTs were synthesized at high yield. The inset in Figure 8 (b) shows the cross-section Raman spectra of low frequency region (I, II, III), whose locations are indicated in Figure 8 (c) as well. The RBM peaks are clearly identified near the top lay of CNTs (I). However, if observation point approaches to the substrate the RBM peak becomes weaker, indicating that as the deposition process continues, the catalyst is sintered and subsequently SWCNTs are changing to MWCNTs. This also indicates that SWCNTs grow in the base-growth regime as shown in Figure 9. In general, CNTs grow in the base-growth. In contrast, CNFs grow in the tip-growth regime due to delamination of catalyst particles from the substrate (e.g. Figure 7 (c)). At the tip-growth regime, CNT growth process is remarkably influenced by incoming reactive species including ions that promote catalyst sintering. Also, selective catalyst heating by plasma occurs. It is likely the reason why structural defects of graphite are pronounced in the tip-growth regime.

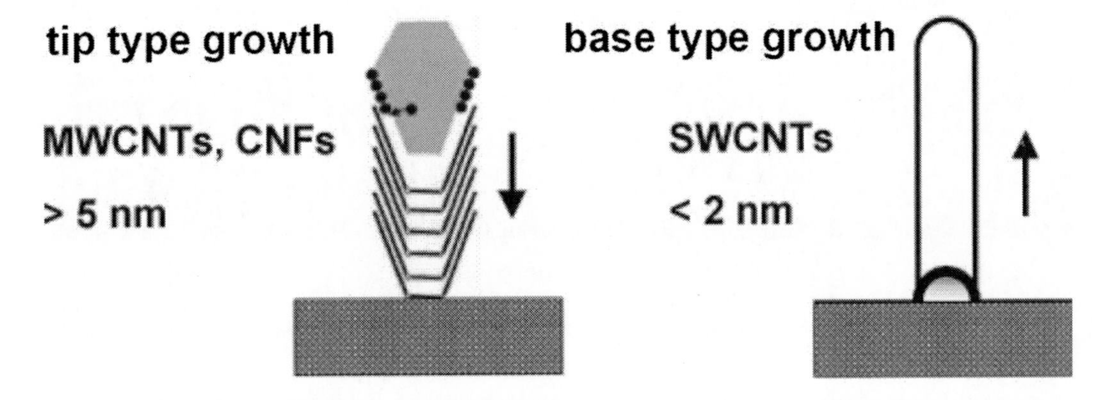

Figure 9. Growth regime of carbon nanotubes [25].

2.4. Effect of Input Power

The effect of input power was investigated, which is one of the important parameters to understand the role of PECVD. Figure 10 shows Raman spectra of CNTs at substrate temperature of 700 °C for 5 minutes with the input powers of 0, 40, 60 and 80 W. At the input power of 40, 60 or 80 W, the G-band peak splits into two G^+ and G^- peaks, while the D-band peak is sufficiently weak; SWCNTs were synthesized with high crystallinity and high selectivity. A width of G-band peak is narrow in Figure 10, which is different from that of CNFs in Figure 7. Figure 11 shows SEM micrographs of those SWCNTs. Carbon deposition is not identified without plasma irradiation, that is, 0 W. Therefore when CH_4 is used as a carbon source, and the substrate temperature is 700 °C, plasma-generated reactive species are indispensable for SWCNT growth. As the input power increased, the growth rate increased, indicating that increasing radical flux accelerates SWCNT growth.

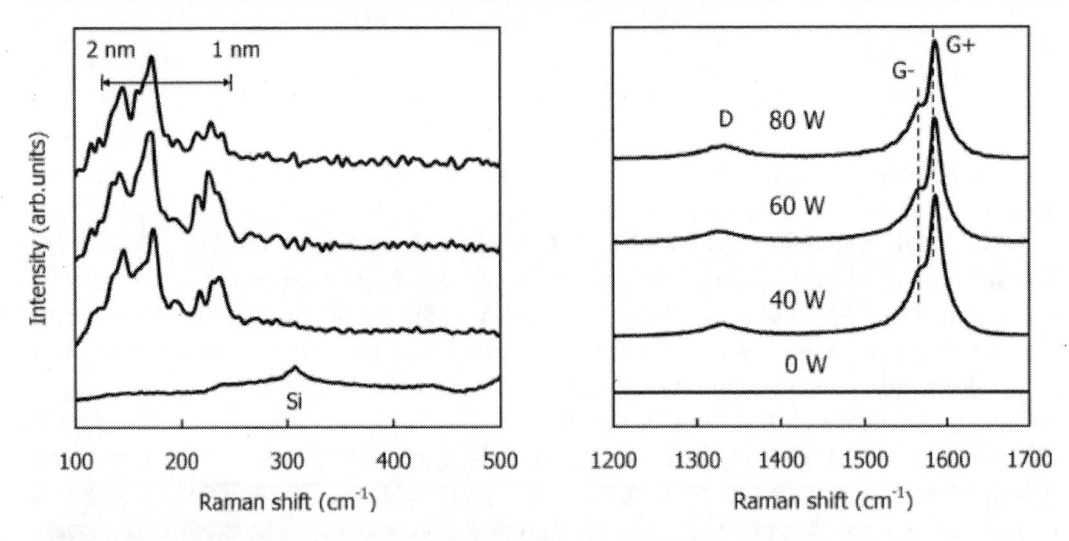

Figure 10. Raman spectra of SWCNTs synthesized at various powers.

(a) 0 W (b) 40W, 356V-147mA (c) 60W, 377V-156mA (d) 80W, 398V-180mA

Figure 11. SEM micrographs of SWCNTs synthesized at various powers. See Table 1 for the details of the sheath parameters.

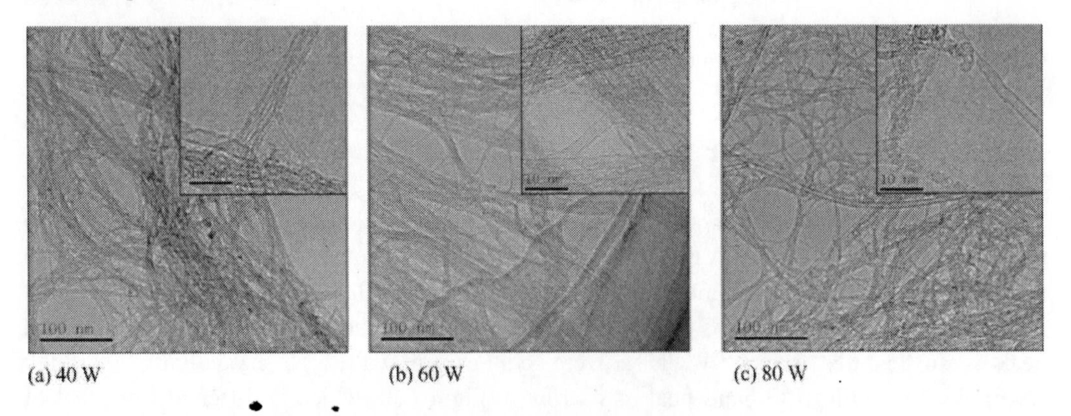

(a) 40 W (b) 60 W (c) 80 W

Figure 12. TEM micrographs of CNTs shown in Figures 10 and 11.

Figure 12 shows the TEM (JEOL JEM-3010) micrographs of those CNTs. SWCNT bundles are observed due to van der Waals force. Isolated 2-3 nm diameter SWCNTs can be observed when they were synthesized at the input power of 40 or 60 W. On the other hand at 80 W, DWCNTs and MWCNTs are frequently observed. Generation of reactive species in atmospheric pressure plasma is indispensable for synthesis of SWCNTs; however, excess supply of reactive species reduces the selectivity of SWCNTs. It is noteworthy that in each sample, CNFs containing bamboo structure were hardly observed. Sheath parameters presented in Table 1 was estimated under the same conditions with Figure 11. Even when the input power is doubled, the sheath electric field did not increase markedly. In addition G/D ratio remained almost unchanged. Ion induced damage due to the increase in the input power is negligible.

2.5. Effect of Substrate Temperature

The effect of substrate temperature is investigated at the input power of 60 W for 5 minutes. As the substrate temperature decreases, the G/D ratio markedly decreased as shown in Figure 13. At the substrate temperature of 600 °C, SWCNTs were hardly synthesized, while amorphous carbons as well as MWCNTs and CNFs were favorably synthesized. At the same time, the deposition rate drastically decreased to 0.7 μm/min. In order to synthesize SWCNTs in this particular setup, the substrate temperature of at least 650 °C must be maintained. In the case of PECVD, methane activation is promoted in the gas phase independently of catalytic reaction [45]. Subsequently, solid carbon is formed on the catalysts, then carbon diffuses across catalyst nanoparticles and extract those carbon as SWCNTs. This process is likely the rate-determining step of CNT growth. Namely, if methane dehydrogenation is promoted at low temperature (600 °C), carbon diffusion can not catch up, and solid carbon covers the catalysts before the SWCNT growth. The extreme example of this case is shown as the TEM micrograph in Figure 7 (c). When the catalyst is covered with carbon film, it prevents carbon species contacting catalyst particle; eventually, CNT growth is terminated. As the size of the catalyst particles decreases and the substrate temperature increases, carbon diffusion across catalyst particle is accelerated and thus CNT growth rate increases. However, if the substrate temperature is too high, both the growth rate and the yield decrease [46, 47].

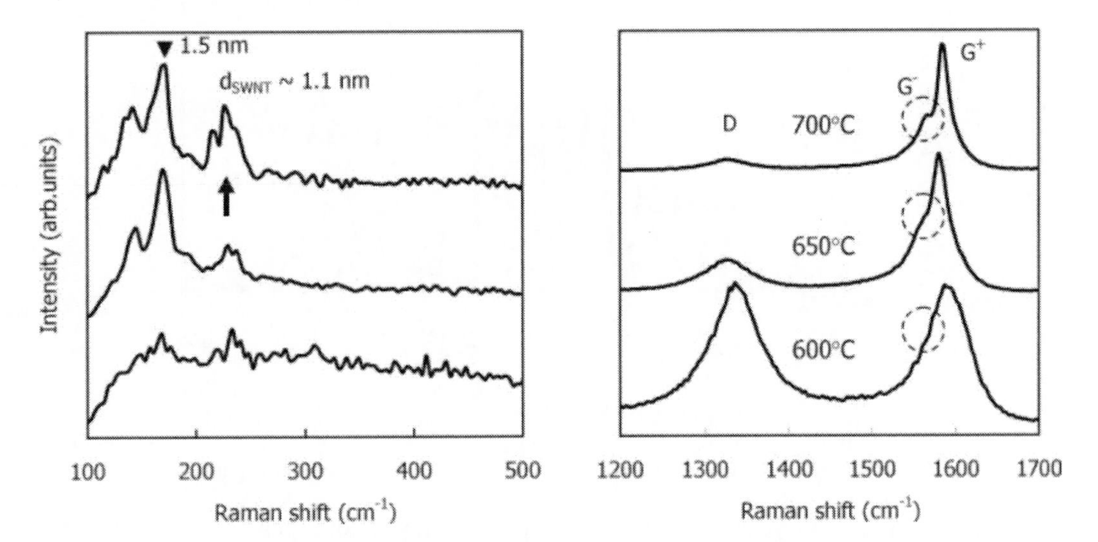

Figure 13. Raman spectra of SWCNTs synthesized at various substrate temperatures.

2.6. Gas Heating Effect by APRFD

According to the previous results, APRFD accelerates methane activation in the gas phase, but would not directly influence the reactivity of catalysts. In addition, it is highly possible that excess supply of reactive species degrades yield and properties of SWCNTs, while effect of ion bombardment is insignificant. On the other hand, apart from supply of radicals, plasma-induced heating such as RF induction heating and recombination of charges on substrate may lead to selective catalyst heating that contributes CNT growth enhancement. In particular, for the synthesis of CNTs in tip-growth regime, since catalytic nanoparticles are exposed to the reactive plasma, CNTs can be synthesized at the substrate temperature less than 200 °C because selective catalyst heating occurs [48,49]. Meanwhile, in the case of APRFD driven at the α-mode, the charge density in the sheath is low, and thus selective catalyst heating due to APRFD seldom occurs. In addition, in the case of the base-growth regime with which the catalyst sticks to the substrate, even if the catalyst is selectively heated, it is estimated that the temperature of the catalyst would not become much higher than that of the substrate due to the heat release to the substrate. Here, the gas temperature was estimated from the rotational temperature of CH, and the effect of input power to the gas heating by APPRFD is discussed.

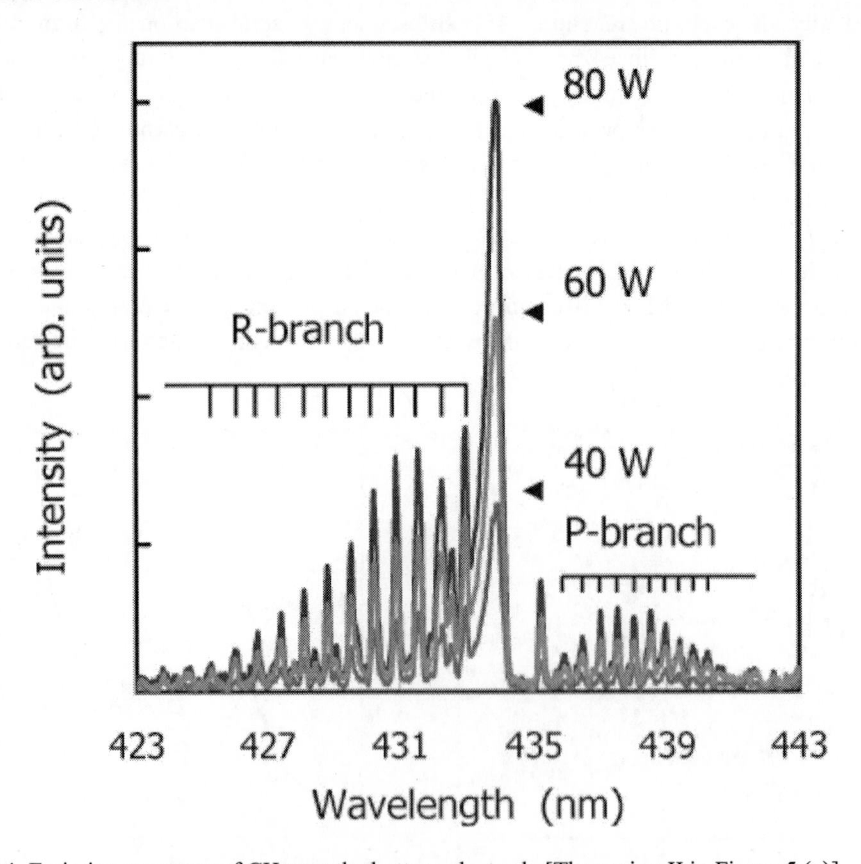

Figure 14. Emission spectrum of CH near the bottom electrode [The region II in Figure 5 (a)].

Figure 14 shows the CH rotational band spectra with various input power. Using the relative intensity of the R-branch peaks, the rotational temperature can be estimated using

Boltzmann plot [50]. Because excited CH radicals will reach the equilibrium with the gas temperature via 20-50 time collisions, the rotational temperature can be used for evaluating gas temperature. For further details, see Refs. [16, 17]. The rotational temperatures are shown in Figure 15. The gas temperature was initially 20 °C, but it is heated up to 400 °C by the negative glow near the upper electrode. After that, it is further heated to 900 °C by the negative glow near the bottom electrode. Here the substrate temperature was 700 °C. Joule heating occurs even in a bulk plasma due to the voltage gradient in it. However due to the small mobility of ions and metastable helium at atmospheric pressure, their contribution should be small. Eventually energy transfer is in majority due to inelastic collisions of electrons. This is most pronounced in the negative glow. The emission intensity increases as the increase in power, but the gas temperature remains almost constant independent of the power. The increase in the power results in the decrease in the yield of SWCNTs, but the gas temperature is almost unchanged. Therefore effect of plasma heating on the degradation of SWCNTs is insignificant. However, as the power increases, the radical density becomes higher, and excess radicals can be generated, which degrade of SWCNTs.

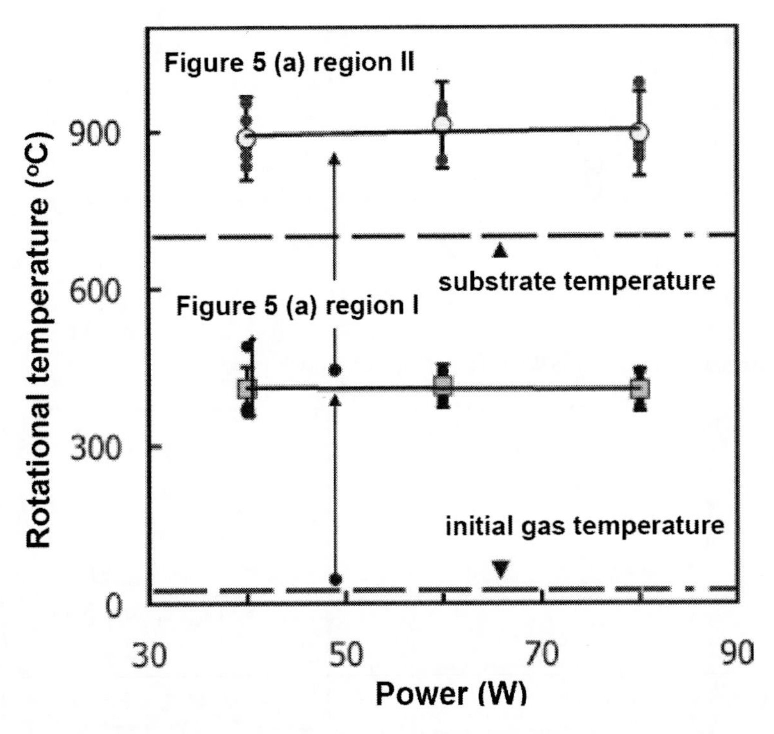

Figure 15. Rotational temperature of the negative glow. [region I and II in Figure 5 (a)]. Bottom electrode temperature: 700 °C, gap: 5 mm, pressure: 108 kPa, He:H_2:CH_4 = 1000:30:16 cm^3/min.

2.7. Growth Rate and Inactivation of Catalyst

In order to estimate the growth rate with the input power of 60 W at the substrate temperature of 700 °C, AP-PECVD was performed with various deposition time as shown in Figure 16. At the initial stage of the growth, deposition rate is approximately 4.0 μm/min. It gradually decreases, and the catalyst was degraded in 20 minutes. As discussed in the previous section, APRFD is not the major factor to shorten the catalyst lifetime. In order to investigate the degradation mechanism, catalyst-coated substrate is kept at 700 °C for relatively long time without generating APRFD, and then CNTs were analyzed to see whether excess annealing would degrade the catalysts.

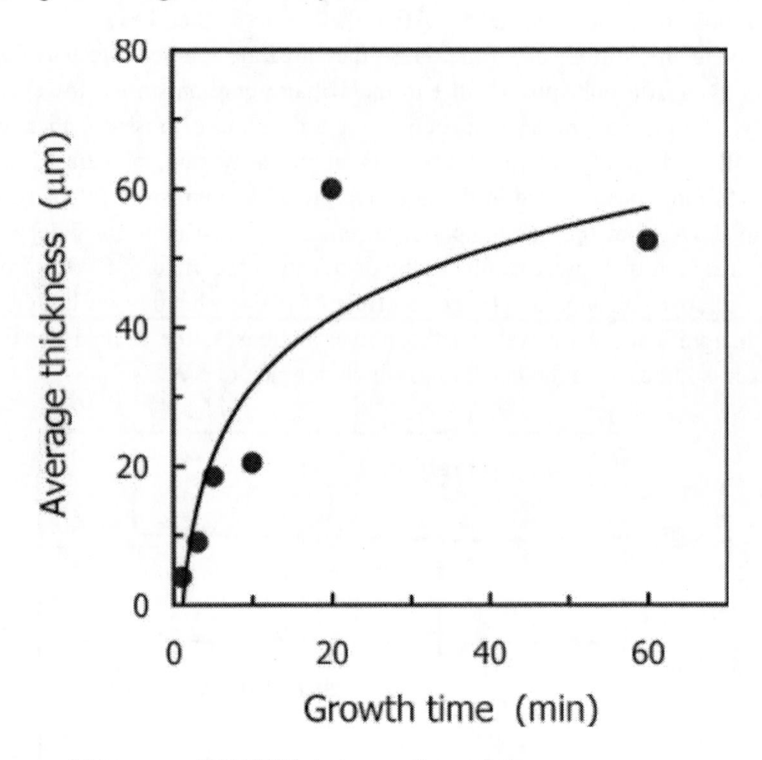

Figure 16. Average thicknesses of SWCNTs in terms of growth time.

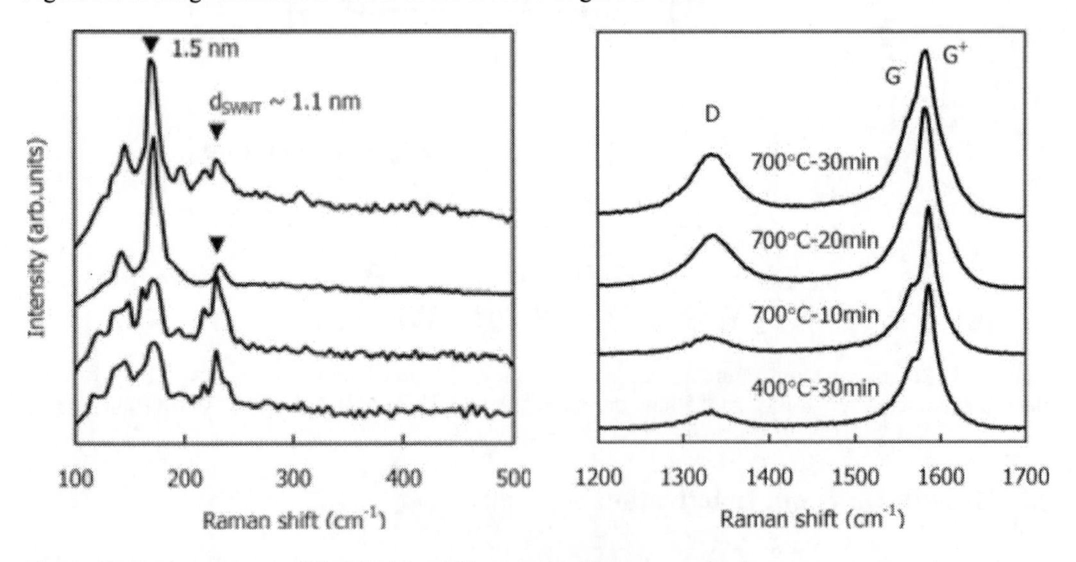

Figure 17. Raman spectra of SWCNTs at different annealing time and substrate temperature.

Figures 17 and 18 show the Raman spectra and the SEM micrographs of CNTs synthesized at 700 °C with the input power of 60 W for 5 minutes. According to the Raman spectra, when the substrate was preheated at 400 °C for 30 minutes before AP-PECVD, there is no remarkable change in the growth rate and the G/D ratio. On the other hand, if the substrate was heated at 700 °C and maintained for more than 20 minutes, the RBM peak corresponding to thinner SWCNTs (d_{SWCNT} ≈ 1.1 nm) disappears, while the peak corresponding to d_{SWCNT} ≈ 1.5 nm increases. The experiment demonstrated that excess annealing at high temperature leads to catalyst sintering and resulting CNT becomes thicker.

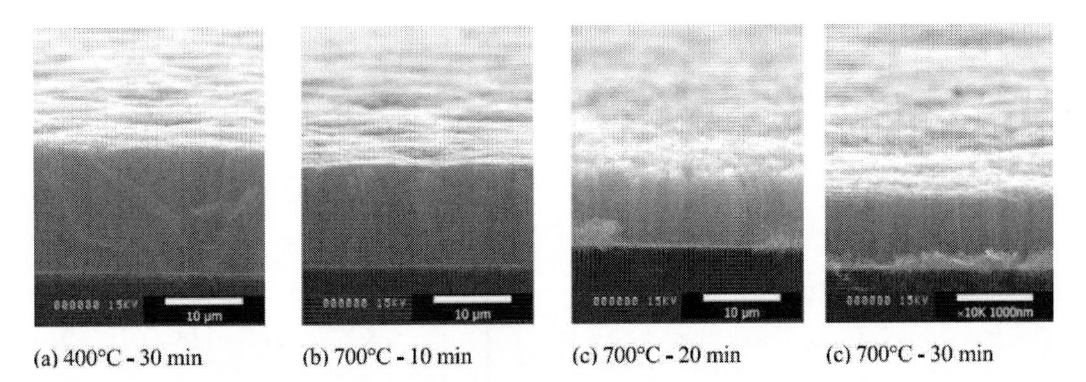

(a) 400°C - 30 min (b) 700°C - 10 min (c) 700°C - 20 min (c) 700°C - 30 min

Figure 18. SEM micrographs of SWCNTs at different annealing conditions.

In addition, Figure 18 indicates that high temperature treatment degraded the alignment and the growth rate. In summary there are two dominant causes for the inactivation of catalysts. One is directly related to the PECVD; namely catalytic reactivity can be degraded by poisoning the catalysts due to excessive deposition of solid carbon. In both cases, the lifetime of the catalyst decreases, and SWCNTs are no longer synthesized. The other is the sintering of catalysts at high temperature. If alumina underlayer is substantially reduced during CNT growth, fine catalyst particles may react with aluminium; subsequently catalysts lose their activity. Even if APRFD does not provide serious damage to catalysts and SWCNTs, high temperature situation induces catalyst poisoning which eventually degraded SWCNT growth rate and yield.

3. Prospects

Figure 19 shows the Raman spectra of CNTs synthesized at 700 °C for 5 minutes at different gas pressures. SEM observation indicates no specific differences in the growth rate and morphology of CNTs. However, the intensity of D-band peak slightly increased at 50 kPa. When the pressure is further decreased to 20 kPa, the intensity of the D-band peak becomes significantly high. At the same time, the intensity of RBM peaks tend to decrease as the pressure decreases. Considering that the ion energy is substantially low even at 20 kPa, the increase in the radical flux to the substrate may increase the intensity of the D-band peak. Figure 20 schematically illustrates SWCNT synthesis in the collisional sheath. It is already indicated in Figures 7 [6] and 11 [7] that without applying APRFD neither SWCNTs nor CNFs were synthesized. It is therefore quite clear that the negative glow formed near the substrate promotes CNT growth. However, it is unknown whether the reactive species such as CH_i and ions generated in the negative glow are actually transported through the 900-μm thick collisional sheath, 1-20 μm dense SWCNT layer, and then arrive at the substrate surface where catalysts are supported. The experimental fact implies that most radicals are readily de-excited or recombined during the transportation through plasma sheath; therefore, emission intensity at the sheath is significantly low. Also, as discussed in section 2.6, the selective substrate heating by APRFD is negligible.

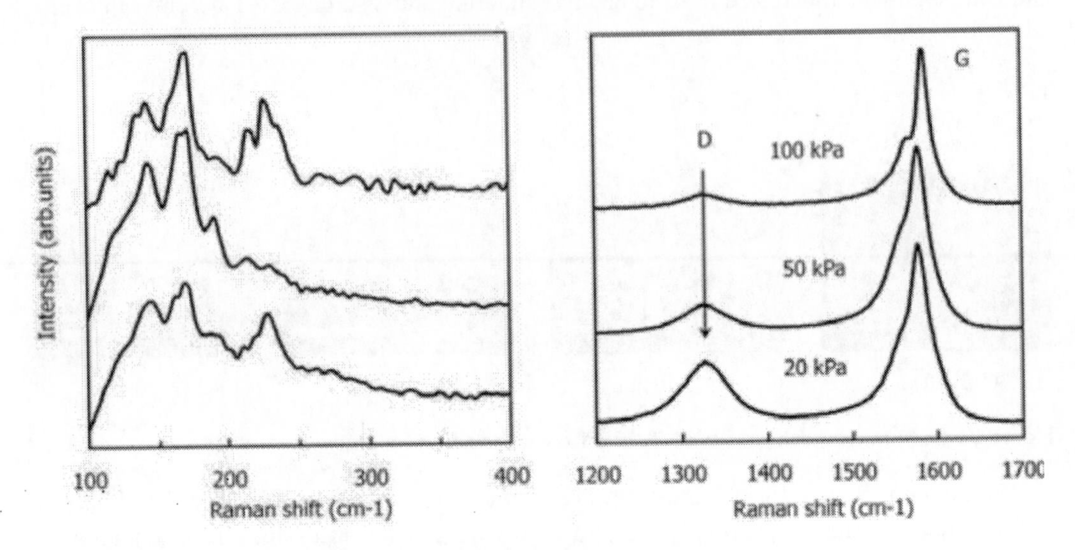

Figure 19. Raman spectra of SWCNTs synthesized at different pressures.

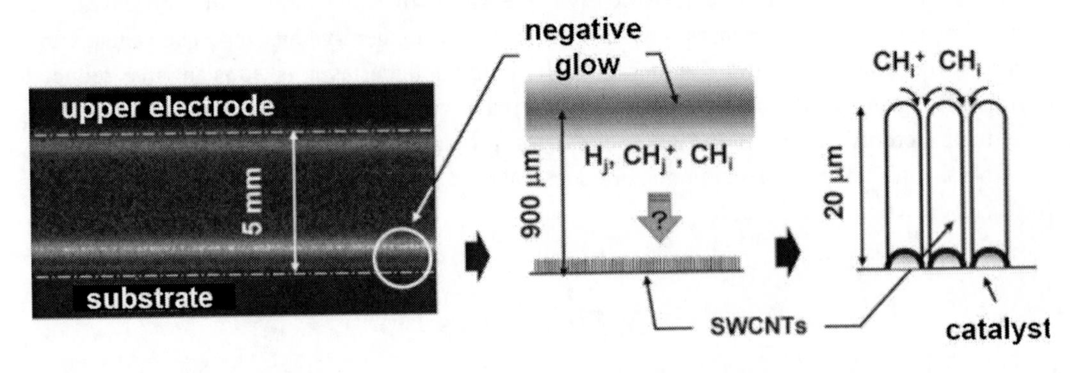

Figure 20. Schematic diagrams of SWCNTs synthesis in the collisional sheath.

In CNT synthesis using low pressure PECVD, it is generally thought that radicals generated in the gas phase promote the CNT growth. On the other hand, in atmospheric pressure plasmas, the reason for high yield SWCNT synthesis is the suppression of excess supply of the radicals to the substrate owing to the function of the sheath as the resistant layer for transporting radicals. The random radical supply at low pressure might rather be the reason for depositing amorphous carbons or reducing reaction selectivity. Further development of APRFD in the field of thin film deposition can be expected with the better understanding of the plasma chemistry under the high density medium such as atmospheric pressure gas.

REMARKS

In this chapter, the basic properties of AP-PECVD and its application to SWCNT synthesis is introduced. Although it was considered to be difficult to synthesize SWCNTs by low pressure PECVD until 2003, development of catalysts and progress of plasma processing have made this technique well-known in these days. However, since plasma is highly reactive, the drawback of simultaneous synthesis of a substantial amount of amorphous carbon is hardly overcome yet with conventional low pressure PECVD. As shown in Figure 19, the yield (G/D ratio) of SWCNTs markedly decreases even at 20 kPa. On the other hand, AP-PECVD enables synthesis of vertically-aligned SWCNTs at lower temperature than thermal CVDs.

As was already mentioned, APG generally relies on helium dilution. Plasma, in general, generates a reactive environment in oxidizing, reduction, or inert atmosphere. If plasma is generated only when helium is used, it may be a disadvantage. More recently, it is reported that a glow-like discharge can be generated in Ar, N_2, or air at atmospheric pressure by using a high frequency power source whose frequency ranges from RF to microwave, or generating a jet-like after-glow plasma. However, many atmospheric pressure plasma processes including APRFD are eventually discussed with specific phenomena obtained by the reactors which researchers developed individually. Understanding of more general phenomena and accumulation of data are indispensable for the further progress of this research field.

ACKNOWLEDGMENTS

Synthesis of SWCNTs was performed by Kuma Ohnishi, former graduate student of Tokyo Institute of Technology. The APRFD was developed by the authors collaborating with Professor Joachim Heberlein and Professor Uwe Kortshagen at the Department of Mechanical Engineering, University of Minnesota. Professor Shigeo Maruyama at the Department of Mechanical Engineering, The University of Tokyo supported catalyst preparation by dip coating method. Professor Naoto Ohtake at Nagoya University supported Raman spectroscopy. Satoshi Genseki at the Centre for Advanced Materials Analysis, Tokyo Institute of Technology performed TEM analysis. The present work was partly funded by the grant by the MEXT Japan (18686018).

REFERENCES

[1] Kogelschatz, U. *Plasma Chem. Plasma Proc.*, 2003, 23(1), 1-46.

[2] Kanazawa, S.; Kogoma, M.; Moriwaki, T.; Okazaki, S. *J. Phys. D Appl. Phys.* 1988, 21(5), 838-840.

[3] Nozaki, T.; Okazaki, K. *Pure Appl. Chem.* 2006, 78(6), 1157-1172.

[4] Nozaki, T.; Kimura, Y.; Okazaki, K. *J. Phys. D Appl. Phys.* 2002, 35, 2779-2784.

[5] Nozaki, T.; Kimura, Y.; Okazaki, K.; Kado, S.; *Plasma Proc. Polym.*, Wiley-VCH-Publisher. 2004, 477-487.

[6] Nozaki, T.; Goto, T.; Okazaki, K.; Ohnishi, K.; Mangolini, L.; Heberlein, J.; Kortshagen, U. *J. Appl. Phys.* 2006, *99*, 024310-1-024310-7.

[7] Nozaki, T.; Ohnishi, K.; Okazaki, K.; Kortshagen, U. *Carbon* 2007, *45*, 364-374.

[8] Kogoma, M. *Jpn. Soc. Plasma Sci. Nuclear Res.* 2003, *79(10)*, 1000-1032.

[9] Hatakeyama, R. *Jpn. Soc. Plasma Sci. Nuclear Res.* 2005, *81(9)*, 651-685.

[10] Kakiuchi, H.; Ohmi, H.; Kuwahara, Y.; Matsumoto, M.; Ebata, Y.; Yasutake, K.; Yoshii, K.; Mori, Y. *Jpn. J. Appl. Phys.* 2006, *45(4B)*, 3587-3591.

[11] Kitabatake, H.; Suemitsu, M.; Kitahara, H.; Nakajima, S.; Uehara, T.; Toyoshima, Y. *Jpn. J. Appl. Phys.* 2005, *44(22)*, L683-L686.

[12] Kondo, Y.; Saito, T.; Terazawa, T.; Saito, M.; Ohtake, N. *Jpn. J. Appl. Phys.* 2005, *44(52)*, L1573-L1575.

[13] Mangolini, L.; Orlov, K.; Kortshagen, U.; Heberein, J.; Kogelschatz, U. *Appl. Phys. Lett.* 2002, *80*, 1722-1724.

[14] Massines, F.; Segur, P.; Gherardi, N.; Khamphan, C.; Ricard, A. *Surf. Coat. Technol.* 2003, *174-175*, 8-14.

[15] Zhang, P.; Kortshagen, U. *J. Phys. D Appl. Phys.* 2006, *39*, 153-163.

[16] Nozaki, T.; Miyazaki, Y.; Unno, Y.; Okazaki, K. *J. Phys. D Appl. Phys.* 2001, *34(23)*, 3383-3390.

[17] Nozaki, T.; Unno, Y.; Okazaki, K. *Plasma Sources Sci. Technol.* 2002, *11*, 431-438.

[18] Yuan, X.; Raja, L.L. *IEEE Trans. Plasma Sci.* 2003, *31(4)*, 495-503.

[19] Raizer, Y.P. Gas Discharge Physics; 2nd Ed; Springer: Berlin, DE, 1997, Chapter 13.

[20] Lieberman, M.A.; Lichtenberg, A.J. *Principles of plasma discharges and materials processing*; Wiley: New York, US, 1994, Chapter 11.

[21] Ellis, H.W. *Atomic data and nuclear data tables*. 1984, *31*, 113-151.

[22] Jeong, G-H.; Satake, N.; Kato, T.; Hirata, T.; Hatakeyama, R.; Tohji, K. *Appl. Phys. A.* 2004, *79*, 85-87.

[23] Huang, Z.P.; Wang, D.Z.; Wen, J.G.; Sennett, M.; Gibson, H.; Ren, Z.F. *Appl. Phys. A.* 2002, *74*, 387-391.

[24] Täschner, C.; Pacal, F.; Leonhardt, A.; Spatenka, P.; Bartsch, K.; Graff, A.; Kaltofen, R. *Surf. Coat. Technol.* 2003, *174-175*, 81-87.

[25] Melechko, A.V.; Merkulov, V.I.; McKnight, T.E.; Guillorn, M.A.; Klein, K.L.; Lowndes, D.H.; Simpson, M.L. *J. Appl. Phys.* 2005, *97*, 041301/1-39.

[26] Meyyappan, M.; Delzeit, L.; Cassell, A.; Hash, D. *Plasma Sources Sci. Technol.* 2003, *12*, 205-216.

[27] In *Carbon Nanotubes, Synthesis, Structure, Properties and Applications*; Dresselhaus, M.S.; Dresselhaus, G.; Avouris, P.; Ed.; Springer-Verlag; Berlin, DE, 2001.

[28] In *Handbook of Microscale and Nanoscale Heat and Fluid Flow*; Maruayma, S.; Ed.; NTS Inc.; Tokyo, JP, 2006, Chaper 8, Section 1.

[29] Kato, T.; Jeong, G.H.; Hirata, T.; Hatakeyama, R. *Thin Solid Films* 2004, *457*, 2-6.

[30] Kato, T.; Jeong, G.H.; Hirata, T.; Hatakeyama, R.; Tohji, K.; Motomiya, K. *Chem. Phys. Lett.* 2003, *381*, 422-426.

[31] Kato, T.; Jeong, G.H.; Hirata, T.; Hatakeyama, R.; Tohji, K. *Jpn. J. Appl. Phys.* 2004, *43(10A)*, L1278-L1280.

[32] Kato, T.; Hatakeyama, R.; Tohji, K. *Nanotechnol.* 2006, *17*, 2223-2226.

[33] Zhong, G.; Iwasaki, T.; Honda, K.; Furukawa, Y.; Ohdomari, I.; Kawarada, H. *Chem. Vap. Dep.* 2005, *11*, 127-130.

[34] Iwasaki, T.; Zhong, G.; Aikawa, T.; Yoshida, T.; Kawarada, H. *J. Phys. Chem. B Lett.* 2005, *109*, 19556-19559.

[35] Wang, Y.Y.; Gupta, S.; Nemanich, R.J. *Appl. Phys. Lett.* 2004, *85(13)*, 2601-2603.

[36] Hiramatsu, M.; Nagao, H.; Taniguchi, M.; Amano, H.; Ando, Y.; Hori, M. *Jpn. J. Appl. Phys.* 2005, *44(22)*, L693-695.

[37] Hiramatsu, M.; Taniguchi, M.; Nagao, H.; Ando, Y.; Hori, M. *Jpn. J. Appl. Phys.* 2005, *44(2)*, 1150-1154.

[38] Matsushita, A.; Nagai, M.; Yamakawa, K.; Hiramatsu, M.; Sakai, A.; Hori, M.; Goto, T.; Zaima, S. *Jpn. J. Appl. Phys.* 2004, *43(1)*, 424-425.

[39] Jašek, O.; Eliáš, M.; Zajíčková, L.; Kudrle, V.;Bublan, M.; Matějková, J.; Rek, A.; Buršík, J.; Kadlečíková, M. *Mater. Sci. Eng. C.* 2006, *26*, 1189-1193.

[40] Lee, J-S.; Chandrashekar, A.; Park, B.M.; Overzet, L.J.; Lee, G.S. *J. Vac. Sci. Technol.* B. 2005, *23(3)*, 1013-1017.

[41] Kyung, S.; Lee, Y.; Kim, C.; Lee, J.; Yeom, G. *Thin Solid Films,* 2006, *506-507*, 268-273.

[42] Murakami, Y.; Miyauchi, Y.; Chiashi, S.; Maruyama, S. *Chem. Phys. Lett.* 2003, *377*, 49-54.

[43] Hu, M.; Murakami, Y.; Ogira, M.; Maruyama, S.; Okubo, T. *J. Catal.* 2004, *225*, 230-239.

[44] Jorio, A.; Saito, R.; Hafner, J.H.; Lieber, C.M.; Hunter, M.; NcClure, T.; Dresselhaus, G.; Dresselhaus, M.S. *Phys. Rev. Lett.* 2001, *86(6)*, 1118-1121.

[45] Hash, D.B.; Meyyappan, M. *J. Appl. Phys.* 2003, *93(1)*, 750-752.

[46] Puretzky, A.A.; Geohegan, D.B.; Ivanov, I.N.; Eres, G. *Appl. Phys. A.* 2005, *81*, 223-240.

[47] Futaba, D.N.; Hata, K.; Yamada, T.; Mizuno, K.; Yumura, M.; Iijima, S. *Phys. Rev. Lett.* 2005, *95*, 056104-1-056104-4.

[48] Boskovic, B.O.; Stolojan, V.; Khan, R.U.; Hag, S.; Silva, S.R.P. *Nat. Mater.* 2002, *1*, 165-168.

[49] Lee, K.Y.; Katayama, M.; Honda, S.; Kuzuoka, T.; Miyake, T.; Terao, Y.; Lee, J.G.; Mori, H.; Hirao, T.; Oura, K. Jpn. *J. Appl. Phys.* 2003, *42*, L804-L806.

[50] Nozaki, T.; Unno, Y.; Miyazaki, Y.; Okazaki, K. *J. Phys. D Appl. Phys.* 2001, *34*, 2504-2511.

In: Generation and Applications of Atmospheric Pressure Plasmas ISBN: 978-1-61209-717-6
Editors: M. Kogoma, M. Kusano and Y. Kusano

Chapter 9

PREPARATION OF Si-BASED THIN FILMS USING ATMOSPHERIC PRESSURE PLASMA CHEMICAL VAPOUR DEPOSITION (CVD)

Kiyoshi Yasutake, Hiroaki Kakiuchi and Hiromasa Ohmi
Graduate School of Engineering, Osaka University, Osaka, Japan

INTRODUCTION

Atmospheric pressure plasma chemical vapour deposition (CVD) is expected to be an industrially low-cost deposition method, since it enables both high rate deposition at low temperature and simplification of vacuum system. Deposition of Si-based thin films using atmospheric pressure plasma CVD has been studied, including a-Si:H thin films for solar cells, poly-crystalline Si, a-SiC thin films, and SiO_x and SiN_x thin films for gas barrier coatings of plastics and for anti-reflective coatings [1-6]. Atmospheric pressure plasma CVD enables not only low-temperature high-rate deposition but also preparation of high quality films due to suppressed ion damage at the film surfaces because of low kinetic energy of ions in the plasma. This chapter focuses on the high quality of the atmospheric pressure plasma CVD films, and shows low temperature growth of mono-crystalline Si films which is applicable for semiconductor devices that require the highest crystallinity.

1. REQUIREMENT FOR LOW TEMPERATURE Si EPITAXIAL GROWTH TECHNIQUE

A Si epitaxial wafer (epi wafer) is a mono-crystalline Si wafer on which a Si mono-crystalline thin film is epitaxially grown. Comparing with a CZ-Si (mono-crystalline Si produced by the Czochralski process) wafer, the epi wafer has a lower defect density in the surface layer, and can fabricate a p/p^+ structure (low boron density epi layer / high boron density substrate). Consequently, epi wafers have been used as major wafers for the Si-device manufacturing due to the requirement of high integration and high performance of semiconductor devices [7]. A current epi wafer manufacturing process is a thermal CVD using $SiHCl_3$ or SiH_4 as a source gas. This is a high temperature process operated at around 1100 °C. Such a high temperature process induces several problems including redistribution of dopant-impurities, auto-doping, reduction of intrinsic gettering ability of CZ-Si substrates by dissolution of oxygen precipitation nuclei, contamination by heavy metals, generation of slip dislocations, and increase of the cost due to large consumption of heat. In order to solve these problems, the temperature of the epitaxial growth process must be as low as less than 900 °C [8,9].

Recently, the low temperature Si epitaxial growth techniques draws considerable attention as inline epitaxial techniques which are applicable to semiconductor device manufacturing process as well as the wafer epitaxial techniques for epi wafer fabrication [10]. For example, it becomes very important for the future semiconductor device technology to develop low temperature doping and epitaxial growth techniques for formation of p^+ or n^+ layers after fabricating main parts of devices, and low temperature selective epitaxial growth techniques for fabricating p or n type microscopic structures. In the case of inline epitaxial techniques, low temperature growth (< 750 °C) is required at which diffusion of the dopant impurities is insignificant in the device region fabricated in the previous processes. Furthermore, if it is applied after Al patterning process, low temperature growth less than 550 °C is required. If such a low temperature inline epitaxial technique is established, novel device structures can be realized like three dimensionally integrated ULSI devices which can not be fabricated with the existing processes.

Since above mentioned techniques for the low temperature epitaxial growth are demanded, low temperature epitaxial growth at a temperature below 600 °C using low pressure plasma CVD is recently studied [11-17]. Among them, remote plasma CVD which aims for damage reduction at the surface by ions, and remote ECR plasma CVD at high plasma density are studied. However, the low temperature epitaxial films obtained by these studies contain high density defects, and cannot demonstrate desirable electrical properties. Presently without increasing the film growth temperature up to approximately 800 °C, epitaxial films with high crystallinity would not be obtained [16, 17].

2. Objectives of Low Temperature Epitaxial Growth Techniques

When the temperature for Si epitaxial growth is lowered, significant problems such as the cleanliness of the atmosphere, reduction of growth rate, and degradation of crystalline quality appear. Each problem based on the data reported on thermal CVDs is discussed below.

2.1. Sustaining the Cleanliness of the Atmosphere

If the atmosphere during deposition contains water vapour or oxygen, Si crystalline surfaces are oxidized and subsequently epitaxial growth is inhibited. Therefore, the cleanliness of the substrate surfaces just before the epitaxial growth is the most important. When oxidizing species exist in the deposition atmosphere, whether SiO_2 is formed due to the oxidation of the Si surfaces or clean Si surfaces are sustained by the surface etching due to the formation of volatile SiO depends on the partial pressure of O_2 or water vapour and the specimen temperature [10, 18]. For example, in order to perform epitaxial growth by thermal CVD at 800 °C, it is enough to decrease the water vapour pressure less than 100 ppb in the deposition atmosphere. However, this critical value drastically decreases at lower temperatures. In fact at 600 °C, a very severe condition of the water vapour pressure less than 0.2 ppb is necessary. In the case of growth process of epi wafer by high temperature thermal CVD, the cleanliness of the Si surfaces is usually secured by annealing in H_2 atmosphere at approximately 1200 °C. On the other hand, when low temperature Si epitaxial growth is performed at the temperature between 700 and 800 °C, it is common to apply hydrogen termination of Si substrate surfaces by diluted hydrofluoric acid.

2.2. Growth Rate

The dependence of Si epitaxial growth rate by thermal CVD on the temperature is discussed next. At high temperature, the dissociation rate of the source gases and crystalline growth reaction at the surface are high, and thus the epitaxial growth rate is limited by the supplying process of the sources gases to the surface. When the growth temperature is lower, the surface reaction limits the growth rate. In this case, the growth rate does not depend on the flow rate of the source gas, but exponentially decreases as the temperature decreases [19]. Although the epitaxial growth rate depends on the choice of the source gas, even if SiH_4, which can be relatively easily dissociated, is used as a source gas, the growth rate becomes less than 5 nm/min at 600 °C, and the eventual growth is not observed at 500 °C. In such a low temperature region, it is thought that the desorption of hydrogen atoms which terminate surface Si atoms limits the growth rate.

2.3. Crystallinity

The degradation of the quality of crystallinity with low temperature growth is the most important problem to solve. It is thought that the following can affect the degradation; influence of impurities such as O and C which remain at the substrate surfaces, relative increase of impurities taken in the thin film due to the low growth rate, insufficient dissociation of the source gas, and reduction of the diffusion length of the surface atoms. When low pressure plasma CVD is used, it is considered to be necessary to take sufficient time for the migration of the surface atoms for establishing high quality single crystalline growth. Therefore, it is necessary to decrease the growth rate substantially. The problem is that the quality of the crystal is degraded if the growth rate increases.

In the case of atmospheric pressure plasma CVD, the above mentioned problem due to the low temperature growth is solved. As will be described in the following sections, high quality Si epitaxial growth can be realized using atmospheric pressure plasma CVD at low temperature and at sufficiently high growth rate.

3. EXPERIMENTAL METHOD

3.1. Atmospheric Pressure Plasma CVD Using Rotary Electrode

We have developed atmospheric pressure plasma CVD equipment using cylindrical rotary electrode for the purpose of high rate growth of functional thin films. The structure is based on the plasma chemical vapourization machining (CVM) equipment described in the chapter 2.13, adding a heater for heating substrates, and a load lock chamber for exchanging substrate. In details, see chapter 2.13 and a previous report [20]. Especially for Si epitaxial growth, cleanliness of the atmosphere is required. For this purpose atmospheric pressure plasma CVD equipment with high cleanliness is developed, considering electrolytic polishing of the inner chamber wall, local evacuation of organic impurity generated at the bearing of the rotary electrode, magnet coupling of the electrode shaft and driving motor, and the use of the magnetic fluid seal, and choice of the material with low contamination of impurities around the susceptor [21,22].

High speed rotary electrodes at over 1000 rpm are useful in that reactive gas is efficiently supplied to the small gap between the electrode and the specimen (< 1 mm) for the generation of an atmospheric pressure plasma, and that particles generated in the plasma is promptly exhausted. In addition, this method has an advantage of introducing high power to the plasma without thermal damage to the electrode since the electrode is efficiently cooled down. Such a rotary electrode is very useful for ultra precision machining by the plasma CVM and treatment of material surfaces by an atmospheric pressure plasma. In fact, high rate deposition of a-Si is established by using atmospheric pressure plasma CVD [1, 2]. However, as it will be discussed in the experimental results below, when the rotary electrode is used for the Si epitaxial growth, it is reported that epitaxial growth is disturbed at the inlet and outlet of the source gas to the atmospheric pressure plasma. In this sense, it is not necessarily the best structure of the electrode for the high rate Si epitaxial growth [23].

3.2. Atmospheric Pressure Plasma CVD Using Porous Carbon Electrode

In order to solve the disturbance of the epitaxial growth at the interface between the atmospheric pressure plasma and the surrounding atmosphere when the rotary electrode is used, a porous carbon electrode is developed [24]. Its concept is summarized below;

i. The source gas is introduced to the plasma through the porous carbon electrode and thus the interface between the plasma and atmosphere does not contact the Si substrate surface,
ii. Only a high purity gas is introduced to the plasma while the used gas will not return to the plasma, and
iii. The whole wafer is exposed in the plasma in order to avoid the adhesion of particles at the surfaces.

Owing to application of such a structure the cleanliness can be easily sustained, and thus it is expected that design of the equipment using a simple vacuum chamber would be possible. Figure 1 (a) shows a cross section of the porous carbon electrode. The diameter of the electrode for a 4-inch Si-wafer and the thickness of the porous carbon are 105 mm and 3 mm, respectively. Figure 1 (b) and (c) show the overview image of the electrode and the SEM image of the electrode surface. The porous carbon material (NISSHINBO ACP-101X) was manufactured by sintering of amorphous carbon powders at 1000 °C. Its density and porosity are 1.1 g/cm^3 and 31.2 %, respectively. Since the porous carbon material has a large surface area, sufficient baking is necessary when it is used for film deposition. In this work, the atmospheric pressure plasma baking at 800 °C for 20 minutes was carried out while a high purity He gas is fed through the electrode at the flow rate of 20 L/min, and adsorbed materials such as water was completely removed.

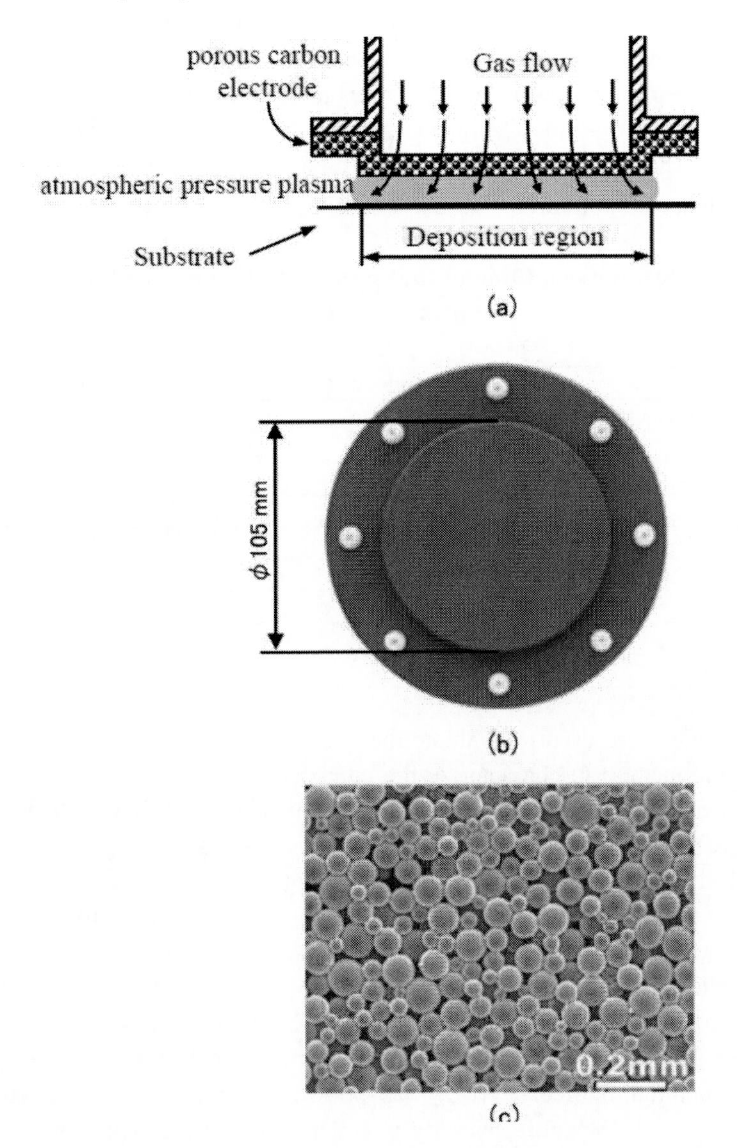

Figure 1. The porous carbon electrode. A schematic diagram of the cross sectional view (a), the overview image (b) and the SEM image of the surface (c).

3.4. Deposition and Characterization Methods

A 4-inch diameter B-doped p-type (001) CZ-Si wafer with the resistivity of 1-15 Ωcm was used as a substrate. A highly clean atmospheric pressure plasma CVD equipment was used as the deposition chamber. The rotary electrode was replaced with the porous carbon electrode. The atmospheric pressure plasma was generated at the 0.8 mm gap between the porous carbon electrode and the Si substrate using a 150 MHz very high frequency (VHF) power supply. A He gas mixed with 0.07 % H_2 was used as a plasma gas. Si deposition was initiated by feeding SiH_4 in to the plasma as a source gas. The deposition was terminated by cut-off of the plasma after a predetermined time. Table 1 shows the condition of the deposition for the epitaxial Si growth. The substrate temperatures indicated in Table 1 are the estimated values at the wafer surface during deposition. The wafer surface temperature just after cutting off the plasma was measured by using a pyrometer. The wafer temperature during deposition was 50 – 60 °C higher than the susceptor temperature measured by a thermocouple. This showed good agreement with the rotational temperature estimated by the analysis of H_2 Fulcher-α band ($d^3\Pi_u^- - a^3\Sigma_g^+$) emission spectra.

Table 1. The deposition condition of the epitaxial Si film by atmospheric pressure plasma CVD.

Electrode type	Rotary electrode	Porous carbon electrode
Rotation speed of the electrode [rpm]	2000	0
Substrate temperature [°C]	200 – 800	400 – 600
He flow rate [L/min]	72	75
H_2 concentration [%]	0 – 30	0.07
SiH_4 concentration [%]	0.1 – 1.0	0.07
Process gas pressure [Torr]	760	760
Plasma gap [mm]	0.7	0.8
VHF power [W]	300 – 3000	1800 – 2500
Deposition time [min]	1.5	4 – 5

Various methods were used to evaluate the quality of the epitaxial Si layers. The most prompt and simple method to characterize the epitaxially grown layer is the observation of film surfaces under an intense illumination from a halogen lamp at 150W. When haze was not observed, RHEED observation confirmed excellent mono-crystalline growth. Atomic force microscopy (AFM) and scanning electron microscopy (SEM) were used for the observation of morphology of the film surfaces, and reflection high energy electron diffraction (RHEED) method, cross sectional transmission electron microscopy (cross sectional TEM) and the selective etch pit method using Secco etchant [26] were used for the observation of crystallinity. Secondary ion mass spectrometry (SIMS) and Fourier transform infrared spectroscopy (FTIR) were used for the analysis of impurities in the films. In addition, the evaluation with photoluminescence (PL) was performed, which enables measurement of defects and impurities with high sensitivity in order to investigate applicability of the low temperature grown Si epitaxial films for the active layer in the semiconductor devices.

4. Low Temperature Si Epitaxial Growth by Atmospheric Pressure Plasma CVD Using Rotary Electrode [22, 23, 27]

In order to investigate the optimum growth condition of low temperature Si epitaxial films by atmospheric pressure plasma CVD using rotary electrode, the crystallinity and surface morphology of the central part of the film (10 mm x 10 mm) are investigated for the Si films grown under various conditions, and then whole part of the film is characterized.

4.1. Influence of H_2/SiH_4 Gas Composition

Figure 2 shows AFM images and RHEED observation results of the Si films grown under the condition of various H_2 concentrations at the fixed SiH_4 concentration of 0.1 % in a He gas, at the substrate temperature of 700 °C with the VHF power of 1100 W. When H_2 concentration was 1 %, the RHEED image indicates the mono-crystalline growth as shown in Figure 2 (b). The epitaxial film surface is significantly smooth, and the surface roughness of the film (R_{PV} = 0.72 nm) is smaller than that of the substrate CZ-Si (R_{PV} = 1.40 nm). On the other hand, when the film was grown without adding H_2 gas into the plasma, the surface of the Si film was rough (R_{PV} = 18.8 nm), and the RHEED image shows the ring pattern indicating poly-crystalline nature (Figure 2 (a)). It is thought that when H_2 gas is not added, residual water vapour of 20 – 30 ppb in the chamber degrades the cleanliness of the Si surfaces, and excellent epitaxial growth would be impeded. He atmospheric pressure plasma treatment mixed with H_2 gas has an effect of cleaning at the Si substrate surface before deposition. It is therefore thought that the addition of 1 % H_2 gas into the plasma is useful for sustaining cleanliness of the Si surface before and during the epitaxial growth.

When H_2 concentration was further increased, the Si film became poly-crystalline and its surface became significantly rough as shown in Figure (c) and (d). The IR spectra of these poly-crystalline Si films include the absorption band at 2000 cm^{-1}, corresponding to stretching modes of Si-H bonding, indicating that many insufficiently dissociated SiH_x molecules would be involved in the films, and that the films would become poly-crystalline. When H_2 concentration was higher, a part of the VHF power was consumed for the vibrational excitation of hydrogen molecules, and thus energy to be used for dissociation of SiH_4 molecules and epitaxial growth reactions would become insufficient. As the above mentioned results and the experimental results with various SiH_4 and H_2 concentrations and the VHF power, it is found that the optimum VHF power exists at each combination of the SiH_4 and H_2 concentrations.

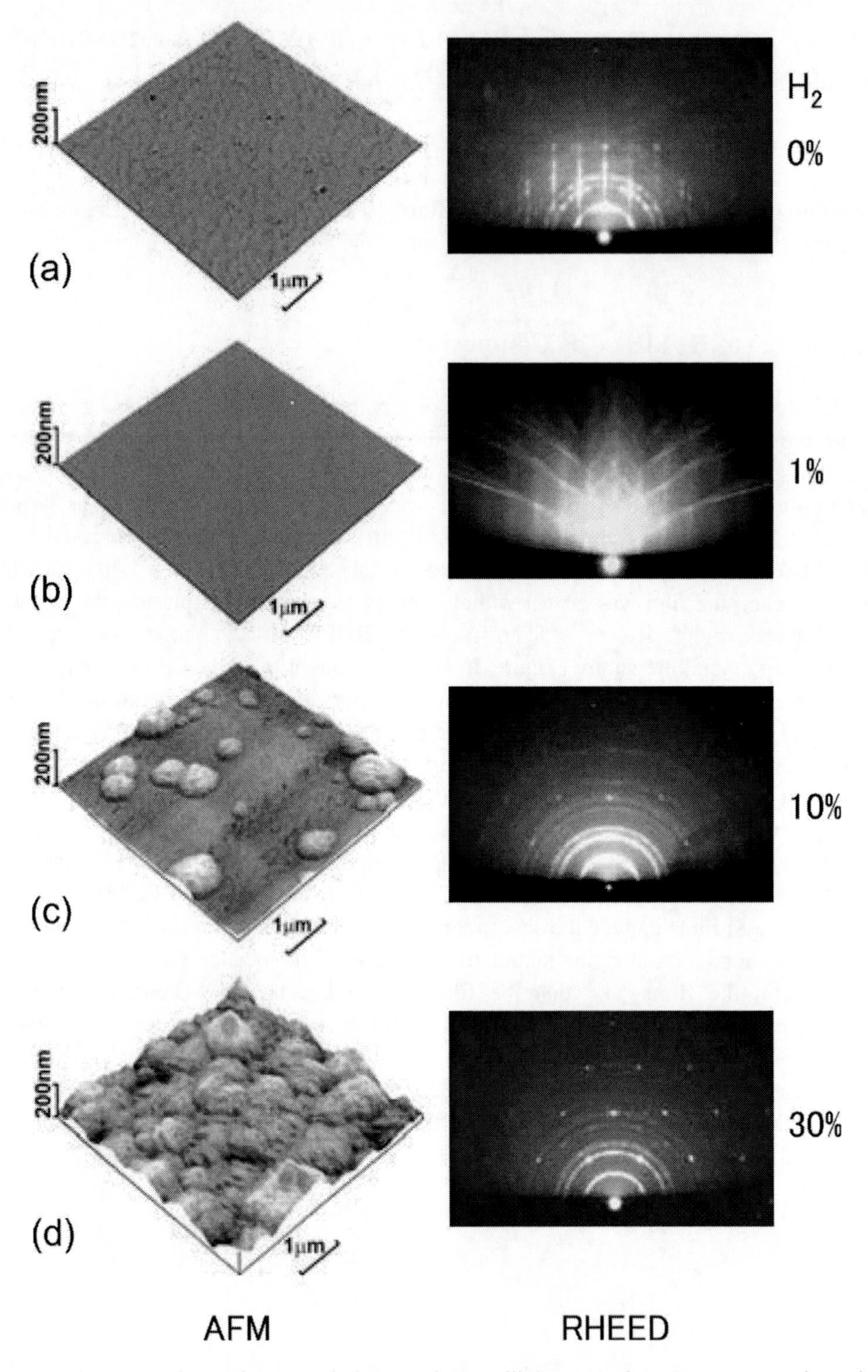

Figure 2. Change of the surface morphology and crystallinity at various H_2 concentrations during deposition. H_2 concentrations: (a) 0 %, (b) 1 %, (c) 10 %, and (d) 30 %. Substrate temperature: 700 °C, VHF power: 1100 W, SiH_4 content in He gas: 0.1 %.

4.2. Influence of Substrate Temperature and VHF Power

As shown previously, the optimum concentration ratio of H_2 and SiH_4 is 10 : 1. The influence of the substrate temperature and VHF power on the crystallinity of Si films was investigated at the concentrations of 1 % H_2 and 0.1 % SiH_4. Figure 3 shows the relation of the crystallinity of the Si films evaluated by RHEED observation, the substrate temperature and the VHF power. It indicates that amorphous or poly-crystalline Si is grown when both the substrate temperature and the VHF power are low. It is noted that epitaxial growth can be established by increasing the VHF power even if the substrate temperature is low. In this experiment, the epitaxial growth was observed even at 500 °C by supplying 3000 W VHF power. On the other hand, although it can be thought that epitaxial growth can be easier at higher substrate temperature, the film became poly-crystalline at 700 °C, if the VHF power was too high (2000 W). It can thus be said that, optimum VHF power must be supplied according to each deposition condition in order to achieve Si epitaxial growth using atmospheric pressure plasma CVD.

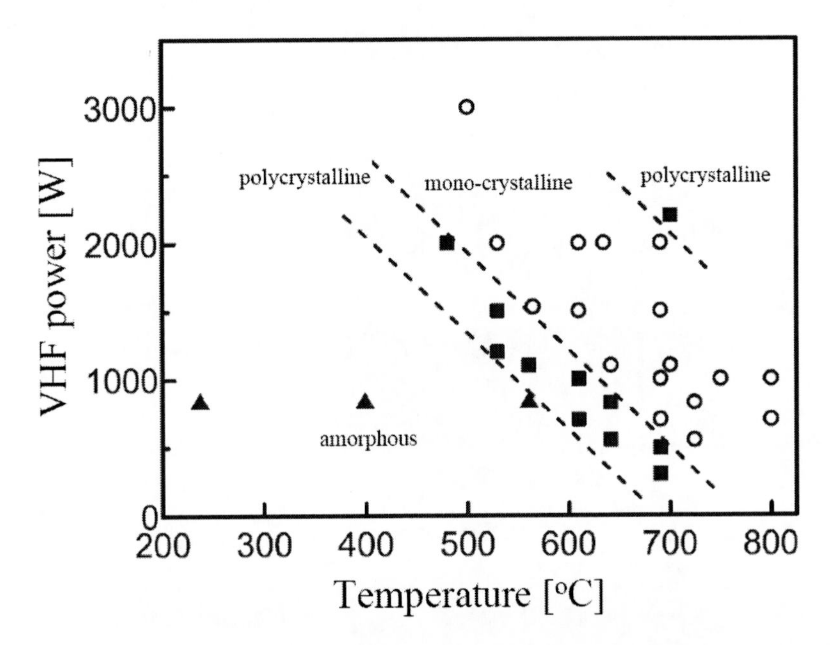

Figure 3. VHF power, substrate temperature and crystallinity of Si films. H_2 content: 1 %, SiH_4 content: 0.1 %.

Figure 4 shows cross sectional TEM images of Si epitaxial films grown at different conditions. Many dislocation lines are observed in the epitaxial film grown at 800 °C with 700 W power as shown in Figure 4 (a). They decreased substantially by increasing the VHF power as high as 1000 W as shown in Figure 4 (b). However, residual contrast at the interface between the epitaxial film and the substrate, and propagation of defects from the interface to the films are observed. Further increase of the VHF power is thought to be necessary in order to grow perfect crystalline films. Figure 4 (c) shows the cross sectional TEM image of the Si film grown at 3000 W VHF power at 500 °C, which is the lowest substrate temperature in this experiment. Here, no contrast by defects is observed both in the film and at the interface between the epitaxial film and the substrate surface, and thus the grown film is free from defects within the level of the TEM observation. When excellent epitaxial growth is achieved, the grown film is mono-crystalline with the same orientation of the substrate. Therefore, the Si substrate and the epitaxial film are not identified at all as shown in Figure 4 (c). The location of the interface indicated in Figure 4 (c) was obtained by the depth analysis of the impurities by SIMS, which will be presented in the next section.

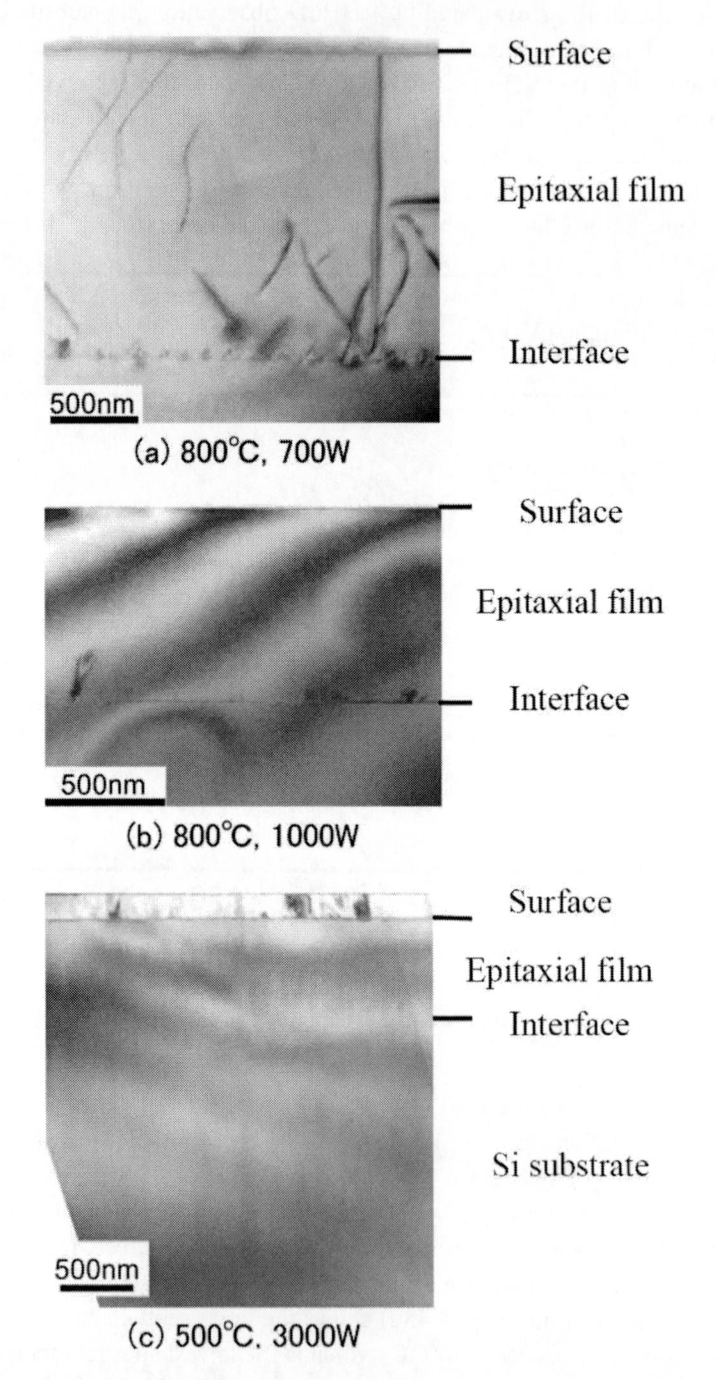

Figure 4. Cross sectional TEM image of the epitaxial Si film grown by atmospheric pressure plasma CVD using rotary electrode. (a) 800 °C, 700 W, (b) 800 °C, 1000 W, and (c) 500 °C, 3000 W.

4.3. Disturbance of Epitaxial Growth at the Edge of the Film [23]

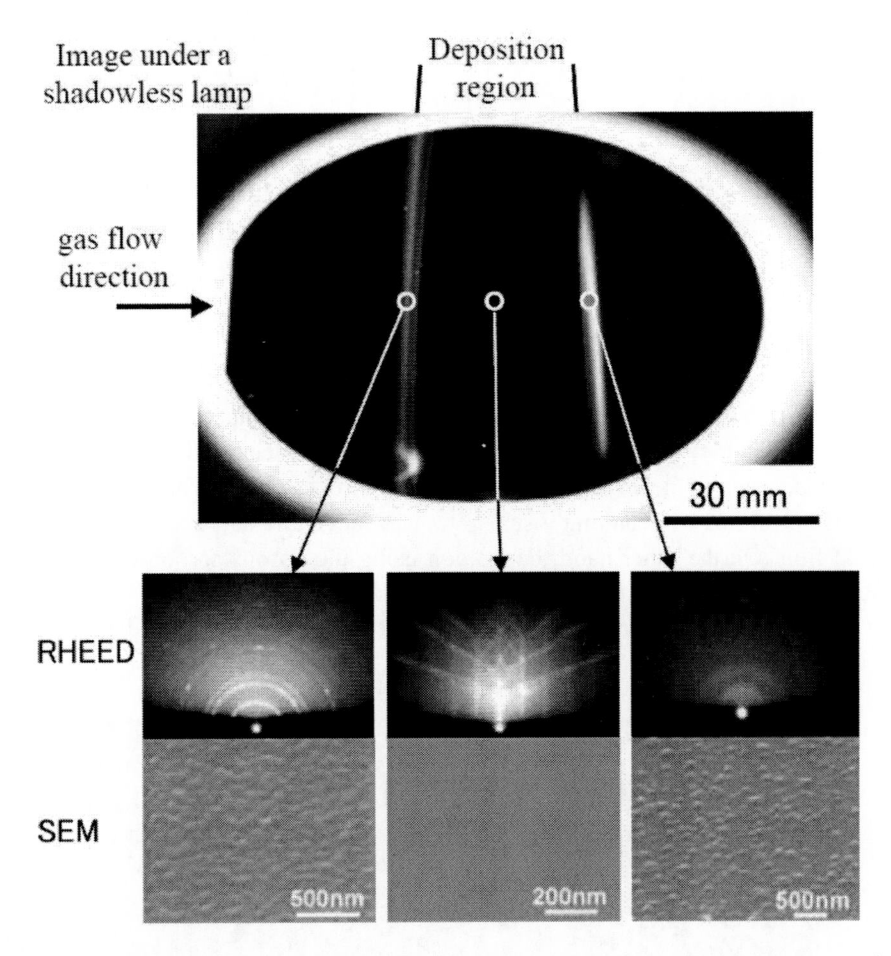

Figure 5. The surface of the Si film grown by atmospheric pressure plasma CVD using rotary electrode. The image under the focused halogen lamp, RHEED, and SEM images.

Figure 5 shows RHEED and SEM images of the typical Si epitaxial film grown with using the rotary electrode. A rotary electrode with 30 cm diameter and 20 cm wide was used. The epitaxial film growth was observed with approximately 40 mm wide at the centre of the 4 inch (001) Si wafer. The observation under the intense illumination from a halogen lamp in Figure 5 (a) shows two white line like regions due to the scattering induced by the surface roughness. The region between the two white lines (approximately 40 mm wide) was exposed in the atmospheric pressure plasma. In this region, high quality Si epitaxial film was grown indicated by the RHEED and SEM images. The two white line regions correspond to the interface between the atmospheric pressure plasma and the surrounding atmosphere. Poly-crystalline and amorphous Si were grown at the white lines of the upstream and the downstream, respectively. The reason of disturbance of epitaxial growth at these white lines is thought to be as follows; at the upstream, the deposition rate becomes too high since a significant amount of precursor for deposition (SiH_x) is generated as the SiH_4 gas is rapidly dissociated when they entered in the plasma. At the downstream, the agglomeration of SiH_x after passing the atmospheric pressure plasma induces generation of a-Si particles. The generation of the particles at the downstream can be solved by optimizing the deposition condition so that most of SiH_x is consumed for the deposition in the plasma. However, the generation of poly-crystalline Si at the upstream remains unsolved as an essential problem if the rotary electrode is used in the atmospheric pressure plasma CVD.

5. Low Temperature Si Epitaxial Growth by Atmospheric Pressure Plasma CVD Using Porous Carbon Electrode [24, 28]

5.1. Evaluation of Crystallinity

In the previous section, it was shown that defect-free Si epitaxial growth is possible at 500 °C by atmospheric pressure plasma CVD using rotary electrode. However, the poly-crystalline Si is grown at the edge when the rotary electrode is used. Therefore, if a large area deposition is performed by scanning a substrate, it is inevitable that defects are introduced in an epitaxial film. On the other hand, deposition using the porous carbon electrode does not show such a problem. Figure 6 shows the optical and RHEED images of the epitaxial film grown at 600 °C, at the VHF power of 1800 W with the concentrations of SiH_4 and H_2 at 0.07 % and 0.07 %, respectively.

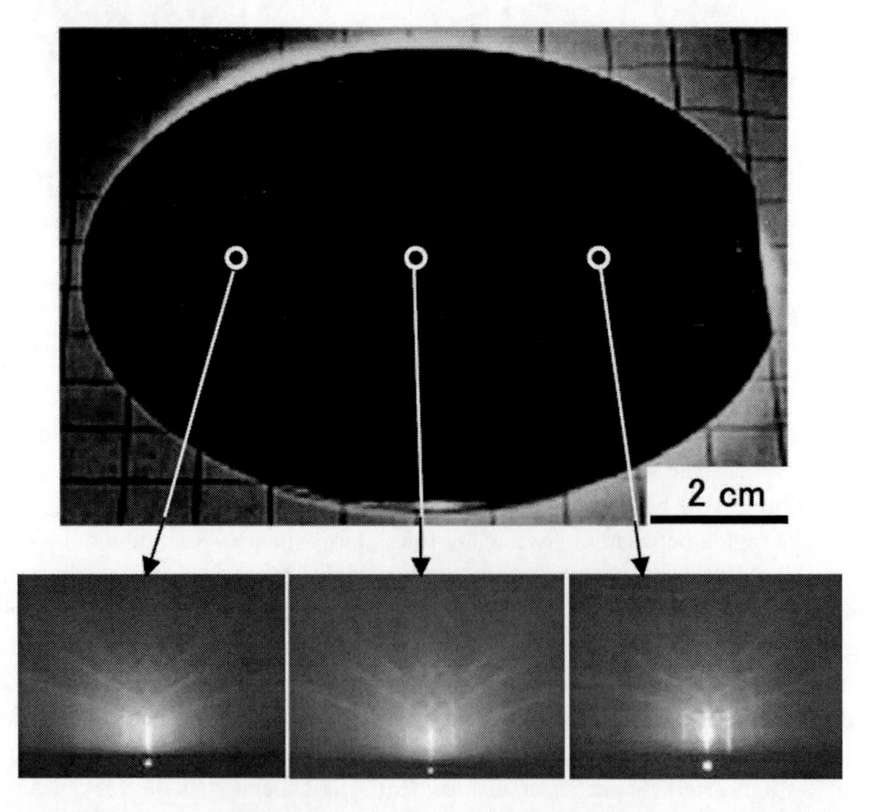

Figure 6. The surface of the Si film grown by atmospheric pressure plasma CVD using porous carbon electrode. The image of the wafer under the focused halogen lamp, and RHEED images.

It is seen that in the overall area of the 4-inch wafer, high quality haze free monocrystalline Si was grown. The averaged deposition rate is 0.35 – 0.4 μm/min, which is sufficient for practical applications. Figure 7 shows the cross sectional TEM and AFM images of this specimen. The cross sectional image of TEM indicates that no contrast is seen at the interface and in the film, and that defect-free epitaxial growth was carried out. The surface roughness was 0.8 nm in PV and 0.07 nm in RMS, corresponding to approximately two atomic layers. Since significantly smooth surface can be obtained by atmospheric pressure plasma CVD, it is expected that surface atoms can sufficiently migrate even at low substrate temperature.

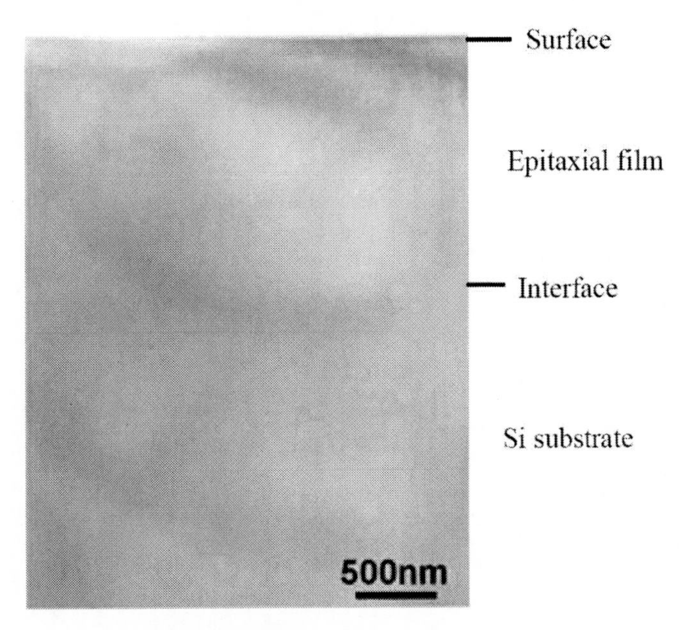

(a) Cross sectional TEM image

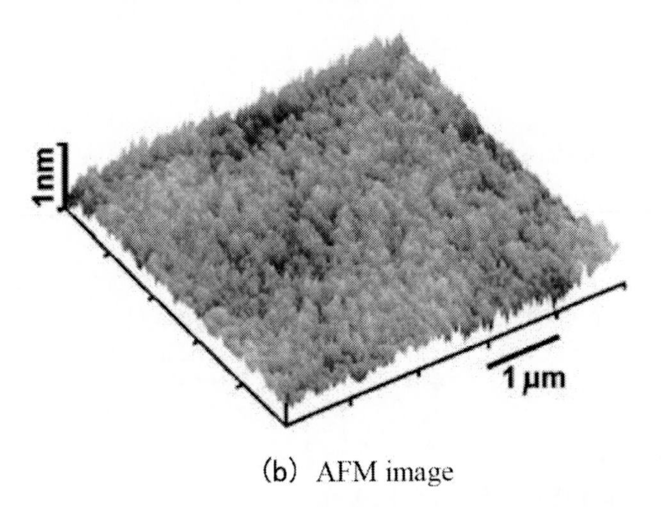

(b) AFM image

Figure 7. The epitaxial Si film grown by atmospheric pressure plasma CVD using porous carbon electrode. (a) Cross sectional TEM image, and (b) AFM image. Substrate temperature: 600 °C, VHF power: 1800 W.

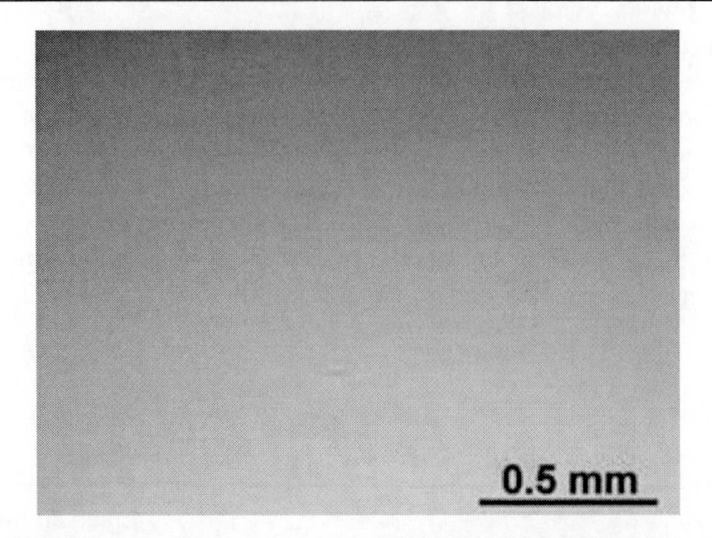

Figure 8. Optical microscopic image of the specimen shown in Figure 7 etched by Secco etchant.

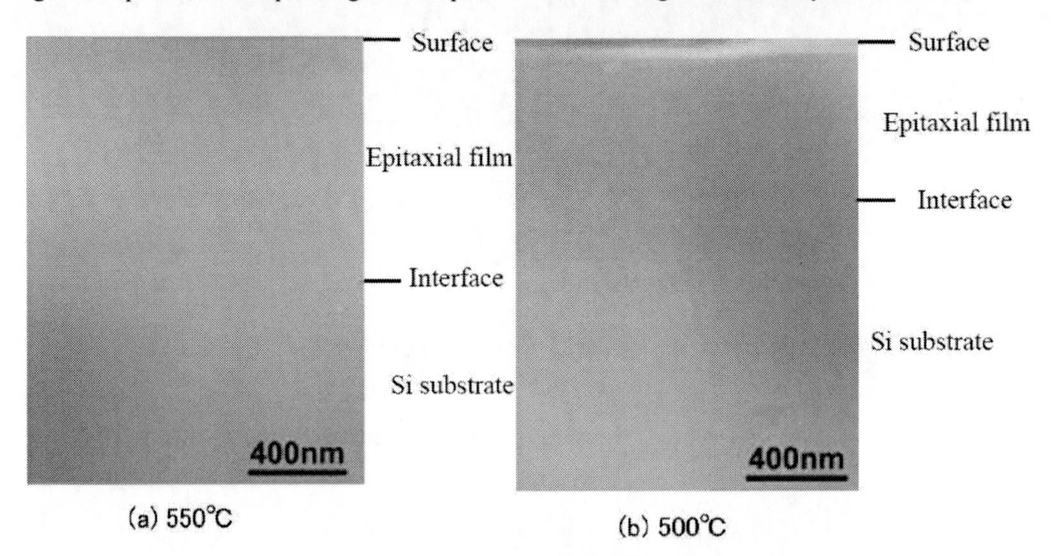

Figure 9. Cross sectional TEM images of the epitaxial Si film grown by atmospheric pressure plasma CVD using porous carbon electrode. Substrate temperature 550 °C (a), and 500 °C (b).

Figure 8 shows the optical microscopic image of the epitaxial film surface after 20-second selective etching in Secco etchant. Neither pits nor mounds were found at all at the surface, indicating that it is defect-free even by the wide area observation.

Figure 9 shows the cross sectional TEM image of the epitaxial film grown under the same conditions but at lower substrate temperatures. It is seen that defect-free growth was achieved at the substrate temperatures of 550 and 500 °C as well.

5.2. Impurity Analysis of Si Epitaxial Layer

SIMS was used for analysing impurities in the Si epitaxial film grown at 600 °C. Figure 10 (a) and (b) shows the depth profiles of the concentrations of O, C and metallic impurities.

As the concentration of each impurity was below detection limit of SIMS, it is confirmed that high purity epitaxial film was grown. On the other hand, approximately 10^{18} cm^{-3} interstitial oxygen was contained in the CZ-Si substrate. Therefore the location of the interface between the epitaxial film and the Si substrate can be identified by the depth profile of the oxygen concentration.

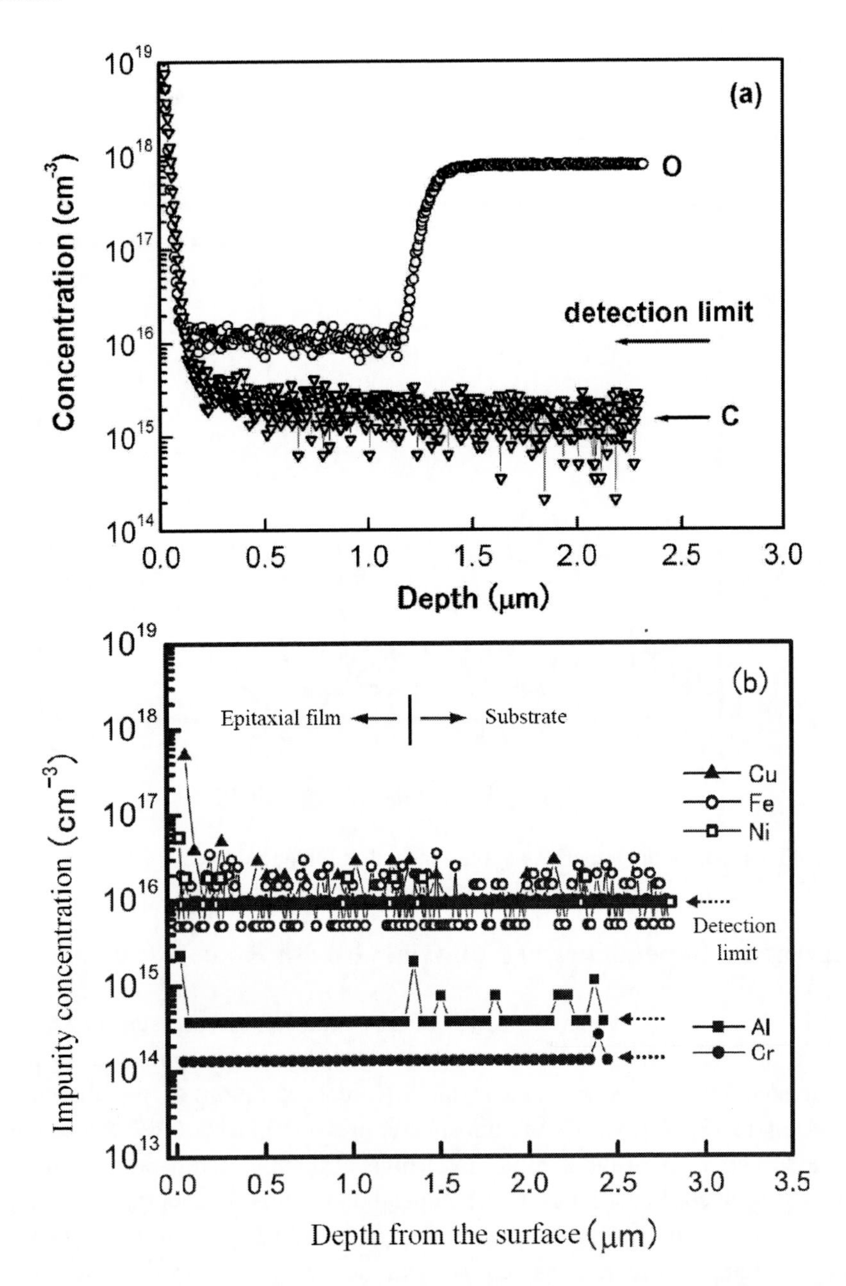

Figure 10. SIMS analysis of impurities in the epitaxial Si film. Depth profiles of light impurities (O, C) (a), and metallic impurities (Cu, Fe, Ni, Al, Cr) (b).

The p/p$^+$ structured epitaxial film was grown next, and sharpness of B concentration profile was estimated using SIMS measurement as shown in Figure 11. It is expected that the transition width of B concentration of the next generation p/p$^+$ epi wafer for ULSI is less than 0.5 μm. The transition width of the epitaxial film grown at 600 °C by atmospheric pressure plasma CVD is approximately 80 nm as shown in Figure 11, satisfying this requirement. Theoretical calculation of B concentration profile was performed, simulating diffusion of B from the substrate into the epitaxial film during the epitaxial growth by using the B diffusion coefficient in Si at 600 °C. The result indicates that the transition width is less than 1 nm. During the SIMS measurement, the film was sputter etched for the depth profiling. It is thought that the measured transition width became larger than the theoretical width due to the atomic mixing effect during the sputtering.

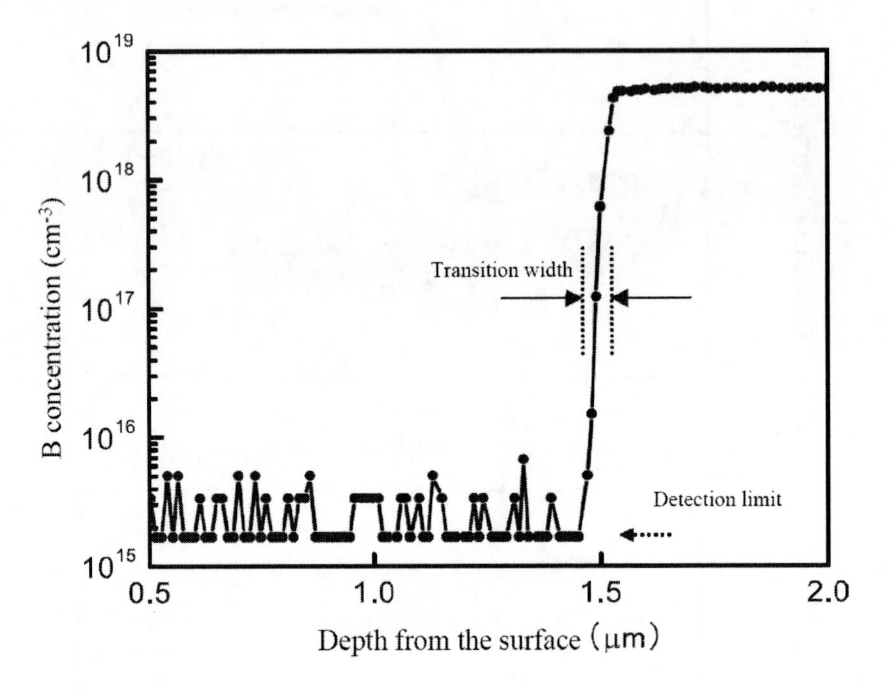

Figure 11. SIMS analysis of the impurities in the epitaxial Si film. The depth profile of B concentration.

5.3 Temperature Dependence of Epitaxial Growth Rate

Figure 12 shows the Si epitaxial growth rate by atmospheric pressure plasma CVD and thermal CVD at various temperatures. The growth rate was determined from the average thickness divided by the growth time. The SiH_4 concentration of the thermal CVD is indicated in Figure 12, while those of atmospheric pressure plasma CVD are summarized in Table 1. The growth rates of the thermal CVD films at the high temperature region over 900 °C seem to approach constant values, which depend on the SiH_4 concentration. It is suggested that the growth rate is limited by the transport process of SiH_4 gas in the high temperature region. In low temperature regions, the growth rate does not strongly depend on SiH_4 concentration, but significantly decreases as the temperature decreases, indicating that the process is surface reaction-limited. The growth rate of the atmospheric pressure plasma CVD films also decreases as the temperature decreases, but not so significant as that of the thermal CVD films. In addition, as the temperature decreases, the decrease of the growth rate tends to saturate, suggesting that the epitaxial growth would be proceeded mainly using the energy of the atmospheric pressure plasma.

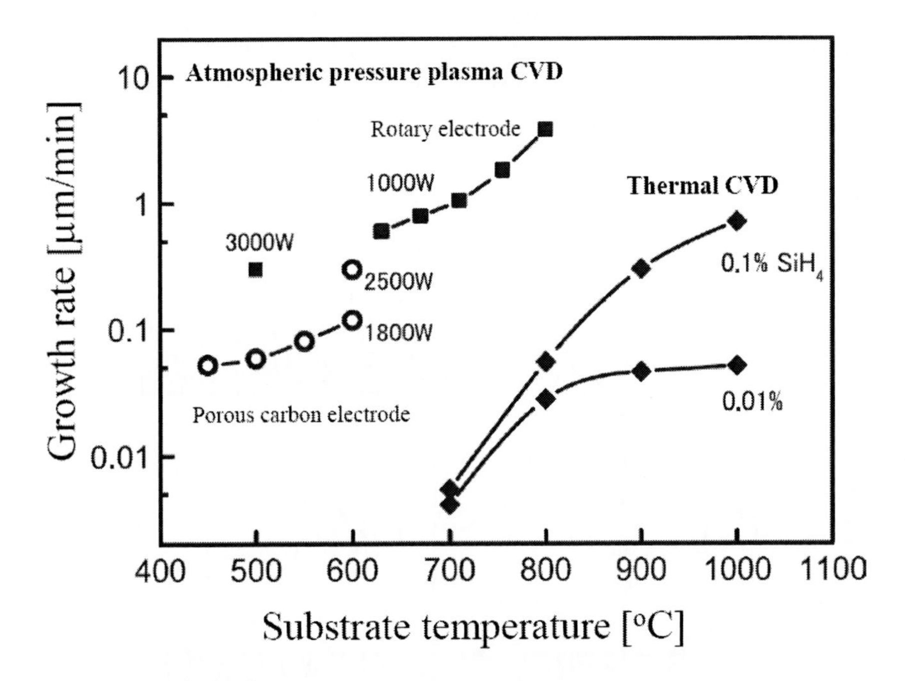

Figure 12. Epitaxial growth rate at various substrate temperatures.

The Si epitaxial growth rate of atmospheric pressure plasma CVD is approximately 0.3 μm/min at 500 °C with 3000 W power using the rotary electrode. It is comparable to that of the thermal CVD films grown at 900 °C. When the porous carbon electrode was used, the growth rate was slightly lower than that using the rotary electrode. It is because the deposition area with the porous carbon electrode is larger than that of the rotary electrode. The growth rate can be increased by increasing the VHF power.

5.4. Photoluminescence of Si Epitaxial Layer [28]

As was already discussed before, the Si epitaxial film grown at 500 – 600 °C using atmospheric pressure plasma CVD is defect-free by the levels of the TEM and the etch-pit observations, and is the high quality crystal with low impurity concentration less than the detection limit of SIMS measurement. In order to use the film as the active layer of the semiconductor device, sensitive technique for evaluation of the concentrations of impurities and defects less than SIMS detection limits is necessary. For example, as heavy metal impurities with deep energy levels in a band gap and point defects give harmful influences on the electrical properties of Si devices, it is said that these densities must be in general much less than 10^{10} cm^{-3}. The measurement of the carrier recombination lifetime is the most effective in order to evaluate the defects with deep energy levels with the highest sensitivity [29-31].

Then the Si epitaxial films grown at low temperature by atmospheric pressure plasma CVD was evaluated by photoluminescence (PL) method [30,32,33] which enables non-destructive and highly sensitive evaluation of impurities and defects, and which can also provide information related with the carrier recombination lifetime [28]. Ar^+ ion laser was used for excitation of PL. The photo-absorption length of Ar^+ ion laser is approximately 1.25 μm, calculated from the photo-absorption coefficient of the Si crystal. Si thin films grown at the temperatures of 500, 550, and 600 °C were used as the epitaxial specimens. The thicknesses of the epitaxial layers are 0.5, 1.0, and 1.5 μm, respectively. As-received CZ-Si wafer was used as a reference specimen.

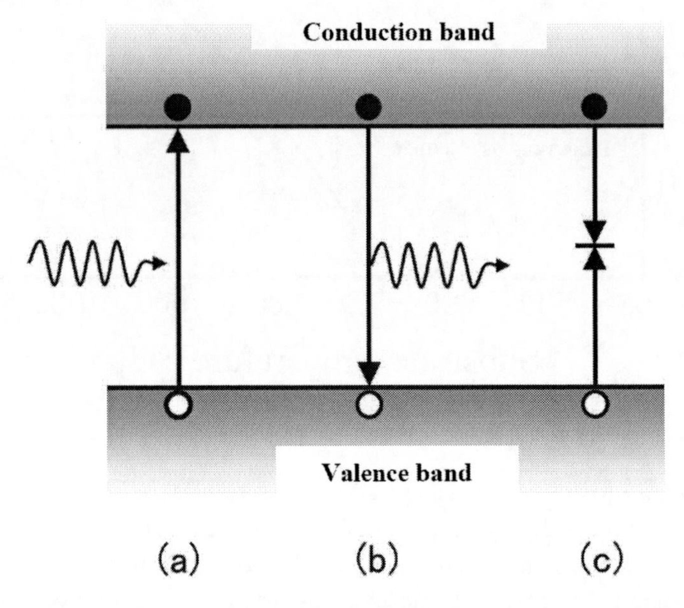

Figure 13. PL emission processes at room temperature. (a) Excitation of electron-hole pair by laser beam, (b) band edge emission by the recombination of carriers, and (c) recombination without photo-emission by defect levels (SRH process [30]).

The PL measurement at the liquid He temperature (4.2 K) can provide the information of radiative impurities and defects. The photoluminescence due to the defect levels was not observed from the epitaxial specimens as well as the reference CZ-Si specimen. The PL measurement at room temperature was performed for evaluating the non-radiative defect levels. The PL emission process at room temperature can be schematically presented as shown in Figure 13. The minority carriers excited by the laser in the process of (a) recombine with majority carriers, and emit PL of the energy corresponding to the band gap (b). If many defect levels exist in a crystal, minority carriers can follow the non-radiative recombination pathway like the case of (c), and thus PL intensity by the process of (b) decreases. The PL intensity is nearly proportional to the effective lifetime of the specimen (τ_{eff}) [33], and thus can be used for evaluation of non-radiative defect levels. However, a bare Si surface shows very high surface recombination velocity. As τ_{eff} is significantly affected by the surface recombination velocity, passivation treatment of the surface was performed on each specimen by forming a thermal oxidation film.

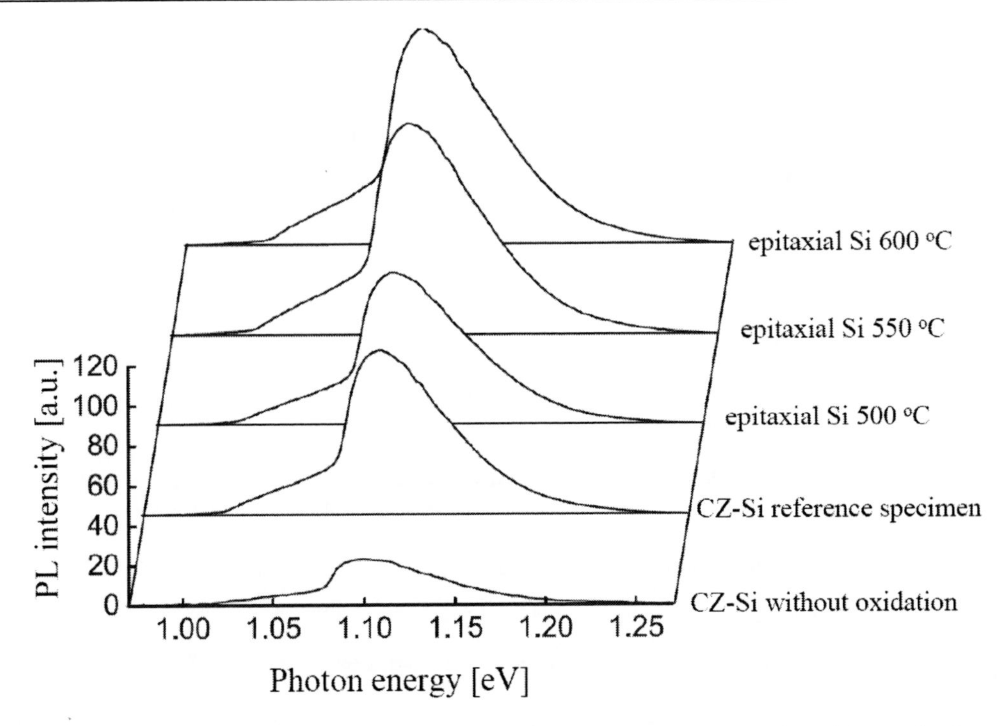

Figure 14. PL emission spectra at room temperature. (a) CZ-Si without oxidation, (b) CZ-Si reference specimen, (c) epitaxial Si film grown at 500 °C, (d) epitaxial Si film grown at 550 °C, and (e) epitaxial Si film grown at 600 °C. Specimens (b) – (e) were surface passivated using oxide films.

Figure 14 is the room temperature PL spectrum of the Si epitaxial specimen whose surface was passivated by the thermal oxide film. The oxide film passivation was performed by the formation of 15 nm thick oxide film by the thermal oxidation at 900 °C for 30 minutes, and subsequent annealing in H_2 gas at 400 °C for 30 minutes. PL spectrum of a bare CZ-Si specimen was compared with that of the thermally oxidized CZ-Si specimen which was used as the reference. These spectra are similar figures, but the intensity is several times higher by the oxide film passivation. It is due to the reduction of the surface recombination velocity. All the PL spectra of the epitaxial specimens show as same as or higher than that of the reference CZ-Si specimen.

Figure 15 summarises the emission intensity at the band edge obtained from the PL spectrum of each specimen measured at room temperature. Each point shows the value averaged over 12 different locations at each specimen surface and the standard deviation. It is indicated that the emission intensity of the epitaxial specimen tends to be higher than that of the reference specimen, and that the PL intensity of the epitaxial specimen grown at higher temperature tends to be higher. As the excess carriers excited by the laser can diffuse into the material and penetrate into the bulk regions which are deeper than the laser absorption length of 1.25 μm, PL spectra contain information not only from the Si epitaxial film but also from the Si substrate. However, the PL intensity increases only when the carrier lifetime in the epitaxial film is longer than that of the substrate. It is therefore confirmed that the Si epitaxial film grown at low temperature by atmospheric pressure plasma CVD at least has a longer carrier lifetime than that of the CZ-Si substrate. As the applications for devices are considered, the film shows excellent electrical properties.

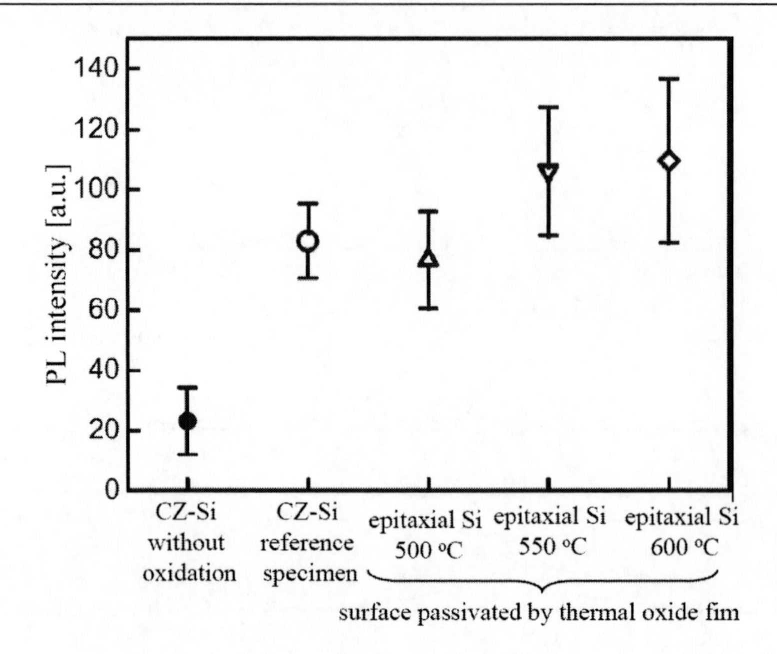

Figure 15. PL emission intensities at room temperature for the specimens shown in Figure 14.

SUMMARY

The technique for growing a high quality epitaxial Si film on the whole area of 4-inch (001) Si wafer at the low temperature region down to 500 °C was established by atmospheric pressure plasma CVD using the porous carbon electrode. It was confirmed that the epitaxial Si film shows no defects observed by TEM and the etch-pit method, that concentrations of O, C and metallic impurities are below SIMS detection limits, that there is no radiative defects measured by the PL method at 4.2 K, and that its carrier recombination lifetime is longer than that of the CZ-Si substrate. The epitaxial growth rate is comparable to that by thermal CVD at approximately 900 °C, which is high enough for practical applications.

The mechanisms is not still sufficiently understood for growing high quality Si epitaxial film at high rate at low temperature using atmospheric pressure plasma CVD. Further study is necessary for its deeper understanding [34-36].

REFERENCES

[1] Kakiuchi, H.; Matsumoto, M.; Ebata, Y.; Ohmi, H.; Yasutake, K.; Yoshii, K.; Mori, Y. *J. Non-Cryst. Solids* 2005, *351(8-9),* 741-747.

[2] Kakiuchi, H.; Ohmi, H.; Kuwahara, Y.; Matsumoto, M.; Ebata, Y.; Yasutake, K.; Yoshii, K.; Mori, Y. *Jpn. J. Appl. Phys*. 2006, *45(4B)*, 3587-3591.

[3] Ohmi, H.; Kakiuchi, H.; Yasutake, K.; Nakahama, Y.; Ebata, Y.; Yoshii, K.; Mori, Y. *Jpn. J. Appl. Phys*. 2006, *45(4B)*, 3581-3586.

[4] Mori, Y.; Kakiuchi, H.; Yoshii, K.; Yasutake, K.; Ohmi, H. *J. Phys. D Appl. Phys.* 2003, 36, 3057-3063.

[5] Kakiuchi, H.; Ohmi, H.; Aketa, M.; Yasutake, K.; Yoshii, K.; Mori, Y. *Thin Solid Films* 2006, *496(2)*, 259-265.

[6] Kakiuchi, H.; Nakahama, Y.; Ohmi, H.; Yasutake, K.; Yoshii, K.; Mori, Y. *Thin Solid Films* 2005, *479(1-2)*, 17-23.

[7] In *MOS Device Epitaxial Wafer*; Tsuya, H.; Shimizu, H.; Yamamoto, H.; Ed.; Realize Corp: Tokyo, JP, 1998.

[8] Imai, M.; Inoue, K.; Mayusumi, M.; Gima, S.; Nakahara, S. *J. Electrochem. Soc.* 2000, *147(2)*, 1568-1572.

[9] Imai, M.; Nakahara, S.; Inoue, K.; Mayusumi, M.; Gima, S. *Electrochem. Solid-State Lett.* 2001, 4(3), G23-G25.

[10] Suzuki, T.; Inoue, H.; Miyauchi, A.; Nakada, M. *CHO-SEIMITSU (Super precision).* 1997, *7*, 73.

[11] Varhue, W.J.; Andry, P.S.; Rogers, J.L.; Adams, E.; Kontra, R.; Lavoie, M. *Solid State Technol.* 1996, *39(6)*, 163-167.

[12] Hattangady, S.V.; Posthill, J.B.; Fountain, G.G.; Rudder, R.A.; Mantini, M.J.; Markunas, R.J. *Appl. Phys. Lett.* 1991, *59(3)*, 339-341.

[13] Rogers, J.L.; Andry, P.S.; Varhue, W.J.; McGaughnea, P.; Adams, E.; Kontra, R. *Appl. Phys. Lett.* 1995, *67(7)*, 971-973.

[14] Platen, J.; Selle, B.; Sieber, I.; Brehme, S.; Zeimer, U.; Fuhs, W. *This Solid Films* 2001, *381*, 22-30.

[15] Schwarzkopf, J.; Selle, B.; Bohne, W.; Röhrich, J.; Sieber, I.; Fuhs, W. *J. Appl. Phys.* 2003, *93(9)*, 5215-5221.

[16] Ohi, S.; Burger, W.R.; Reif, R. *Appl. Phys. Lett.* 1988, *53(10)*, 891-893.

[17] Comfort, J.H.; Reif, R. *J. Electrochem. Soc.* 1989, *136(8)*, 2398-2405.

[18] Ghidini, G.; Smith, F.W. *J. Electrochem. Soc.* 1984, 131(12), 2924-2928.

[19] Eversteyn, F.C. *Philips Pes. Rep.* 1974, 29, 45.

[20] Mori, Y.; Yoshii, K.; Kakiuchi, H.; Yasutake, K. *Rev. Sci. Instrum.* 2000, *71(8)*, 3173-3177.

[21] Mori, Y.; Kakiuchi, H.; Yoshii, K.; Yasutake, K. *SEIMITSUKOUGAKKAISHI (Jpn. Soc. Precision Eng.)*, 2000, *66(11)*, 1802-1806.

[22] Mori, Y.; Yoshii, K.; Yasutake, K.; Kakiuchi, H.; Ohmi, H.; Wada, K. *Thin Solid Films* 2003, *444(1/2)*, 138-145.

[23] Yasutake, K.; Ohmi, H.; Kakiuchi, H.; Wakamiya, T.; Watanabe, H. *Jpn. J. Appl. Phys.* 2006, *45(4B)*, 3592-3597.

[24] Ohmi, H.; Kakiuchi, H.; Tawara, N.; Wakamiya, T.; Shimura, T.; Watanabe, H.; Yasutake, K. *Jpn. J. Appl. Phys.* 2006, *45(10B)*, 8424-8429.

[25] Oshikane, Y.; Yamamura, K.; Kakiuchi, H.; Oda, A.; Western, C.; Nagao, A.; Endo, K. In *Proc. 6th Int. Conf. Reactive Plasmas and 23rd Symp. Plasma Proc.* Sendai, JP, 2006, *23*, 743-744.

[26] Secco d'Aragona, F. *J. Electrochem. Soc.* 1972, *199(7)*, 948-951.

[27] Yasutake, K.; Kakiuchi, H.; Ohmi, H.; Yoshii, K.; Mori, Y. *Appl. Phys. A.* 2005, *81(6)*, 1139-1144.

[28] Yasutake, K.; Tawara, N.; Ohmi, H.; Terai, Y.; Kakiuchi, H.; Watanabe, H.; Fujiwara, Y. *Jpn. J. Appl. Phys.* 2007, *46(4B)*, 2510-2515.

[29] Takahashi, H.; Maekawa, T. *Jpn. J. Appl Phys.* 2000, *39(7A)*, 3854-3859.

[30] Schroder, D.K. *Semiconductor Material and Device Characterization*; 2nd ed; John Wiley & Sons: New York, US, 1998.

[31] Schroder, D.K., Choi, B.D.; Kang, S.G.; Ohashi, A.; Kitahara, K.; Opposits, G.; Pavelka, T.; Benton, J. *IEEE Trans. Electron Devices* 2003, *50(4)*, 906-912.

[32] Katsura, J.; Nakayama, H.; Nishino, T.; Hamakawa, Y. *Jpn. J. Appl. Phys*. 1982, *21(5)*, 712-715.

[33] Tarasov, I.; Ostapenko, S.; Feifer, V.; McHugo, S,: Koveshnikov, S.V.; Weber, J.; Haessler, C.; Reisner, E.U. *Phys. B*. 1999, *273-274*, 549-552.

[34] Yasutake, K.; Ohmi, H.; Kirihata, Y.; Kakiuchi, H. *Thin Solid Films* 2008 *517 (1)*, 242–244

[35] Kirihata, Y.; Nomura, T.; Ohmi, H.; Kakiuchi, H.; Yasutake, K. *Surf. Interf. Anal.* 2008, *40(6-7)*, 984-987.

[36] Inagaki, K. Hirose, K.; Yasutake, K. *Surf. Interf. Anal.* 2008, *40(6-7)*, 1088-1091.

In: Generation and Applications of Atmospheric Pressure Plasmas ISBN: 978-1-61209-717-6
Editors: M. Kogoma, M. Kusano and Y. Kusano

Chapter 10

DEPOSITION OF OXIDES BY ATMOSPHERIC PRESSURE PLASMA CHEMICAL VAPOUR DEPOSITION

Noritoshi Saito[1] and Motokazu Yuasa[2]
[1]MEMS Core Co., LTD, Miyagi, Japan
[2]Sekisui Chemical CO., LTD, Kyoto, Japan

INTRODUCTION

Current manufacturing processes of Large Scale Integration (LSI) and Flat Panel Display (FPD) have a lot of problems, since more than 50 % of the processes are performed in vacuum or reactive gas atmosphere. For example, due to the increase of the complexity and the number of layers of the devices, the number of the carrying-in/out time of the wafers and glass substrates in vacuum chamber increases. It is said that carrying in/out to/from a vacuum chamber is complicated, and that it takes 90 % of the total transportation time. As a result, it is difficult to shorten the total treatment time, even if the treatment time in a vacuum chamber is decreased.

In order to solve the problem, we focused on generating a plasma at atmospheric pressure, and succeeded to generate a stable plasma at atmospheric pressure, independent of the gas atmosphere by controlling pulsed electric field [1]. The characteristics of the atmospheric pressure plasma developed here are summarized as follows;

(1). Vacuum-Less Processing

Investment for vacuum pumps and facilities needed for vacuum environment is unnecessary. This affects the choice of the material and the pressure resistance of the chamber. If vacuum is not used, the price and the size for the equipment can be reduced. In addition, it enables the treatment of substrates which contain water and volatile components. This developed technique is preferably applied for the treatment of plastic films.

(2). Continuous Treatment Processing

A continuous dry process such as a roll-to-roll method and a belt-conveyer transportation method can be easily constructed at atmospheric pressure, which can demonstrate high productivity. Low pressure process requires a complicated shielding system between atmospheric pressure and vacuum, while process at atmospheric pressure needs only a very simple system to achieve a continuous process.

(3). High Speed Treatment

In addition to the reduction of pumping time, atmospheric pressure plasma has its high treatment speed due to its high density.

Making the best use of these characteristics, an atmospheric pressure chemical vapour deposition (AP-CVD) technique has been developed as one of the applications of the vacuum-less plasmas. We started with deposition of anti-reflective (AR) coatings on plastic films [2], and have extended the applications in semiconductor industry and liquid crystal display (LCD) process [3]. In this chapter, SiO_2-CVD technique is introduced, which is under development for the Micro Electro-Mechanical System (MEMS) application.

EQUIPMENT QND BASIC PROCESS

(a) automatic type,

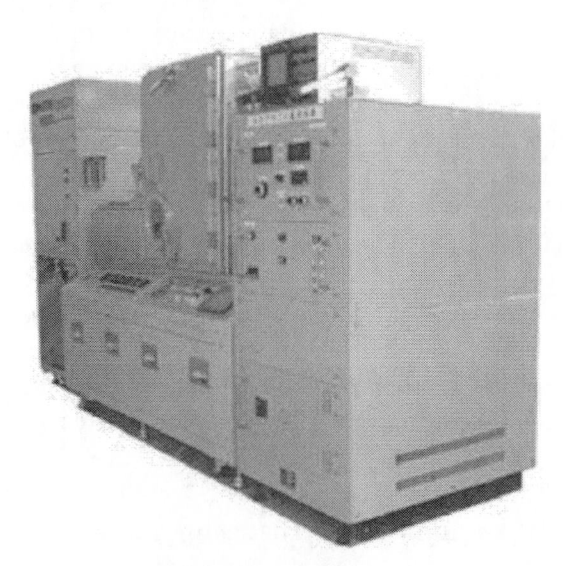

(b) manual type.

Figure 1. AP-CVD for MEMS (AP-01M).

Figure 1 shows the appearance of the AP-CVD system (AP-01M) for the MEMS processing. The schematic diagram and the photo image of the AP-CVD plasma source are shown in Figure 2. This system enables high speed deposition of insulating coatings at low temperature using an atmospheric pressure high density plasma source. This AP-CVD system can form SiO_2 coatings by the reaction of the alkoxysilane type precursor with oxygen plasma which is generated by applying high frequency pulsed voltage at a slit electrode at atmospheric pressure. The generated plasma and the substrate are several mm away. Namely this system is the remote plasma CVD with which excited oxygen in a plasma are transported to the surface by a gas flow. This technique avoids the damage of the substrate by ion bombardment, and is also suitable for deposition after fabricating IC and MEMS devices. Deposition is taken place by placing the substrate on the susceptor which can be heated and placed beneath the linear plasma source, and moving right and left. Treatable substrate size ranges from indeterminate small form to maximum 200 mm square. It can serve the best for research and development, and production of small quantities.

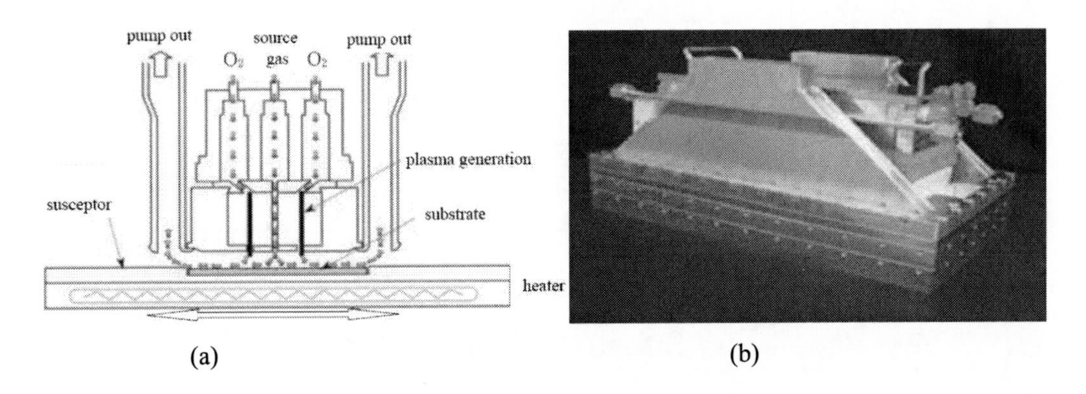

Figure 2. AP-CVD system. (a) Schematic diagram of remote type plasma source. (b) A plasma source for 200 mm wide substrate.

Liquid materials such as various alkoxysilanes represented by tetraethoxysilane (TEOS) can be used as source gases for the deposition. This system can stably supply high flow rate source gases at atmospheric pressure using a uniquely developed vaporization-supply system. It is also possible to dope phosphorus into the SiO_2 film using trimethylphosphate (TMOP ($PO(OCH_3)_3$)).

Table 1 shows the standard deposition condition for SiO_2 coatings using this system for 200 mm ϕ wafer. Using TMOS as the alkoxysilane source, which has lower content of carbon and shows higher reactivity than TEOS, high deposition speed was demonstrated. Furthermore, by adding a small amount of H_2O, the -OH functional group in the coatings are reduced. In such a way, a high quality SiO_2 insulating coating was synthesized, showing improved electrical insulating withstand voltage and high resistance against HF wet etching.

Table 1. Leakage rate with different hole diameters of the feed through electrodes (* Targeted leakage rate < 4×10^{-9} Pa m³ / s).

Hole diameter of electrode [μm]	Leakage rate [Pa m^3 / s]*
No electrode	1.2×10^{-9}
50	$1.1 - 1.5 \times 10^{-9}$
100	$1.4 - 2.0 \times 10^{-9}$
200	$1.0 - 2.8 \times 10^{-9}$

Figure 3 shows the deposition rate of the SiO_2 coatings at various substrate temperatures. The deposition rate obtained here is the coating thickness after scanning an 8-inch wafer beneath the remote plasma system in one minute. The deposition rate increased up to approximately 200 °C but decreased at higher temperatures. It indicates that the process is surface reaction limited deposition between 200 and 350 °C and diffusion limited deposition if > 350 °C, and thus this process is based on surface reactions.

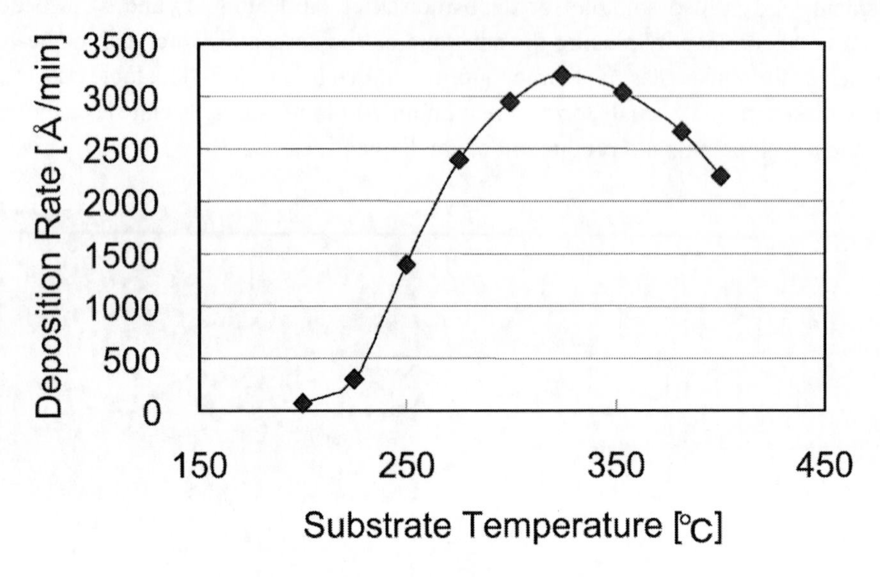

Figure 3. Deposition rate at various substrate temperatures.

It is generally said that alkoxysilane like TEOS is easy to migrate, and shows excellent step coverage as a CVD source material. It is due to the formation of intermediates such as polysiloxiane at around the substrate surface, when alkoxysilane transfers to SiO_2. The present method also relies on reaction of alkoxysilane with activated species such as O* and O_3 in a remote plasma. It is thought that this reaction results in formation of intermediates with relatively large molecular weight at around the surface. They are adsorbed at the surface and forming SiO_2 during their migration (surface reaction). In this case, migration of the intermediates proceeds so that surface energy is at relaxation related to unevenness of the substrate surface. As a result, the coating shows excellent local smoothness and step coverage.

2. Excellent Coverage

Here, fabrication process of piercing interconnections substrate is introduced as an example of formation of insulating coating for the MEMS structure [5]. Figure 4 shows a general cross section of the piercing interconnections. A perforated hole is created on a Si substrate using Deep-reactive ion etching (RIE), and then an oxide coating is deposited on the front, back sides, and side walls. After that conductive copper is embedded by plating. During the processing, the following is important;

[1] Insulation between Si and Cu, (a key issue in the processing is the uniform formation of the insulating coating on the side walls),
[2] High tightness after embedding copper (a key issue in the processing is flattening the side walls before copper plating).

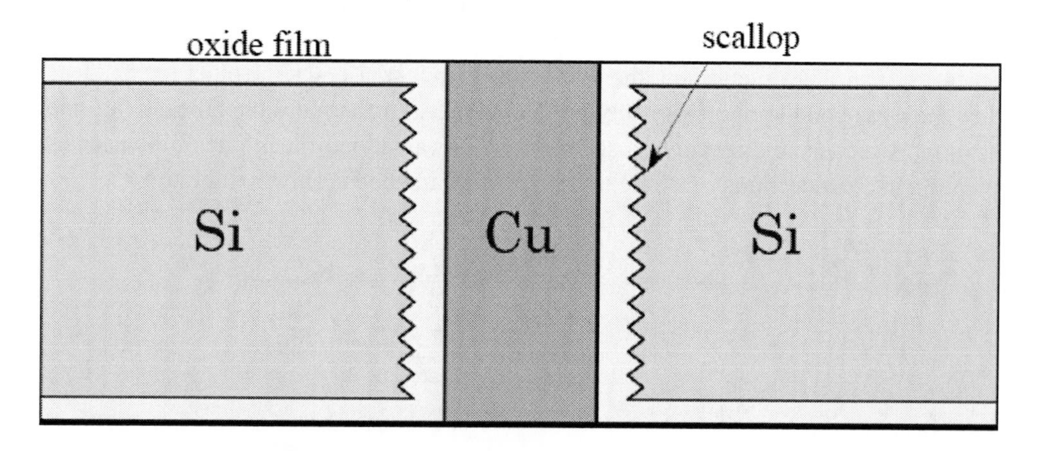

Figure 4. Cross section of through silicon via (TSV).

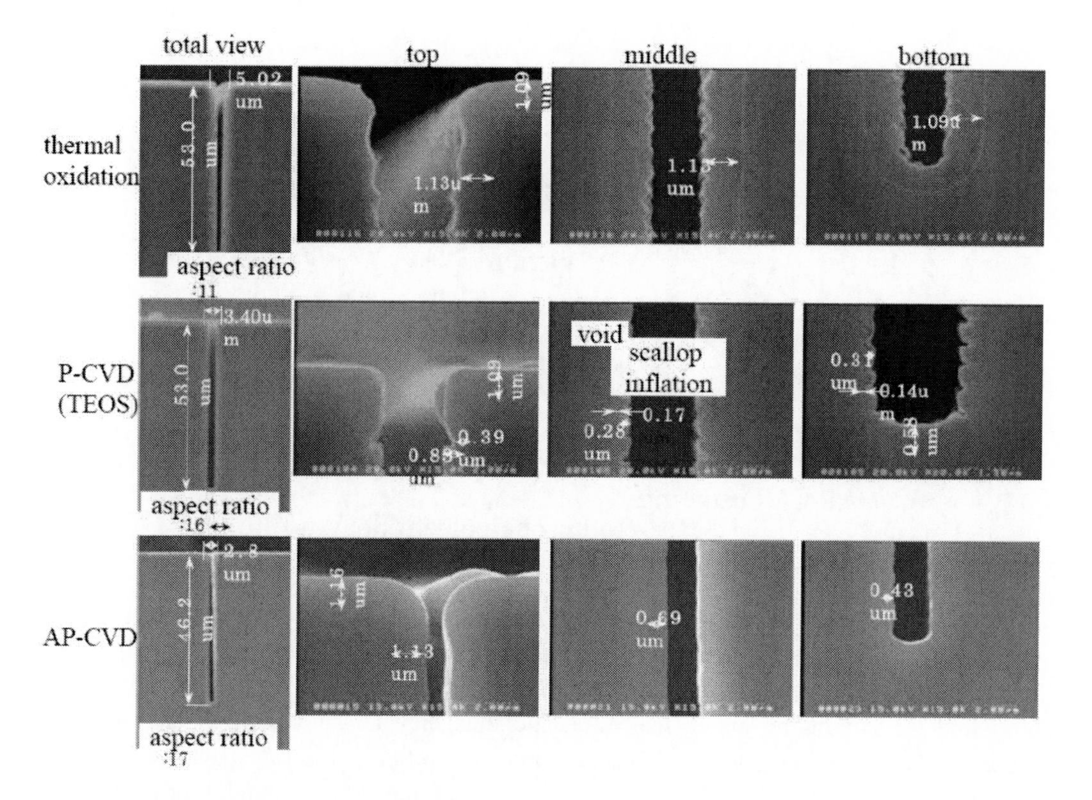

Figure 5. SiO_2 coverage at the feed-through.

Figure 5 shows the scanning electron microscopic (SEM) images of the cross-sections after deposition of insulating coating at the deep ditch which was formed by etching. The oxide coatings were synthesized using thermal oxidation, low pressure plasma TEOS CVD (P-CVD) [6], and AP-CVD at the 50 μm deep and 4 – 5 μm wide ditches. The thermal oxidation method uniformly forms oxide layer up to the bottom of the ditch, while the shape of the scallop at the side walls which was formed using the Deep-RIE remains as it is. P-CVD shows poor coverage of approximately 13 % thick at the side wall near the bottom. Here the coverage is defined as the coating thickness at the side wall divided by the coating thickness at the substrate surface. In addition, the scallop tends to increase. On the other hand, AP-CVD coating shows as good as approximately 37 % coverage. Furthermore the process can smooth the scallop. A schematic diagram of a sample for the evaluation of leak rate of the package of the piercing interconnections substrate using AP-CVD method is shown in Figure 6.

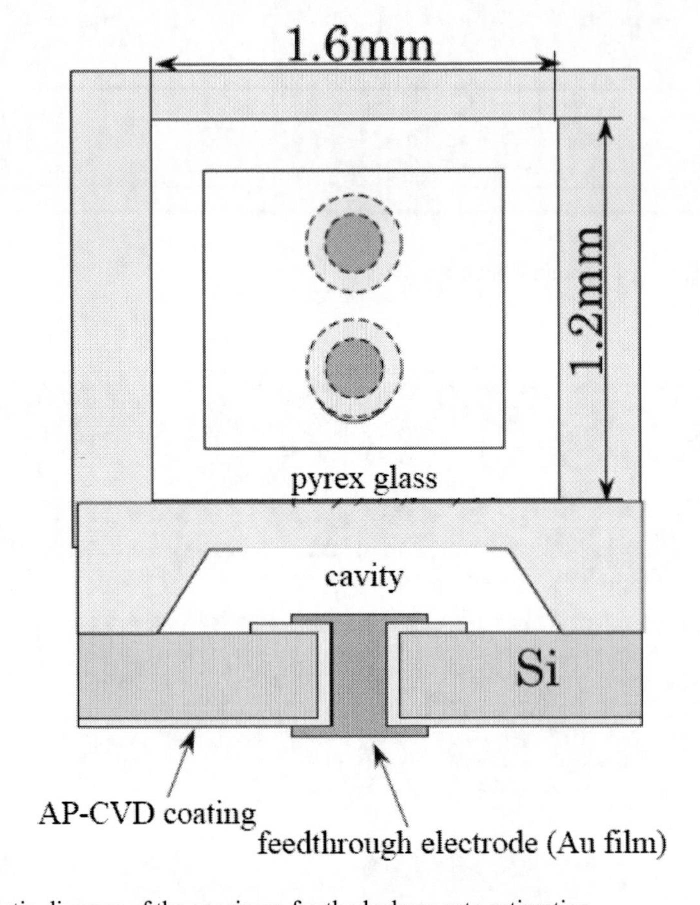

Figure 6. Schematic diagram of the specimen for the leakage rate estimation.

Table 1 summarizes the helium leak rate of the above mentioned samples. As the size of the hole diameter of the electrode increased, the data scattered more. However, they showed approximately half of the aimed leak rate, and thus they all showed high gas tightening properties. On the other hand, the piercing interconnections substrate package fabricated using the thermal oxidation method could not keep the gas tightening property. It is concluded here that the requirements of electrical insulation between Si and Cu, and gas tightening after embedding Cu can be achieved by synthesizing insulating coatings using the AP-CVD system.

Deposition on the backside of the undercut represents one of the examples of its excellent coverage which is the characteristics of the AP-CVD coatings. Figure 7 shows the cross sections of the undercut parts on which insulating films were coated by using AP-CVD and P-CVD. After etching the 15 μm wide Line & Space using Deep-RIE, 5 μm of the sacrifice layer (oxide coating) was side-etched. After that, 1-μm insulating coatings were deposited using AP-CVD and P-CVD. Though P-CVD coating was deposited on the backside of the undercut, there was a gap of maximum 0.54 μm. On the other hand, the gap at the undercut after the AP-CVD was 0.04 μm. It is thus indicated that AP-CVD shows excellent coverage at the backside of the undercut. The structures for MEMS require the formation of driving piezoelectric devices and metallic interconnections on Si surface, and synthesis of insulating coatings at low temperature. It is noted here that AP-CVD can deposit insulating films at the backside of the undercut with excellent coverage at low temperature.

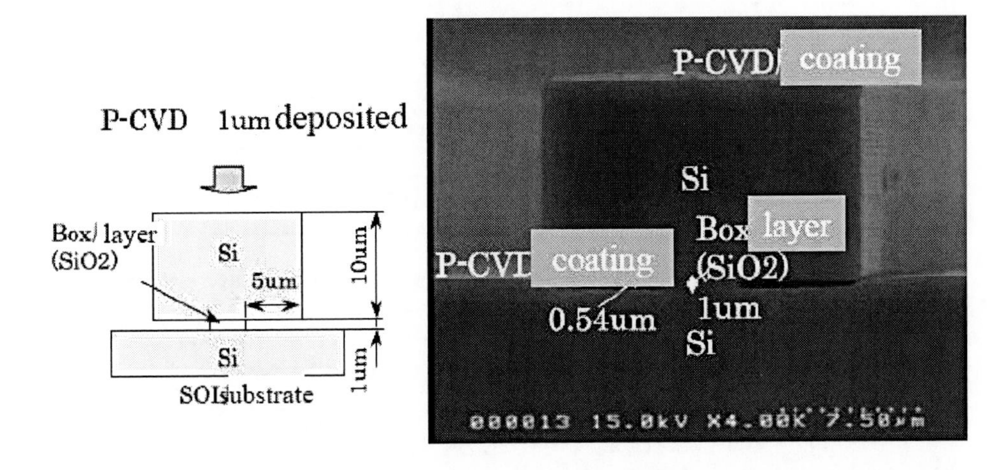

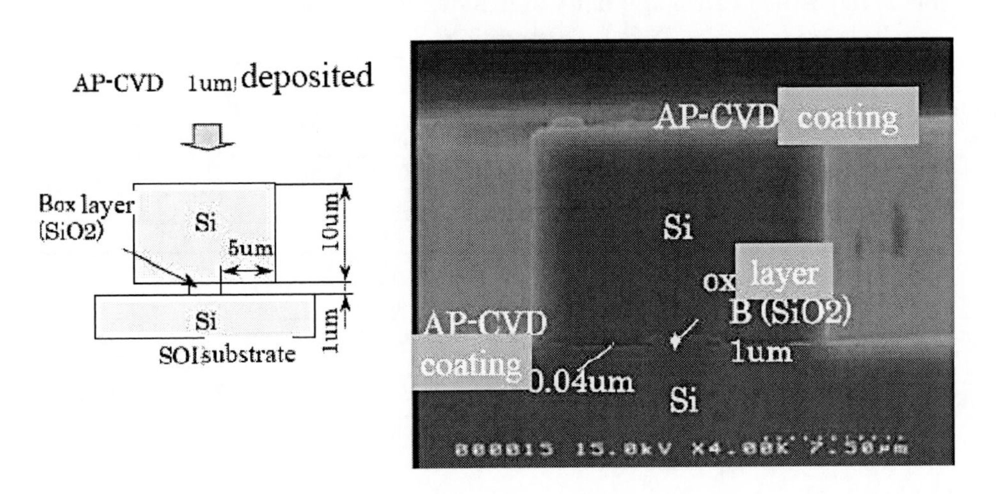

Figure 7. Coating coverage at the backside of the undercut.

3. Application for Sacrifice Layer

Formation of the structures using a sacrifice layer is MEMS's own processing, which does not exist in semiconductor processing. Figure 8 shows the fabrication processes of a diaphragm using a sacrifice layer.

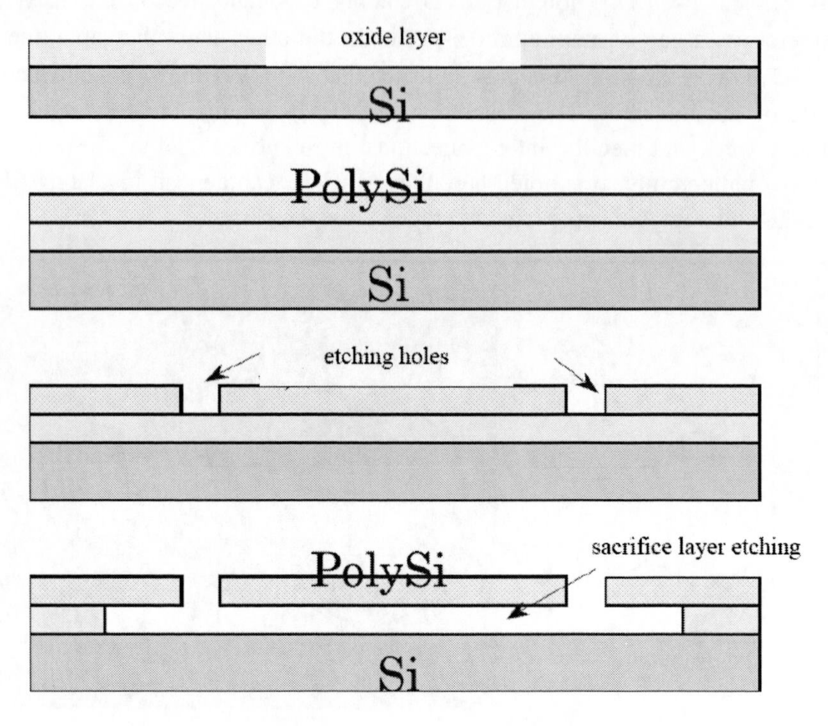

Figure 8. Diaphragm fabrication processes.

Table 2. Deposition rate and quality of the SiO_2 coating deposited by the standard condition.

Plasma gas	O_2	Flow rate: 10 L/min
Substrate temperature		350 °C
Stage transportation speed		200 mm / min
Plasma condition	Power supply	Pulsed
	frequency	20 kHz
	Applied voltage	14 kV
Deposition time	(ϕ 200 mm)	1 minute
Deposition rate	(ϕ 200 mm)	300 nm / min
-OH content in the coating		3 %
WERR (1 % DHF)		3.5 times of thermally oxidized coating
Electric Field	1×10^{-7} A / cm^2	10 MV / cm

By etching the sacrifice layer of the oxide film the poly-Si diaphragm is fabricated. In order to form a desired shape of the diaphragm, the sacrifice layer of the oxide coating must be etched horizontally more than several tens μm. This usually requires long etching time, inducing several problems. That is why the oxide coatings which show high etching rates are required. AP-CVD can synthesize high quality coatings by adding a small amount of H_2O during deposition as shown in Table 2. However, if H_2O is not added, low quality coatings with high etching rate can be synthesized as well.

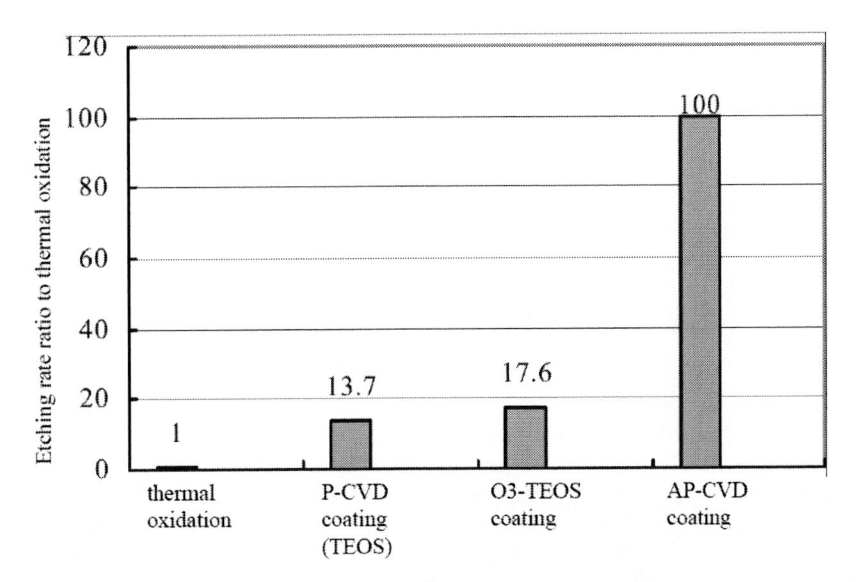

Figure 9. The ratio of sacrifice layer etching rates of various oxide coatings.

Figure 9 shows the ratio of etching rates of various oxide films. Here, the etching rate was evaluated by a dry etching system for sacrifice layers using HF and CH_3OH. When the etching rate of the thermally oxidized layer is 1, those of P-CVD (TEOS) and AP-CVD (O_3-TEOS) are 14 and 18, respectively. On the other hand, that of AP-CVD without adding H_2O is significantly high, as high as 100. Figure 10 shows the ratio of the etching rate at various TMOS flow rates. As the flow rate of TMOS increases, the etching rate of a sacrifice layer increases, exceeding 200 times of that of thermally oxidized layer. Synthesis of Phospho-silicate Glass (PSG) coatings is the attempt to obtain high-etch rate oxides. However, if a PSG coating is dry-etched as a sacrifice layer, products containing phosphorous acid generate, inducing sticking, and resulting in a failure of forming a normal structure. AP-CVD coatings do not generate liquid reactive products, and thus this sticking problem will not occur at the dry etching of a sacrifice layer.

Figure 11 shows an oscillation sensor which has an AP-CVD coating as a sacrifice layer. The interval of the sacrifice layer holes is 140 μm, and thus the etching of more than 70 μm from both sides of the holes is necessary. If it is a thermally oxidized layer, its etching requires more than 700 minutes. When the AP-CVD coating was used, the etching completed within approximately 10 minutes. Namely, by using the AP-CVD coatings as sacrifice layers, the time for etching can be reduced. It subsequently reduces damages to the other materials used.

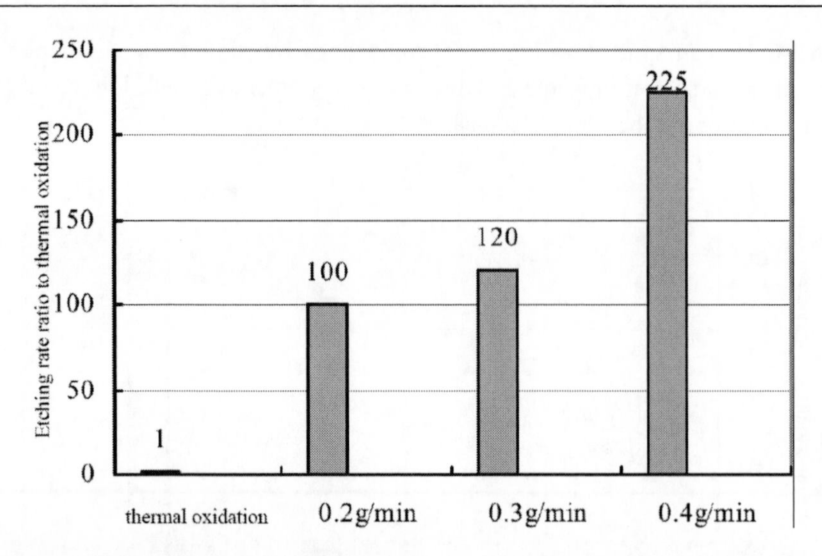

Figure 10. The ratio of sacrifice layer etching rates at different TMOS flow rates.

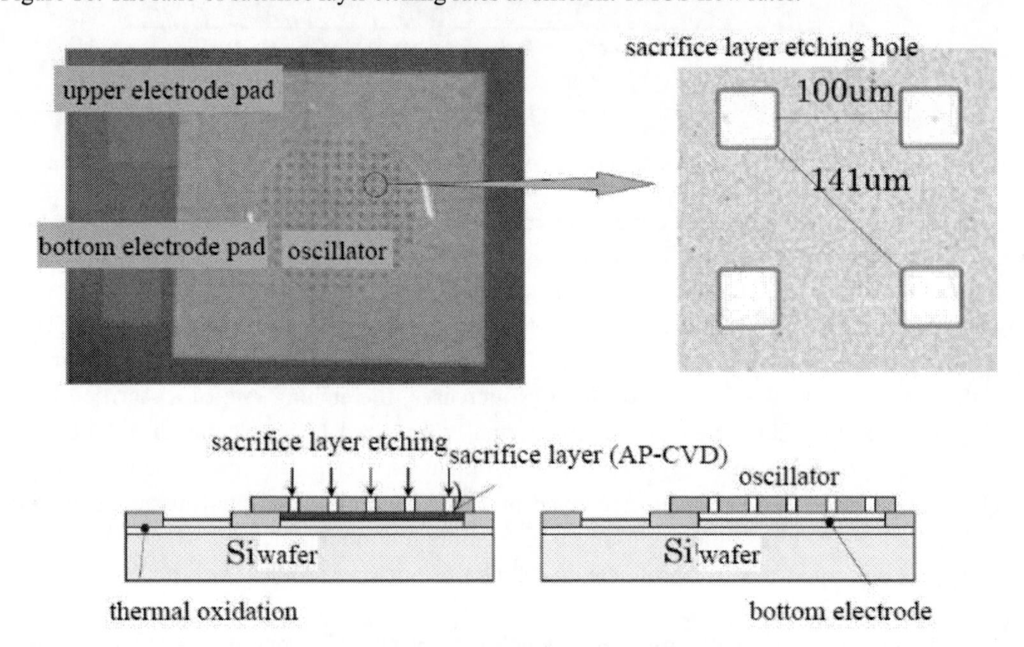

Figure 11. Oscillation sensor using a sacrifice layer deposited by the AP-CVD.

SUMMARY

Although the introduction of this technique is just an example, we hope that the excellent coverage of SiO_2 coatings by the AP-CVD technique is well understood. It is due to the surface reaction process using liquid source of alkoxysilane type. Utilizing this characteristic, its applications for insulating layers and sacrifice layers for Surface-MEMS and Bulk-MEMS will be attempted. At the same time, oxide coatings by the AP-CVD with different characteristics are also under development. The expansion of application areas and inline continuous multi-layer deposition technique will be proposed.

REFERENCES

[1] Yuasa, M. *Nikkei Micro-devices*. In Japanese. 2001, *4*, 139-146.
[2] Yuasa, M. *Monthly Display*. In Japanese. 2002, *8(11)* 29-35.
[3] Yara, T.; Iwane, K.; Yuasa, M. *A monthly publication Jpn. Soc. Appl. Phys. (Oyo Butsuri)* in Japanese. 2006, *75(4)*, 456-460.
[4] Maeda, K. *Surface reaction process for next generation ULSI manufacturing (Jisedai ULSI Seizou no TamenoHyoumen Hannnou Purosesu* in Japanese; Science Forum Inc.: Tokyo, JP, 1994; 111-145.
[5] Bonkohara, M.; Takahashi, K. *J. Inst. Electr. Inform. Commun. Eng*. 2001, *84(11)*, 803-810.
[6] Inoue, Y.; Takai, O. *Jpn. Soc. Plasma Sci. Nucl. Fusion Res. 2000*, *76(10)*, 1068-1073.

In: Generation and Applications of Atmospheric Pressure Plasmas ISBN: 978-1-61209-717-6
Editors: M. Kogoma, M. Kusano and Y. Kusano

Chapter 11

SYNTHESIS OF HIGH GAS BARRIER CARBON COATING BY ATMOSPHERIC PRESSURE PLASMA CVD

Hideyuki Kodama[1] and Tetsuya Suzuki[2]
[1]Aoyama Gakuin University, Sagamihara, Kanagawa, Japan
[2]Keio University, Yokohama, Kanagawa, Japan

INTRODUCTION

Surface treatment is the technique to improve properties of a substrate by treating its surface. In particular, coating techniques are indispensable for the current materials development, since they enable significant improvement of material properties or addition of completely new properties by covering the substrate with a very thin material. Figure 1 categorizes the surface treatment techniques. Among the coating techniques, vapor deposition techniques using plasma have been industrialized in all application areas, as they can control and synthesize nano-level structure of material consisting of more than one element. These chemical vapor deposition (CVD) and physical vapor deposition (PVD) methods are in general low pressure synthesis methods using vacuum equipments. Due to low level of contamination and low collision frequency of ions and radicals in a plasma at low pressure, high quality thin films can be easily synthesized. However, if vacuum equipments are used, the cost is high and a treatable area is limited. Therefore practical use of these methods is difficult for quite a few products. As the areas for the application of the coating techniques have been extended, the interest in atmospheric pressure plasmas which do not require vacuum techniques has been significantly increased for several years. If thin films can be also synthesized at atmospheric pressure, showing similar properties by low pressure synthesis, it would be scientifically interesting as well as industrially useful. We have studied synthesis of thin films using atmospheric pressure plasma, in particular synthesis of amorphous carbon films since 2000, which will be described later. Although thin film synthesis was achieved a few years after starting the research, only rough coatings could be synthesized. However, after improving the structure of the equipment and synthesis conditions, we are now able to synthesize amorphous carbon coatings whose gas barrier property is as high as or higher than that synthesized at low pressure. In this chapter, synthesis of high gas barrier amorphous carbon coatings is described using atmospheric pressure plasmas.

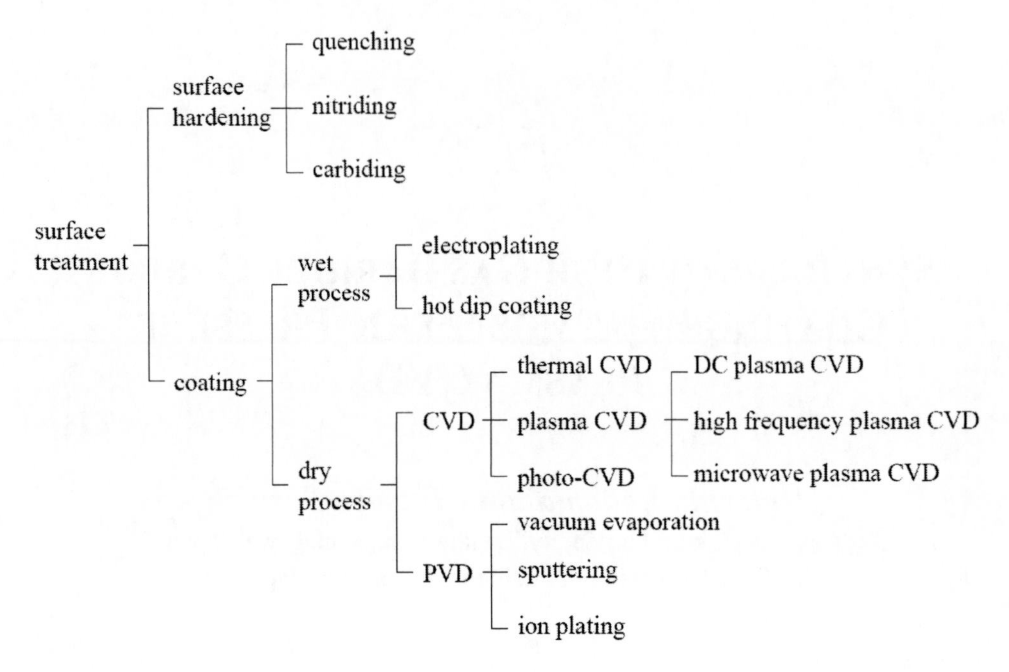

Figure 1. Surface treatment techniques.

1. Amorphous Carbon Film

Many of you might already hear the word "diamond-like-carbon (DLC)". DLC was reported in 1970s as amorphous inorganic materials whose hardness is nearly as high as that of diamond. It turned out that DLC shows not only the high hardness but also various desirable properties such as extreme smoothness, excellent sliding property, wear resistance and chemical stability. It is now expected that the DLC can be applied in a wide range of topics. DLC was at first regarded as the byproduct of artificial diamond, and diamond was of people's interest. However, since DLC can be synthesized more easily with lower cost than diamond, the DLC became widely studied in 1990s, and the name of DLC became used for amorphous carbon films in general. Figure 2 shows microscopic structures of diamond, graphite and DLC. Figure 2 (c) schematically shows the mixed structures of diamond and graphite, but in fact it is unclear in total. Comparing with diamond consisting of sp^3 hybridized carbons only and graphite consisting of sp^2 hybridized carbons only, DLC is amorphous which contains three dimensional structures similar to diamond but does not have long range regularity, and is thought to contain carbon atoms showing characteristics of both diamond and graphite. Initially DLC was regarded as a hard carbon film containing high content of sp^3 hybridized structure. However, DLC does not have fixed structure, and varies according to the synthesis equipment and/or conditions. Consequently even a soft carbon film containing only a small amount of sp^3 hybridized structure is now called DLC. Even now the definition of DLC is not fixed. In a broad way it is called amorphous carbon (a-C) and hydrogenated amorphous carbon (a-C:H). In this chapter it is called amorphous carbon film.

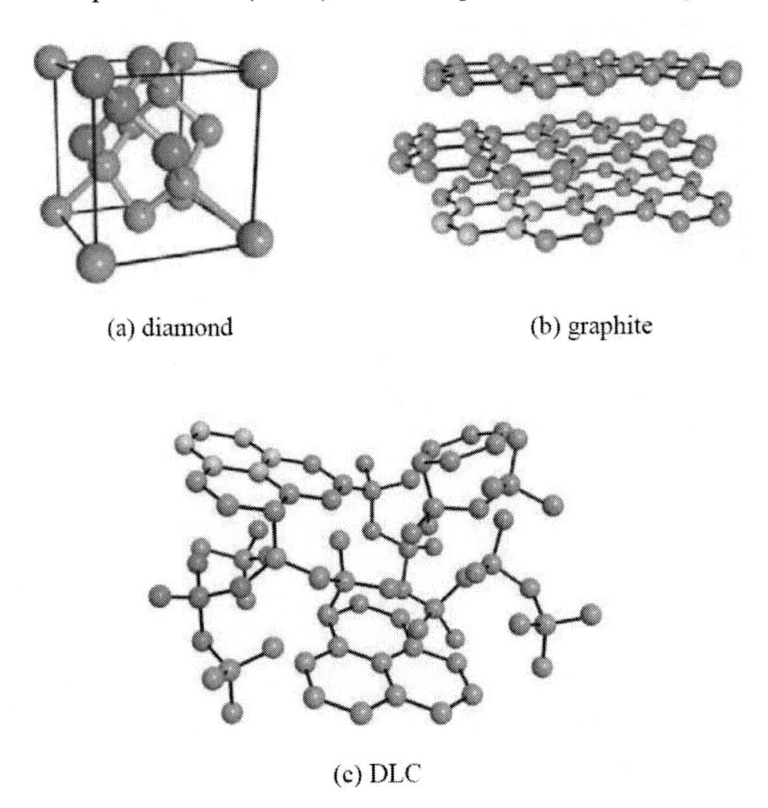

Figure 2. Structures of diamond, graphite and DLC.

The oldest reports on synthesis of amorphous carbons are using ion beam evaporation method with carbon ion beam [1], and ionizing evaporation method with hydrocarbon ion beam [2]. In the first half of 1980s, amorphous carbon films were applied for the oscillating plate of speakers. While research on diamond was declined in 1990s, amorphous carbons were applied for single-lever basin mixers and then familiar products such as frames of watches as well as machine tools. The reason of such a rapid progress of the applications of amorphous carbons is that it shows not only high hardness which is close to that of diamond but also excellent sliding properties. Even at that time there existed techniques for improving sliding properties by using hard ceramic coatings represented by TiN and fluorine processing represented by TEFLON coatings. Then amorphous carbons attracted significant interests for machine parts and tools which require hardness, sliding properties and wear resistance simultaneously. Furthermore, attentions are paid for its applications for food packages and medical devices as it shows not only excellent mechanical properties but also high gas barrier properties and high bio-compatibility such as antithrombogenicity. Therefore the applications of amorphous carbons extend widely. One of the reasons why amorphous carbon shows various excellent properties comes from its structure. As mentioned previously, amorphous carbon consists of amorphous structure mixed with sp^2 and sp^3 hybridized carbons, does not have fixed structures, and its structure changes significantly by synthesis methods and conditions. There are various synthesis methods including radio frequency (RF) plasma CVD, microwave plasma CVD, ion plating, and sputtering as well as the above mentioned ionized evaporation method. Noticeable difference between RF and microwave plasma CVD methods and PVD methods such as ion plating and sputtering is that the CVD mainly uses gas source of hydrocarbons such as methane (CH_4) and acetylene (C_2H_2), while the PVD uses solid carbon. If hydrocarbon gas is used as a source gas, synthesized film contains hydrogen as the source gas does, while use of solid source enables synthesis of hydrogen-free amorphous carbon coatings as the source does not contain hydrogen.

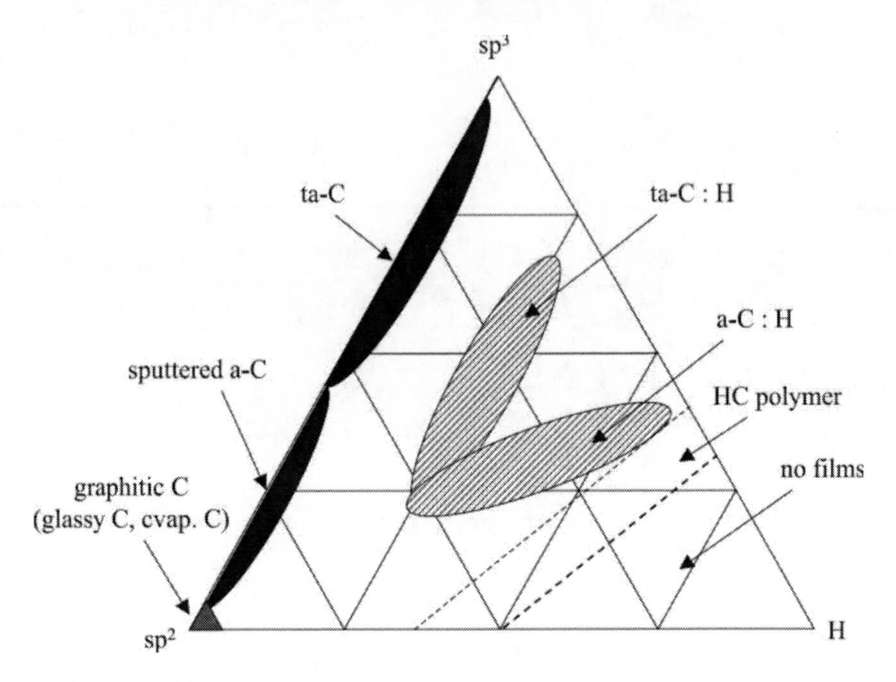

Figure 3. Three-phase diagram of amorphous carbon film.

Figure 3 shows the structure of amorphous carbon summarized by Ferrari and Robertson *et al.* using a three phase diagram of sp^3, sp^2, and H bond [3]. Depending on the ratio of these bonds, properties of the coating change. Although the ratio of sp^2 and sp^3 bonds is of course important (it can be said that the coating will be harder if the ratio of sp^3 bonding is higher.), H bonds play a key role in a variety of properties of amorphous carbon coatings. As the H bond increases, in another word, if the H terminals increase, three dimensional bond decreases, and consequently the hardness decreases. On the other hand, it is possible to give flexibility in a film. In this case, the amorphous carbon can be deposited not only on metallic substrates but also flexible substrates such as plastics. The research on additional functionalities by doping metals or fluorine is also extensively performed.

Amorphous carbon has been noticed as its lowest friction coefficient, excellent wear resistance, and high stability among numerous vapour deposited coatings. Its applications have been drastically progressed for recent ten years. In fact it is almost impossible to know all the products which use amorphous carbon; the examples include a hard disc drive for a computer, house use faucet fittings, a compact zoom camera, a frame and band of a watch, and a blade for a shaver. Here we focus on the high gas barrier property of amorphous carbon and present its application to food containers.

2. Application of Amorphous Carbon for Food Containers

Nowadays most of the food containers are made of polymers such as polyethylene (PE), polypropylene (PP) and polyethylene-terephthalate (PET). Polymers are generally transparent, light, inexpensive, and easy to mold. Due to these advantages, they are used in food industry, changing the style of sales. In particular, as a PET bottle which is used for a container of a beverage is robust against impact and can be reclosed, a beverage bottle becomes portable. In the beverage corner of a convenience store its area is now larger than that of metallic cans. In such a way, polymers are advantageous in usefulness and very suitable for food containers. However, it does not mean that there is no disadvantage. A noticeable disadvantage is its lower gas barrier property than glass or metals. The gas permeability a day is very low, but is not negligible for the application such as a beverage which can be stored for relatively long period. If O_2 included in air outside a bottle penetrates into the bottle, it causes the decrease of nutrition such as vitamin in fruit juice and catechin in tea and change of the color. In addition, CO_2 gas gradually emits out of a bottle and becomes stale. Recently most of beverages are on sales enclosed in PET bottles, but the polymer bottle with beer is not commercialized since the gas barrier property is a big concern. In the case of fizzy drink which is categorized as juice, the storage period can be prolonged to some extent by adding excess amount of carbonic acid. However, as beer is manufactured in a different process, this method is not applicable. In order to sustain a desirable quality of beer in a PET bottle it is necessary to control both introduction of O_2 from outside the bottle and leak of CO_2 from inside the bottle simultaneously. Therefore, the quality of a beverage is degraded gradually once it is closed in a PET bottle, and the expired period of sustaining the quality will be significantly short during the process of distribution and sales without controlling the gas permeability. For example, O_2 permeability of a typical 500-mL PET bottle at 23 ^{o}C is approximately 0.03 cc/bottle/day = 0.09 ppm/day. Under the assumption that permitted O_2 introduction into beer is 1 ppm, the quality of the beer can be sufficiently retained for approximately 12 days. As the gas permeable quantity is proportional to the surface area of a bottle, a large volume bottle is advantageous in quality retention. However, even for 1.5 L bottle, the quality of beer can be retained only for approximately 1 month. Comparing to glass bottles and metal cans in which beverage has a shelf life of 9 months, the quality retention of a PET bottle is significantly low. It is therefore said that at least approximately 10 times improvement of the gas barrier property is necessary if a PET bottle is used to fill beer. In addition, the gas barrier property of a bottle depends on temperature, and is lower at higher temperature. Not only for the beer but also for a bottle for hot beverage and the quality retention in sub-tropical areas the technology for decreasing permeability of a PET bottle, say in another word, for a high gas barrier PET bottle is said to be necessary worldwide. Previously multilayer structure containing high gas barrier materials were developed for a high gas barrier bottle. However, nowadays in terms of recycling it is thought to be the best to use coatings for improving the gas barrier property. Kirin Beer Ltd. successfully developed a high gas barrier PET bottle inside of which is coated with DLC using a plasma CVD for the first time in the world [4]. The PET bottle coated by DLC has 10 – 30 times higher O_2 barrier property than a normal PET bottle, corresponding to the quality retention sufficient for a container of beer or wine. In addition, it has no significant problem in recycling.

Figure 4 shows a schematic diagram of a PET bottle coating equipment using a radio frequency (RF) plasma CVD. A tubular electrode inside the bottle is also used for gas inlet, while the outer electrode of the similar figure covers the bottle. The deposition is carried out by first placing a PET bottle in the outer electrode, evacuating inside the bottle, introducing a source gas (C_2H_2), and applying a radio frequency power to the outer electrode. A plasma is generated in the bottle and DLC is deposited inside the bottle. In this method, by changing conditions such as RF power and gas flow rate, plasma species, density distribution, and ion bombardment in the bottle change. Therefore it is indispensable to optimize the conditions for each size of a bottle. The DLC can be coated in bottles with various shapes and sizes by preparing similarly-figured outer electrode individually.

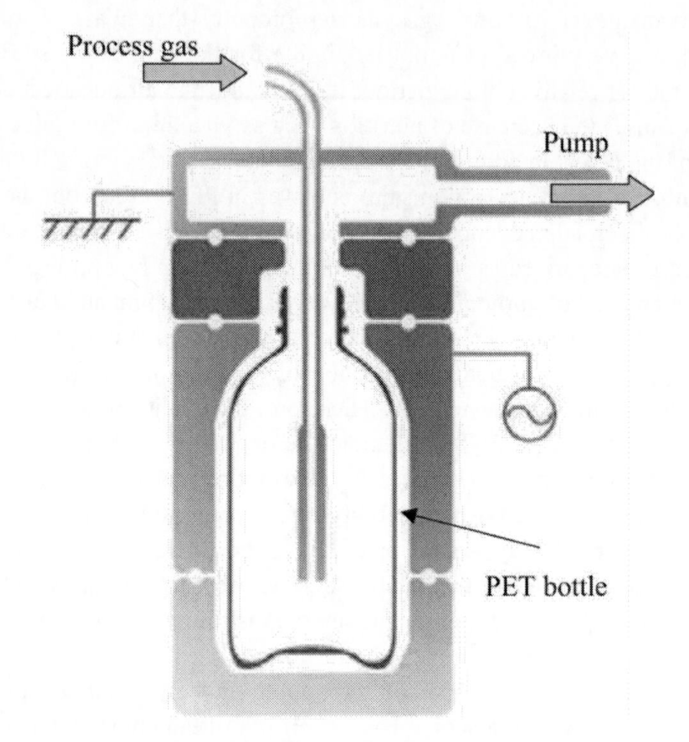

Figure 4. Schematic diagram of PET bottle coating system using RF plasma CVD.

3. Synthesis of Amorphous Carbon Coating by Atmospheric Pressure Plasma CVD and the Evaluation of Gas Barrier Property

The high gas barrier PET bottles coated with amorphous-carbon introduced here are partly commercially applied already. However, as mentioned before, the vacuum technology is very costly for inexpensive products like PET bottles. It is therefore strongly expected that atmospheric pressure plasma technique is developed for this application. We have worked on synthesis of amorphous carbon using atmospheric pressure plasma, and evaluated the coatings particularly for the gas barrier property. The atmospheric pressure plasma technique was reported by Professor Okazaki etc. in Sophia University in 1980s [5]. However, up to now, this technique is not widely applied. It is possibly but stabilization of the discharge is difficult. Demand for plasma treatment is increasing, and the limitation of vacuum technology is apparent in terms of cost and treatment areas, and practical application of the atmospheric pressure plasma technique is desired.

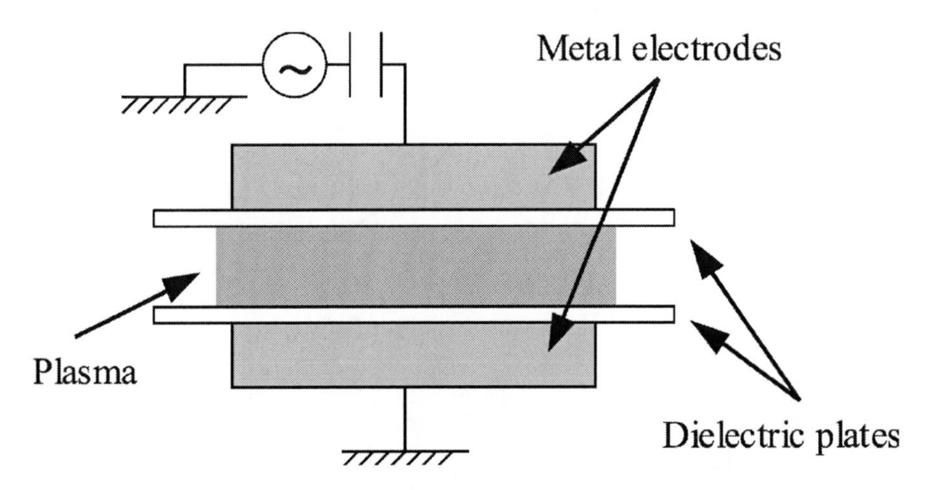

Figure 5. Schematic diagram of the atmospheric pressure plasma setup.

Based on the atmospheric pressure glow plasma technique which Professor Okazaki *et al.* reported, we started with development of equipment, collaborating with Sekisui Chemical. Figure 5 shows the basic principle of the equipment. At least one of the parallel plate electrodes is covered with an insulating plate, and voltage is applied between the electrodes using a pulsed power supply. At first helium, argon or nitrogen discharge was generated, a small amount of acetylene (approximately 1 %), which is general source material for amorphous carbon film, was mixed as a source gas of amorphous carbon films, and subsequently a film was synthesized. In this method, a carbon coating was successfully synthesized. However, the gas barrier property by the oxygen transmission test of the coated specimen shows similar value of O_2 transmission for the untreated PET substrate, indicating that the coating synthesized was very rough with no gas-barrier property. The reason of synthesizing the low gas barrier coating is thought to be formation of linear structure and termination of bonding at the three dimensional carbon network by mixing oxygen in air or nitrogen as a dilute gas. In fact, XPS analysis indicates the existence of oxygen and a small amount of nitrogen in the coating. Another reason is thought to be that due to the insufficient power input, formation of particles in a plasma would be pronounced rather than reactions at the surfaces, and the coating would be deposited by falling them onto the surface. It was necessary to improve the equipment in order to solve the problem. The dissociation energy of a gas varies very much depending on the kind of the gas. Professor Okazaki *et al.* used a huge amount of helium as a dilute gas. It is possibly because helium is relatively easy to be discharged, and has long lifetime of the discharge. However, using a huge amount of dilute gas for the synthesis of amorphous carbon coatings decreases the deposition rate, while impurities like nitrogen can be involved in the coating. When helium gas is used, impurities are not easily mixed, but the use of helium does not show good cost performance. Therefore, it is desirable to establish a discharge of the gas mainly consisting of acetylene. While rare gases such as helium and argon are easy to be ignited, N_2 requires a high voltage to ignite, and acetylene does still higher. Therefore a power supply for higher voltage and high frequency was introduced, and the ratio of dilute gas was reduced by optimizing the material of the electrodes and the discharge area.

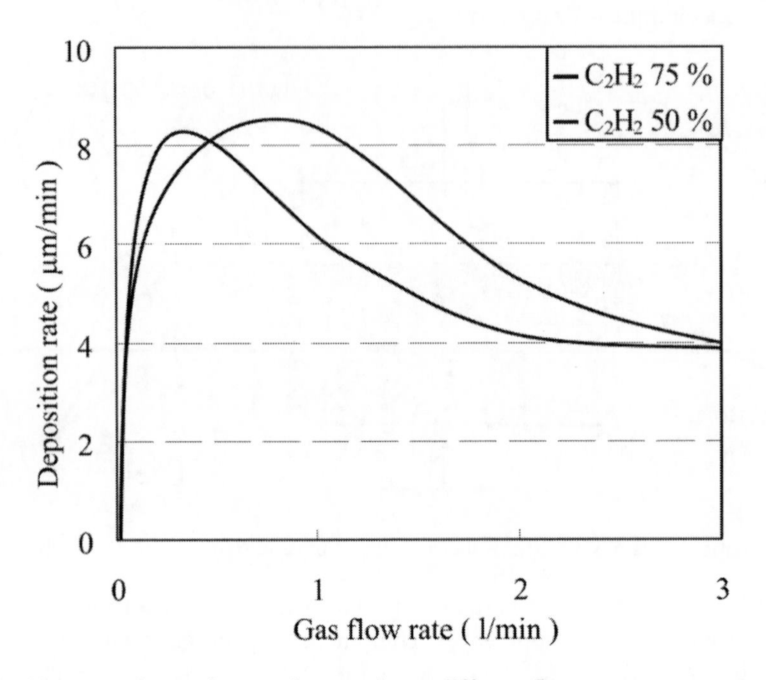

Figure 6. Deposition rate of amorphous carbon coating at different flow rates.

Figure 6 shows the deposition rate at various gas flow rates for the gas mixtures of 50 and 75 % acetylene. At atmospheric pressure, the gas flow rate, in fact the flow speed is the very important parameter. The deposition rate increased as the flow rate increased to a maximum, and then decreased. Comparing the deposition rates for the mixing rates of 50 and 75 %, the maximum deposition rates were both 8-9 μm/min, but the flow rate at the maximum deposition rate was different. The result indicates that not the total flow rate but the flow rate of acetylene determines the deposition rate. However, as mentioned before, the total flow rate significantly affects the synthesis process. When the flow rate of acetylene is significantly reduced, the maximum deposition rate mentioned above is not achieved. The deposition rate varied substantially at different flow rates, but generally high deposition rate of 4-9 μm/min was achieved, which is several tens times higher than that of the conventional low pressure technique. This is probably one of the attractive properties of the atmospheric pressure plasma. In order to utilize this advantage most effectively, taking the flow rate as an optimum with which the highest deposition rate is obtained, synthesis and characterization of the coatings were carried out.

Figure 7 shows the images of transmission electron microscopy (TEM) of the coating at the optimum flow rate in 75 % acetylene. In general, lattice patterns are not observed, confirming that it is amorphous. It shows no significant difference from coatings synthesized at low pressure. Figure 8 shows the gas barrier property by the O_2 transmission test. Although the O_2 transmission rate significantly varies at different flow rates, it is mainly due to the difference of the thickness as the deposition time was same. As the flow rate increases, the gas barrier property is lowered. However, in all conditions, the gas barrier property was improved, compared with the O_2 transmission rate of 27.7 cc/m^2/24h/atm for the untreated PET substrate. It is noted that the O_2 transmission rate of the coating synthesized at the optimum flow rate in 75 % acetylene was less than the detection limit (< 0.01), and that almost perfect gas barrier property is established. This level is difficult to be achieved even with low pressure processing. It is therefore concluded that atmospheric pressure plasma technique is applicable for high gas barrier coatings. Although the deposition rates at the optimum flow rates in 50 and 75 % acetylene were almost same, the gas barrier property of the coating synthesized in 50 % acetylene was not so high as that in 75 % acetylene, possibly due to the influence of the dilute gas as was mentioned before. The XPS analysis indicates that a small amount of nitrogen was detected at the coating when N_2 was used as a dilute gas, and that the nitrogen content in the coating increased as the N_2 content in the gas increased. Consequently in order to synthesize a coating with a high gas barrier property, it is ideal not to use a dilute gas if it is possible, and to generate a discharge with acetylene gas only. We further upgraded the equipment, and are now able to demonstrate very high speed deposition at up top approximately 1 μm/s in a pure acetylene gas. By upgrading equipment and optimizing the synthesis conditions potential of the atmospheric pressure plasma technique is further extending.

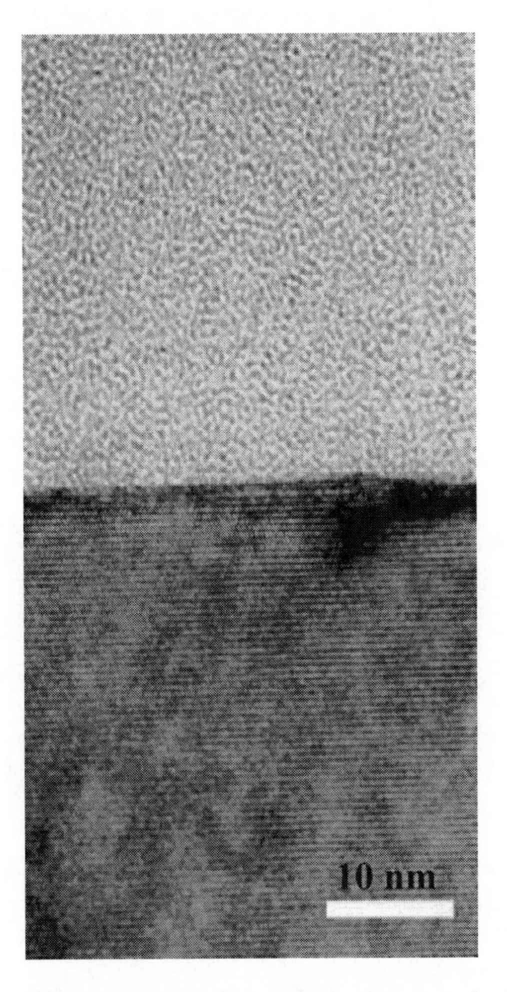

Figure 7. Cross sectional TEM image of the coating synthesized at atmospheric pressure.

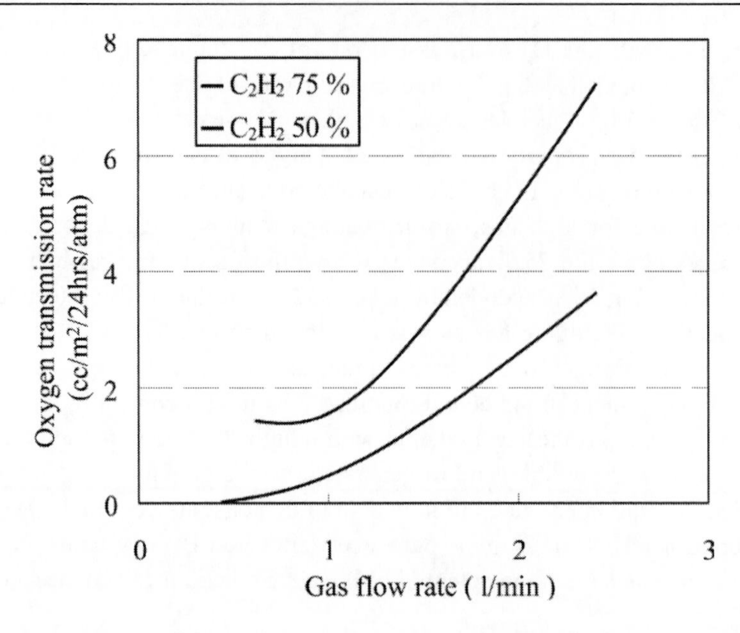

Figure 8. Gas barrier property of the coating synthesized at various gas flow rates.

The advantage of the atmospheric pressure plasma technique is not only the increase of the high treatment speed, but also possible renewal of the existing production line. The conventional low pressure technique uses a vacuum chamber and the process is inevitably a batch method, in which a substrate is set in the camber, and taken away after treatment. In the case of atmospheric pressure plasma, such a vacuum chamber is not necessary, and thus a continuous line-treatment process can be introduced by moving one of the electrodes or by transporting substrates using a belt conveyer. Such a transition of the production method enables large area plasma treatment and use of more than one plasma treatment continuously. We also developed such a line treatment system.

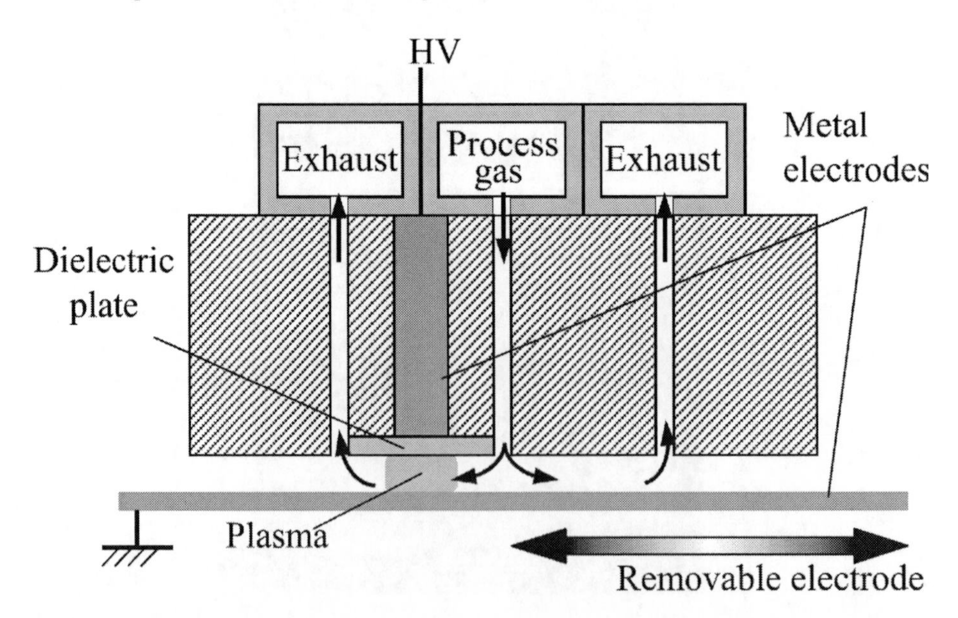

Figure 9. Schematic diagram of line type atmospheric pressure plasma CVD system.

Figure 9 shows schematic diagram of line-type atmospheric pressure plasma CVD system. The materials and structure used here are basically same as those used in the pre-described experiment for the batch type system. However, in this case, moving the bottom electrode enables continuous treatment. As was described before, the influence of the gas flow is significant at atmospheric pressure, and thus the control of the gas flow is important for the construction of stable and reproducible system. At low pressure, once it is introduced in a chamber, a source gas diffuses and uniform condition can be retained to some extent. On the other hand, it is difficult to control to retain uniform gas flow at atmospheric pressure. Uneven flow directly affects the uniformity of a plasma, and the synthesized coating becomes non-uniform. In addition, at low pressure, as inside the chamber is pumped down to vacuum, it is uncommon that the process gas contains impurity coming not from the source gas. At atmospheric pressure, however, due to the existence of atmosphere, coatings containing many impurities will be formed unless the flow of the source gas is properly controlled. In the case of line system, sufficient care must be paid for the control of the gas flow because a substrate or the electrode itself moves. The developed system has a structure of sustaining the gas flow between the electrodes as uniform as possible by introducing the source gas at the centre while exhausting it from both sides. This exhaust gas system also plays a role not to introduce ambient air surrounding the system. The result of the similar experiment using this system is introduced here.

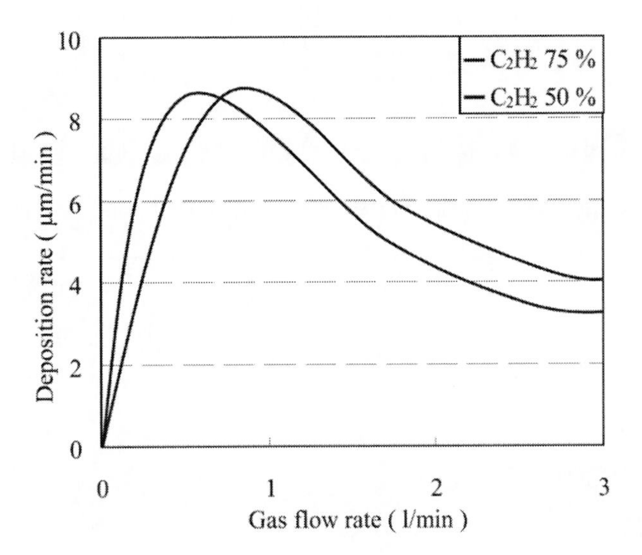

Figure 10. Deposition rate of the coating synthesized at various gas flow rates using the line type system.

Figure 10 shows the deposition rates at different gas flow rates. Though the value of the flow rates differs, it shows similar trends to the batch treatment in Figure 6, and the obtained maximum deposition rates are same. Figure 11 shows the cross sectional SEM image of the synthesized coating. The coating was grown uniformly from the substrate surface, and deviation of the coating by the movement of the substrate was not observed. It is therefore indicated that if the structure of the system is the same, movement of the electrode does not affect the synthesis process significantly. Although adjustment of synthesis conditions is necessary, it can be said that the result obtained using the batch system can be reproduced by the line system. The gas barrier property of the coating synthesized at the optimized flow rate was evaluated as shown in Figure 12. It is confirmed that the synthesized coating shows high gas barrier property. In addition, even in acetylene 50 % condition, almost perfect gas barrier property was achieved, which is different from the result with the batch system. It is therefore concluded that practically applicable amorphous carbon coatings can be synthesized even with line-type atmospheric pressure plasma treatment, and that the application of atmospheric pressure plasma becomes more realistic. We recently started more practical research such as continuous treatment of films using rolls, as premise of applicability of the technique.

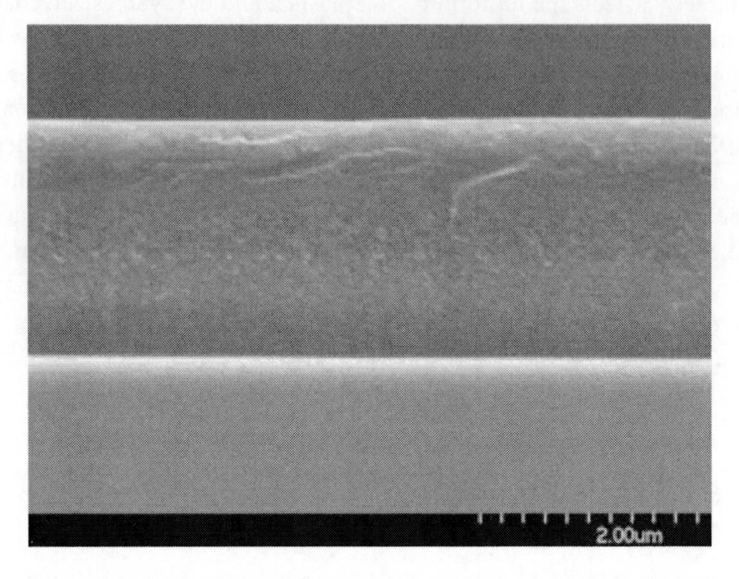

Figure 11. The cross sectional SEM image of the coating synthesized using the line type system.

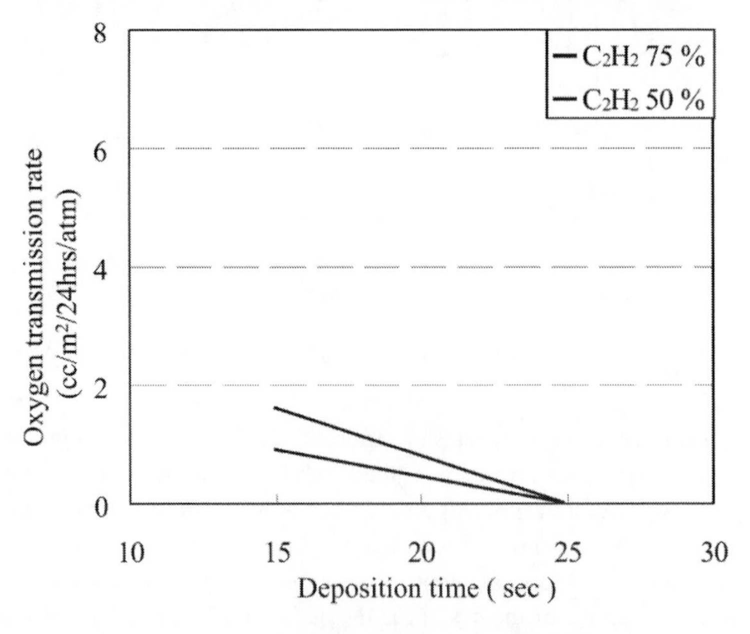

Figure 12. Gas barrier property of the coating synthesized using the line-type system for various deposition times.

REMARKS

It was shown here that practically applicable functional coatings can be synthesized even at atmospheric pressure, if atmospheric pressure plasma technique is used. They are applicable to daily products such as inexpensive food packaging with which application of amorphous carbon coatings is conventionally considered to be difficult because of its high coat and limited treatment area. Furthermore, since materials containing water vapour like paper with which it is impossible to pump down to vacuum can be plasma-treated, it is anticipated that applications of plasma surface modification can extend furthermore. As many researchers already show interest in the atmospheric pressure plasma, it is expected that not only for the amorphous carbon materials but for also many functional coatings which have been synthesized at low pressure, possibilities of synthesis at atmospheric pressure will be clarified. It is anticipated that if the atmospheric pressure plasma technique becomes the alternative of the low pressure plasma technique, it will give an evolutional change in the area of surface treatment.

REFERENCES

[1] Angus, J.C.; Will, H.A.; Stanko, W.S. *J. Appl. Phys.* 1968, *39*, 2915-2922.

[2] Aisenberg, S.; Chabot, R. *J. Appl. Phys.* 1971, *42*, 2953-2958.

[3] Ferrari, A.C.; Robertson, J. *Phys. Rev.* 2000, *B 61*, 14095-14107.

[4] Kazufumi, N. *Jpn. patent.* 1996, JP08053116A2.

[5] Kanazawa, S.; Kogoma, M.; Moriwaki, T.; Okazaki, S. *J. Phys. D Appl. Phys.* 1988, *21*, 838-840.

In: Generation and Applications of Atmospheric Pressure Plasmas ISBN: 978-1-61209-717-6
Editors: M. Kogoma, M. Kusano and Y. Kusano

Chapter 12

ATMOSPHERIC PRESSURE MICROPLASMA AND ITS APPLICATIONS FOR MATERIALS PROCESSING

Kazuo Terashima
Department of Advanced Materials Science, Graduate School of Frontier Sciences, University of Tokyo, Chiba, Japan

INTRODUCTION

Atmospheric pressure plasmas have recently attracted significant attentions as novel plasma sources for materials processing. In addition to the development of conventional thermal plasmas at equilibrium states with high controllability and functionality, the development of various kinds of low temperature non-equilibrium plasmas including atmospheric pressure glow discharges have been carried out by various process controls such as time control, space control, and reaction species control. Accordingly, applications for novel processing have been intensively carried on. One of the examples is an atmospheric pressure microplasma whose size ranges from a few mm to a few μm.

This chapter introduces the generation and physical properties of the atmospheric pressure microplasma and its applications for materials processing, such as thin film deposition process.

1. ATMOSPHERIC PRESSURE MICROPLASMA

A thermal plasma at a thermally equilibrium state [1,2] is easily generated at around atmospheric pressure due to frequent collisions among gas molecules, atoms and electrons (See Figures 1 and 2). This is a high temperature plasma whose gas temperature varies from several thousand K to several tens thousand K. It is expected that such a plasma can be applied for super high-speed materiaals processing by utilizing its high energy and high particle-densities as a high enthalpy fluid. The applications of the thermal plasmas were in fact limited to heat sources in heavy chemical industry, particularly in steel industry. However, the recent improvement in controllability of the process parameters enabled development of nano cluster synthesis for high functional materials in energy and information technology, super high-speed nano-processing for fabrication of nano-materials by deposition, and surface treatment. Successful industrial application includes nano-particle synthesis for cosmetic products.

Figure 1. Image of thermal plasma.

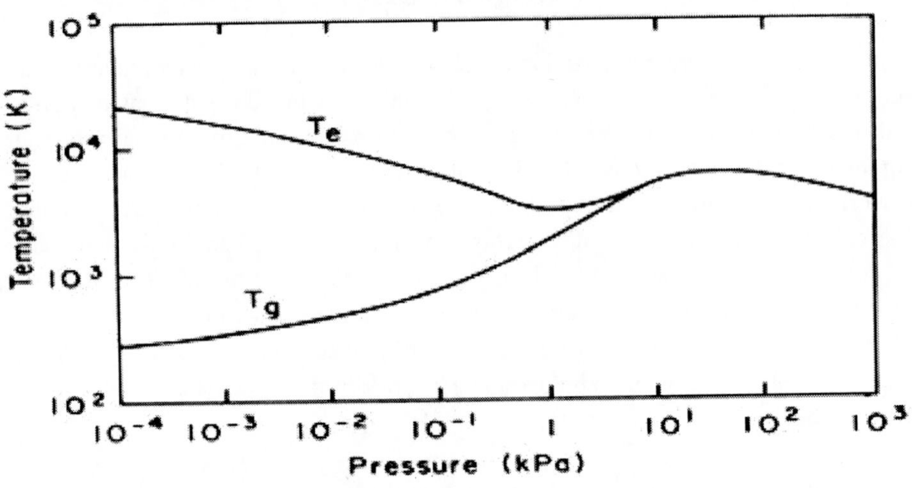

Figure 2. Temperatures of electron (T_e) and gas (T_g) at various gas pressures.

Meanwhile, Prof. Okazaki, Prof. Kogoma and their group invented atmospheric pressure glow discharge (APGD) using a dielectric barrier discharge (DBD) with a Mylar film and metallic mesh [3]. This invention has initiated development of various generation methods of non-equilibrium low temperature atmospheric pressure plasmas. In order to sustain the non-equilibrium state in a process space,

[1] extinguishing a discharge in a time shorter than the energy-dispersion time for collisions of the electron system with gas particles etc. (e.g. pulse excitation, high frequency excitation), and
[2] improvement of heat release processed by a high-speed gas flow and use of a gas which has high heat conductivity

are attempted. Furthermore generation of high density non-equilibrium low temperature microplasma at atmospheric pressure became possible by

[1] eventual reduction of collision frequency of electrons and gas particles,
[2] improvement of heat release by the increase of the interface areas between a plasma and chamber walls (electrodes), and
[3] improvement of heat release by increasing the flow rate of a process gas.

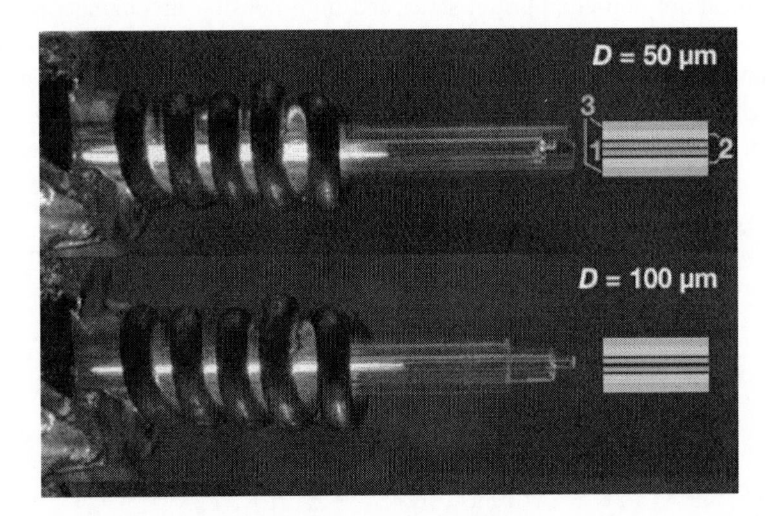

Figure 3. Generation of inductively coupled microplasma.

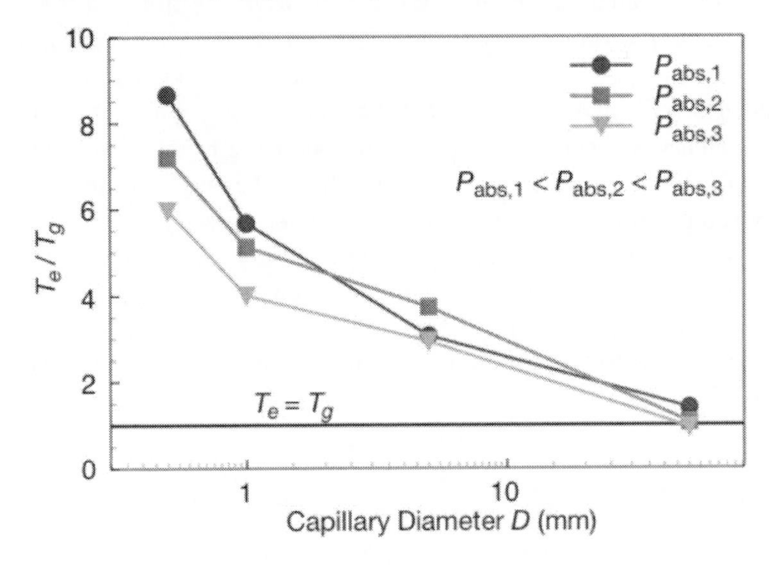

Figure 4. The ratio of temperatures as a function of capillary diameter.

Figure 3 shows the images of an inductively coupled microplasma. The temperature ratio of electrons and gas is numerically calculated at different diameters of a capillary tube as shown in Figure 4. As the diameter of the capillary decreases, the temperature ratio increases, corresponding to a highly non-equilibrium state. The development of microplasmas led to enormous success in the application for plasma display panels (PDPs). The generation of a microplasma generally needs high working gas-pressure in order to ignite a discharge in a very small space [4]. For example, when the characteristic length of a plasma discharge cell is in a range between 100 μm and 1 μm, non-equilibrium plasma is easily generated at high gas pressure between 1 atm (atmospheric pressure) and 100 atm. Plasma phenomena in a new complex system which are caused by clustering phenomena among atoms and molecules, appearance accompanying the physical and chemical properties, and development of new plasma science can be expected at high gas-pressure including atmospheric pressure and a supercritical fluid state which is between gas and liquid states [5]. Furthermore, by utilizing its spatial localization, non-equilibrium state, and high density, its engineering applications include synthesis of nano-materials, development of integrated large area plasma process devices, infinitesimal environmental analysis system, portable health care devises, short wavelength light source between vacuum ultraviolet and extreme ultraviolet regions, and their application for nano-lithography. Among atmospheric pressure non-equilibrium plasmas, microplasmas are studied as the 21st century ecological social technology associated with energy conservation and low environmental influence.

In the following sections, generation and properties of a microplasma and its applications for materials processing are presented.

2. MICROPLASMA [4,5]

2.1. Introduction

Conventionally research targets of “plasmas” mainly range from experimental plasmas (macro-space plasmas) in the region between several hundred μm and mm, to gigantic plasmas (super macro-space plasmas) in the macroscopic region including plasmas in a natural life and space plasmas. On the other hand, almost nothing about plasmas in a micro-space has been understood, particularly a hyperfine space – a nano-space less than a few μm as shown in Figure 5. Therefore a “microplasma” whose size ranges between several mm and approximately 100 μm attracts significant interest as a new research target in plasma material science. Several generation methods of a microplasma have been developed, and subsequently various applications are extensively attempted. For example, micro-light sources for plasma nano-process diagnostics, a portable microplasma for air environmental decontamination, a plasma-operation-mess, and microplasma spray system have been already commercialized.

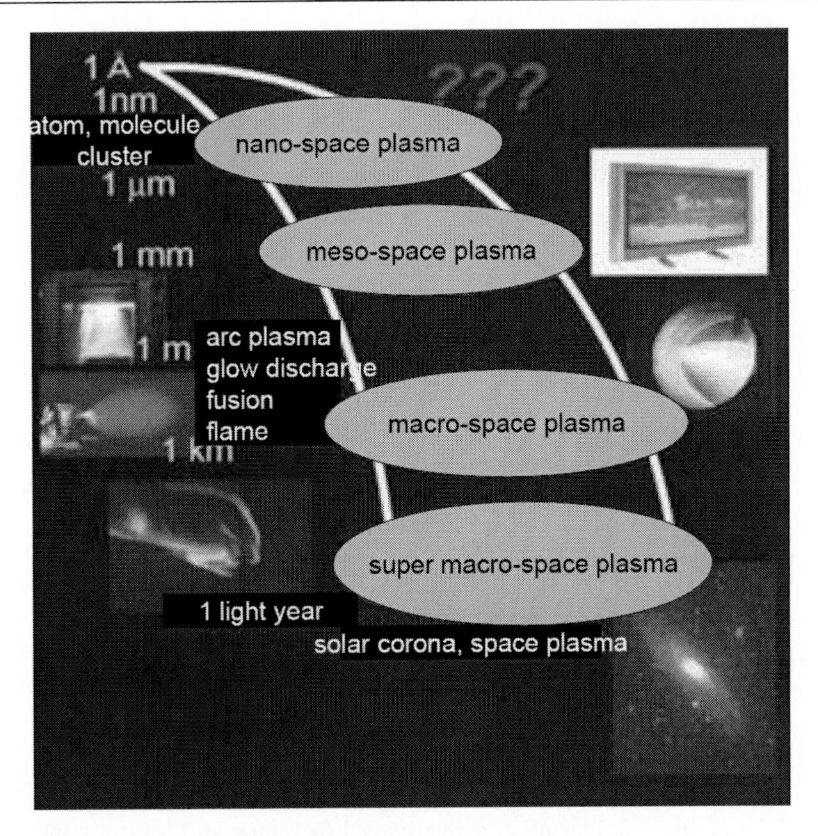

Figure 5. Overview of plasmas as function of their scale.

2.2. Microplasmas

2.2.1. Direct Current (DC) Microplasma [6-11]

Generation of microscopic-scale plasma requires a microscopic space. Our group generated a DC plasma using 0.5 – 100 μm microscopic gap electrodes as shown in Figure 6, which were manufactured using micro-fabrication technology, depositing a metallic material such as W, Au, or Pt on an insulating substrate like SiO_2, and etching. The electrodes have following characteristics; fabrication of highly controllable micro-scale gap (order of 10 nm), flexibility of the shape of electrodes, and easy integration of electrodes for plasma generation. Generation of a microplasma was attempted using the electrodes in various gases (He, H_2, Ar, CH_4, air etc.). The results indicated that Paschen's-like law for a discharge ignition voltage can be extended even in the microscopic region of 5 μm. In addition, electron density was estimated to be approximately 4×10^{21} m^{-3} by the measurement of Stark broadening for a H_2 atmospheric pressure microplasma with 10-μm gap Pt electrodes at the voltage of 450 V. On the other hand, a numerical simulation using PIC-MCC method for a H_2 atmospheric pressure microplasma with 10-μm gap at the voltage of 500 V indicated that the thickness of the plasma region is only about 1.5 μm, and thus most of the gap turned out to be a sheath. Under this condition, the electron density at the bulk plasma is 5.5×10^{21} m^{-3} and Debye length is 0.065 μm, which is substantially smaller than the size of the plasma (1.5 μm). It is therefore concluded that this plasma can be categorized within conventional macro-scale plasmas.

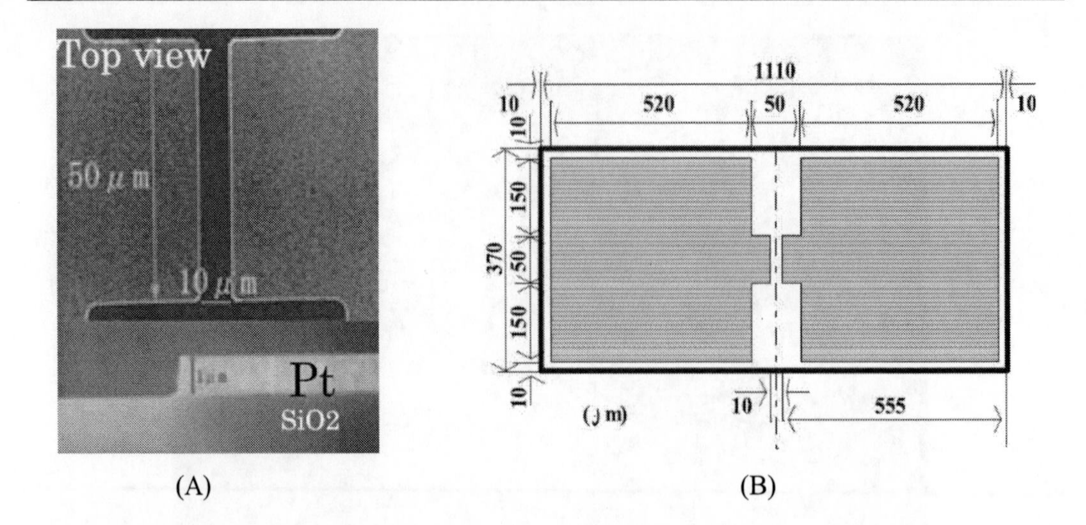

(A) (B)

Figure 6. Microscopic gap electrodes.

2.2.2. Dielectric Barrier Discharge (DBD) Microplasma [12]

Significant damage of the electrode with a DC microplasma due to its high density is more problematic than that with a macroplasma. Consequently, it is difficult to sustain a stable discharge for a long time, and generation time of a plasma must be as short as between several s and several tens s. In order to solve this problem, application of a dielectric barrier discharge (DBD) to the generation of a microplasma is attempted, as a DBD enables generation of a macroscopic plasma at a relatively low temperature. In this method at least one of the electrodes is covered with a dielectric material, alternative voltage is applied at the electrodes, the plasma is extinguished before high current arc plasma is generated, and consequently low temperature pulsed plasma is generated. Our group fabricated a DBD microplasma consisting of a powered electrode of electrically polished 250 μm diameter W wire and a copper-plate ground electrode adhering a cover glass plate with a conductive paste with a distance of the electrodes less than 50 μm. Optical emission measurement of an N_2 DBD microplasma at 4 MPa indicates that a low temperature microplasma at 400 K can be generated even in such a high gas pressure in the super critical fluid phase.

2.2.3. Ultra High Frequency (UHF) Microplasma Jet [13-17]

As microplasmas are in general less stabilized at the surface of the plasma than macro-plasmas, stable generation and sustain of microplasmas requires increase of ionization frequency and rise of electron temperature etc. For such purposes, for example, study on electrode materials with high coefficient of secondary electron emission such as MgO has been carried out for the development of plasma display panel (PDP). On the other hand, our group developed a thermoelectron-enhanced microplasma (TEMP), which enables sustainability of microplasma by supplying hot electrons from outside of the plasma [12-14]. Hot electrons supplied from inductively heated W wires and sustaining inductively coupled plasma easily enabled generation and sustainment of a microplasma even in a capillary with the inner diameter at or less than 1 mm. Generation of a microplasma was not observed with the same condition without a W wire or with a low-temperature W wire, indicating that hot electrons play an important role in the generation of a plasma. Depending on the conditions, both low temperature and thermal plasmas can be generated. So far, a microplasma was successfully generated in a capillary with the minimum inner diameter of 20 μm. In addition, this plasma jest is used for synthesis of various carbon-related nano-materials such as carbon onions, carbon tubes, and carbon walls, and micro- nano-structures of metal oxides.

2.2.4. Microwave Microplasma [18-19]

Furthermore, a strip-line-type microwave microplasma is developed utilizing electrodes fabricated by the strip-line technique which is originally established as a key technique in the area of communication technology. This microplasma has the following advantages due to the use of microwave power supply; a power supply can be inexpensive, loss by irradiation is low, and focusing applied power to a necessary place is possible, electrical matching is easy using simple devises, generation of a plasma at wide range of gas pressures is stable, electrode-less operation is possible, and long lifetime of sustaining a discharge can be shown. Furthermore, the microwave microplasma has advantages due to the application of the strip-line technique. Namely, the strip-line technique is already well-established as a basic technique in the area of communication technology, operation at low power consumption is possible, production of strips is inexpensive, and construction of small (mobile) plasma generation system is possible. In practice, a ground electrode is arranged at one side of a dielectric substrate, and a micro-strip-line electrode is arranged on the other side, introducing microwave. In addition, fine gaps of several hundred μm are fabricated on both sides of the electrodes, in which a microplasma is generated. By designing width and length of the electrodes, a strong electric field can be applied only at the gaps, resulting in generation of a microplasma with high efficiency.

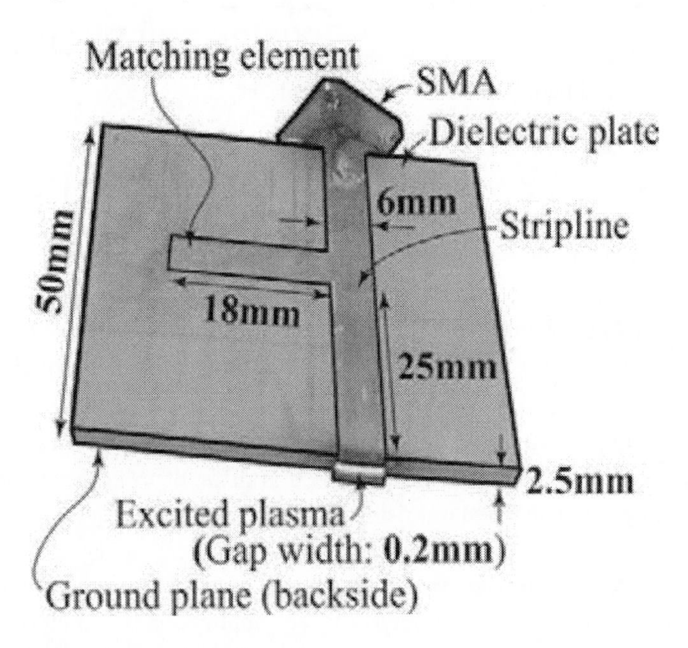

Figure 7. Fabricated microplasma electrode system.

Figure 7 shows the microplasma electrode system which we eventually fabricated. An alumina plate was used as a substrate and copper-tape strip-line was used as an electrode. Utilizing the strong electric field, a low temperature non-equilibrium plasma can be generated without a gas flow at atmospheric pressure without introducing a noble gas such as He and Ar. For example, in a 6 mm long gap and with a 0.2 mm wide electrode, plasma can be sustained at atmospheric pressure with approximately 5 W microwave and at or less than 1 W for self-ignition of a discharge. A gas temperature at this condition is estimated to be approximately 500 K, using the optical emission spectrum of N_2. It is thus indicated that this plasma is an atmospheric pressure non-equilibrium (low temperature) plasma. This plasma can be stably generated continuously for no less than 3 hours. This non-equilibrium plasma at atmospheric pressure is in particular called Strip-line Microwave Micro Atmospheric Plasma (SMMAP). It is extensively applied for atmospheric pressure high speed surface treatment of various materials. Furthermore, a large-scale large-area plasma devise (plasma panel) has been developed by integrating the electrodes horizontally and vertically as shown in Figure 8.

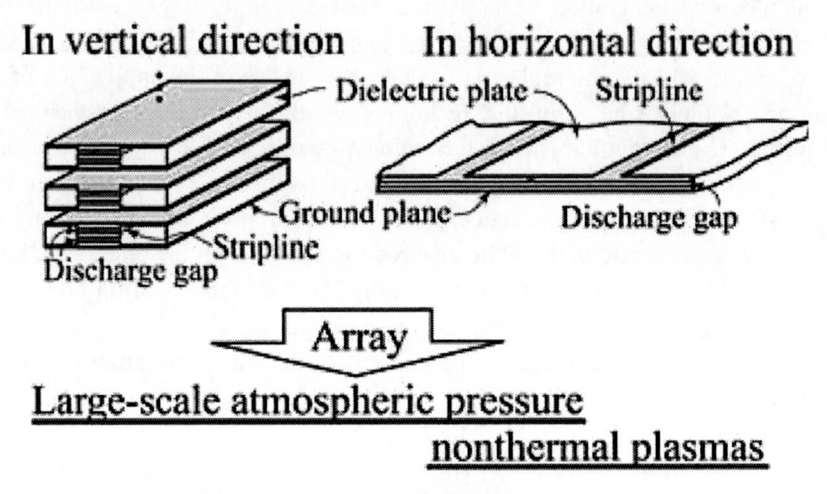

Figure 8. Large-scale atmospheric pressure non-thermal plasmas.

2.3. Applications Of Microplasma For Material Processing [20]

Utilizing the characteristics such as

- high density (no less than 1000 times of a conventional plasma),
- super non-equilibrium state (no less than 1000 times of conventional plasma),
- super localization (μm order),
- easiness of up-scaling (the order of metre) by using multi-arraying,
- development of material processing which is difficult to achieve by a macroplasma has been intensively carried out using microplasmas.

For example, a process device which integrates microplasmas on a substrate, process plasma chip, is proposed and developed for an application of material processing of microplasma material [7-11]. A process plasma chip is the process devise apparatus which enables mass process treatment by arranging and integrating microplasma process devises on a substrate. Figures 9 and 10 show the schematic diagram of the process plasma chip and the image of plasma ignition using a prototype plasma chip (2 x 2 matrix), respectively. It resulted from the increase in plasma density due to the decrease in the size of a plasma.

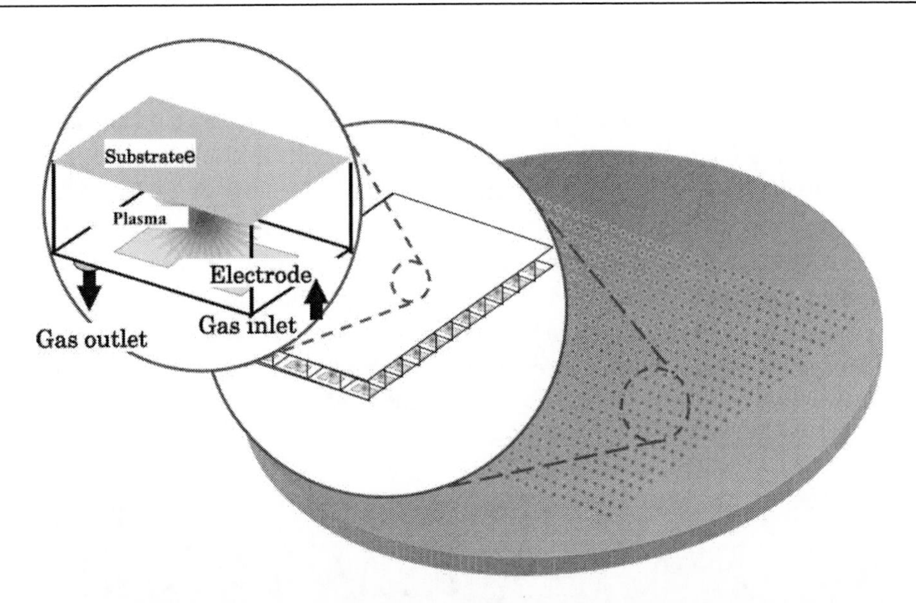

Figure 9. Schematic diagram of the process device system integrating microplasma process units.

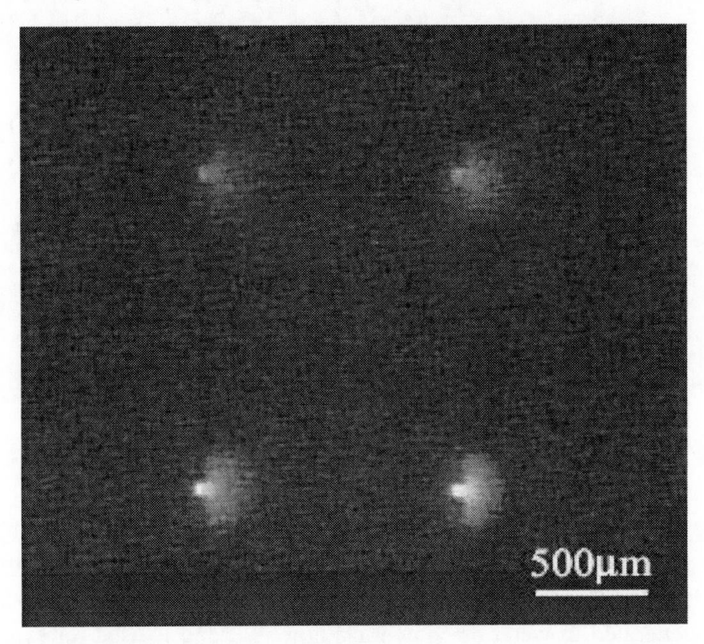

Figure 10. Image of microplasmas of a prototype process device system.

High speed processing due to the increase in plasma density with microplasmas, and drastic improvement of efficiency by simultaneous treatment with multiple plasmas can be expected. Utilizing a prototype 2 x 2 matrix microplasma system with 10 mm gap substrate electrodes, applications for synthesizing nano-structured C and CN materials using CH_4-H_2 and CH_4-N_2 gaseous plasma enhanced chemical vapour deposition (CVD) are attempted. By controlling the experimental parameters, carbon coatings such as graphite, amorphous carbon, and diamond, and CN_x coatings with various nitrogen contents have been synthesized. Measurement of coating thickness indicates that the deposition rate can be no less than 1 μm/min. This experiment successfully demonstrated simultaneous deposition of 4 different specimens within as short as approximately 1 minute, indicating not only the importance of initial development of the device, but also high capability of this plasma chip and high density super microplasma process. In addition, by parallel arrangement of such microplasmas in a same way of plasma TV, metre-scale large-area plasma treatment can be easily achieved.

Furthermore, proposal of basic concepts and development of plasma fibre [15] has been attempted that has a variety of microplasma source at the top of a flexible tube. This devise enables localized plasma process without restricting operation environment and locations including in water, under the ground, in a living body, and in a vacuum, that is similar to laser process using optical fibres. It is expected that the plasma fibres can afford to realize on-demand-plasma-nano-process irradiating a necessary place, at a necessary time, with necessary volume of plasma.

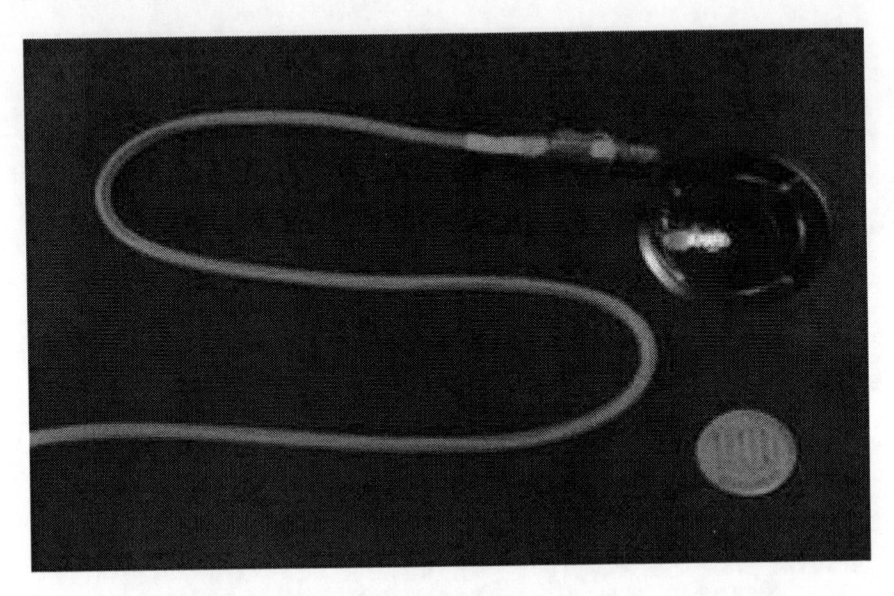

Figure 11. Hot electron assisted microplasma fibre.

An example of hot electron assisted microplasma fibre is shown in Figure 11. This plasma is energy (several – several tens W) and space (no more than several mm) saving. Since decreasing the size of a plasma enables easy generation of a discharge at high gas density, it is possible to generate it even in liquid and a living body. Utilizing this plasma fibre, TEMP is generated in water as shown in Figure 12, and is used for synthesizing nano-carbon materials using rapid cooling in liquid. Applying this technique, urchin-like carbon nano-materials as well as carbon nanotubes, and graphite are synthesized. It is considered that characteristic synthesis of nano-structured materials is influenced by the reactions at the interface between a plasma and liquid, and that applications for new processing such as DNA bio-systems are developed, positively utilizing the reactions at the interface between a plasma and liquid. Apart from the heat-aided-type microplasma fibre, DBD microplasma fibres are also developed. Figure 13 shows the plasma pen in which a DBD microplasma fibre is installed. This plasma pen can be used for hydrophobic and hydrophilic treatment of a plastic surface at atmospheric pressure as shown in Figure 14. Localized treatment of several hundred μm is successfully demonstrated in an open air.

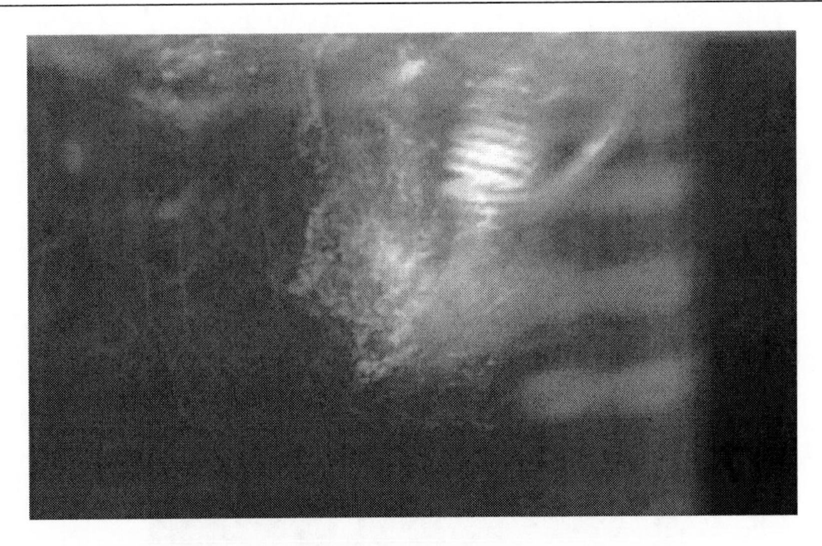

Figure 12. Generation of TEMP in water using the plasma fibre.

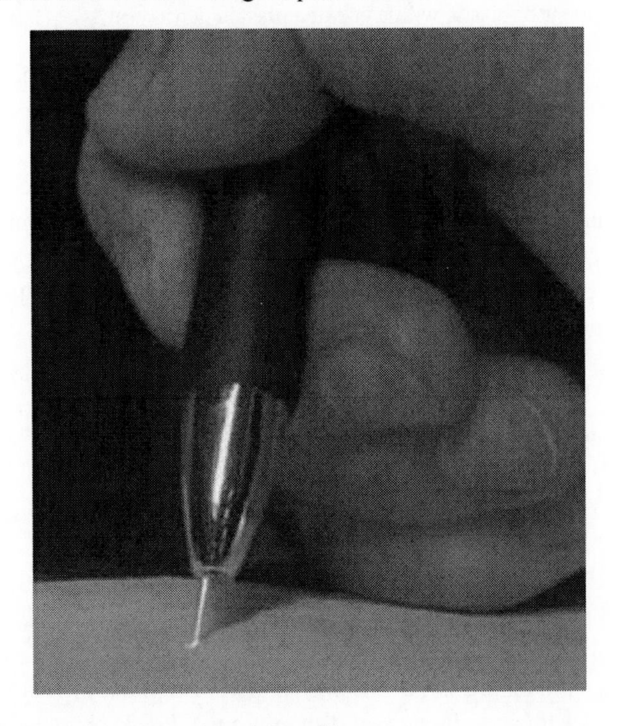

Figure 13. Plasma pen involving DBD microplasma fibre.

As shown in these examples, atmospheric pressure microplasma can successfully realize various characteristics which a normal atmospheric pressure plasma has; free from expensive vacuum system, creation of exotic reaction space such as in water, and high speed processing. Conventional plasma processing has shown enormous mass of waste since only a surface in a region of 1 nm and 1 μm is processed by dipping a specimen in a huge bulk plasma in a huge chamber. On the other hand, application of an atmospheric pressure microplasma can lead a complete shift of process idea which well-suits saving resource and energy as the process irradiates a plasma only at a necessary place, necessary time, and necessary amount.

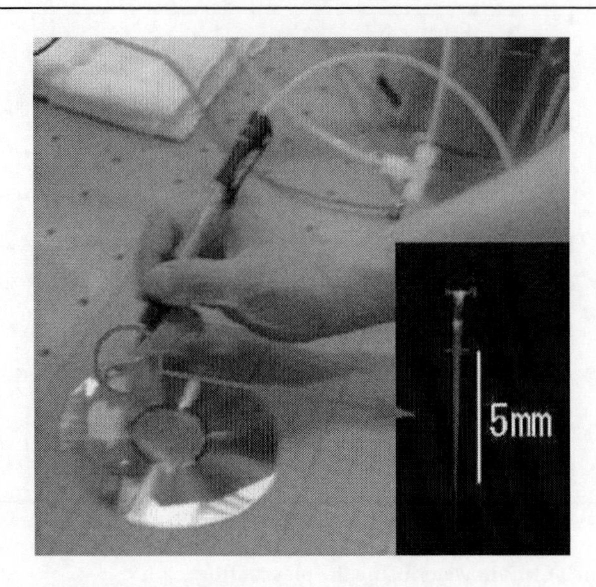

Figure 14. Surface treatment of plastic with a use of a plasma pen.

SUMMARY

Plasma material science plays an important role as a basic scientific technology to solve urgent objectives which Japan now confronts, such as energy and environmental problem in a global scale, construction of new social systems in IT age, and the aging problem. It is highly expected that a new plasma materials science, which atmospheric pressure microplasma, can develop, will play a part of the role for solving these problems.

REFERENCES

[1] for example, in *Proc. JSPS Int. Symp. Thermal Plasma Deposition*; Yoshida, T.; Terashima, K.; Ed., Tokyo, JP, 2002.

[2] Terashima, K.; Yoshida, T. *J. Plasma Fusion Res*. 1996, *72*, 529-535.

[3] Kanazawa, S.; Kogoma, M.; Moriwaki,T.; Okazaki, S. In *Proc. Int. Symp. Plasma Chem. (ISPC-8 (Tokyo))*. 1987, *3,* 1839-1844.

[4] for example, Tachibana, K. *IEEJ Tran. Elec. Elec. Eng.*. 2006, *1*,145-155.

[5] for example, *Microplsma –Basic and Applications–*, Ed. Tachibana, K.; Ishii, S.; Teraashima, K.; Shirafuji, T.; Ed.; Ohbumsha: Tokyo, JP, 2009. in Japanese.

[6] Ito, T.; Kerashima, K. *Appl. Phys. Lett.* 2002, *80*, 2854-2856.

[7] Ito, T.; Izaki, T.; Terashima, K. In *Proc. 14th Symp. Plasma Chem. (ISPC14, Prague).* 1999,967-972.

[8] Ito, T.; Izaki, T.; Terashima, K. *Surf. Coat. Technol.* 2000, *133-134*, 497-500.

[9] Ito, T.; Izaki, T.; Terashima, K. *Thin Solid Films* 2001, *386*, 300-304.

[10] Ito, T.; Terashima, K. *Thin Solid Films* 2001, *390*, 234-236.

[11] Ito, T.; Katahira, K.; Asahara, A.; Kulinich, S.A.; Terashima, K. *Sci. Tech. Adv. Mater.* 2003, *4*, 559-564.

[12] Tomai, T.; Ito, T.; Terashima, K. *Thin Solid Fims* 2006, *506-507*, 409-413.
[13] Ito, T.; Terashima, K. *Appl. Phys. Lett.* 2002, *80*, 2648-2650.
[14] Shimizu, Y.; Sasaki, T.; Ito, T.; Terashima, K.; Koshizaki, N. *J. Phys. D Appl. Phys.* 2003, *36*, 2940-2944.
[15] Ito, T.; Nishiyama, H.; Terashima, K.; Sugimoto, K.; Yoshikawa, H.; Takahashi, H.; Sakurai, T. *J. Phys. D Appl. Phys.* 2004, *37*, 445-448.
[16] Shimizu, Y.; Sasaki, T.; Liang, C.; Chandra Bose, A.; Ito, T.; Terashima, K.; Koshizaki, N. *Chem. Vapor Deposition* 2005, *11*, 244-249.
[17] Shimizu, Y.; Sasaki, T.; Bose, A., C.; Terashima, K.; Koshizaki, N. *Surf. Coat. Technol.* 2006, *200*, 4251-4256.
[18] Kim, J.; Terashima, K. *Appl. Phys. Lett.* 2005, *86*, 191504-1-191504-3.
[19] Ogata, K.; Terashima, K. *J. Appl. Phys.* 2009, *106*, 023301.
[20] for example, Mariotti, D.; Sankaran, R.M. *J. Phys. D Appl. Phys.* 2010, *43*, 323001.

In: Generation and Applications of Atmospheric Pressure Plasmas ISBN: 978-1-61209-717-6
Editors: M. Kogoma, M. Kusano and Y. Kusano

Chapter 13

SI ETCHING

Tomohiro Okumura
Panasonic Corporation, Osaka, Japan

INTRODUCTION

Dry etching technique using low pressure plasmas is an expensive process in that it uses vacuum technology and resist masks. On the other hand, an atmospheric pressure plasma does not require vacuum pumps, and make loading/unloading of work-pieces easy to be treated. Consequently, the processing system can be simplified, and very inexpensive etching equipment can be constructed. In addition, an atmospheric pressure plasma enables high speed processing, since high density ions and radicals can be generated by the atmospheric pressure plasma, but not by a low pressure plasmas. Furthermore, as an atmospheric pressure plasma can be easily generated locally, spot treatment in a microscopic level is possible. Therefore mask-less etching in several tens μm ~ several hundreds mm level is achievable, which is advantageous for constructing low cost processing.

An atmospheric pressure plasma as a mask-less etching processing has advantages of insignificant roughening of the processed surfaces and easily achievable high etching selectivity due to its use of glow discharge and chemical reactions compared with other mask-less micro-machining techniques such as laser machining, micro electric discharge machining (μEDM) and mechanical machining.

This chapter introduces micro-scale Si etching utilizing the desirable properties of an atmospheric pressure plasma mentioned above.

1. SI ETCHING TECHNIQUE USING AN ATMOSPHERIC PRESSURE PLASMA

Development of Si etching using atmospheric pressure microplasmas has been reported in a literature. For example, a beam plasma is generated using a 1 mm diameter stainless-steel cylindrical electrode which is surrounded by a 4 mm inner diameter cylindrical dielectric, feeding a process gas between the electrode and the dielectric, and supplying 13.56 MHz radio frequency (RF) power to the electrode [1]. A Si substrate, also acting as an electrode, is etched by a micro-hollow cathode discharge generated using electrodes manufactured by photo-lithography process [2]. First, VHF power is supplied to a solenoid antenna which surrounds the quartz-glass discharge tube whose inner diameter is 1.0 mm, and SF_6 atmospheric pressure microplasma is generated. Then reactive species are supplied from an outlet of the discharge tube onto the substrate, and a 400-μm diameter, 450-μm deep hole is created on a Si substrate [3]. Line region etching of a Si substrate is performed by perpendicularly arranging plate electrodes [4].

When microplasma etching is applied for manufacturing a device which requires line etching of thin films, generating a microplasma at the line-shaped region is advantageous in terms of treatment speed than generating a microplasma in a spot region and scanning it for line shaped etching. Based on this viewpoint, we developed atmospheric pressure line-shaped microplasma sources which enable high speed linear etching of thin films and substrates. One example of a devise manufacturing process which requires such a kind of treatment is a dicing process of a Micro-Electro-Mechanical Systems (MEMS) substrate [5] as shown in Figure 1. The technique is thought to be applicable to a dicing process for thin Si substrates which are necessary for mounting a multi-chip.

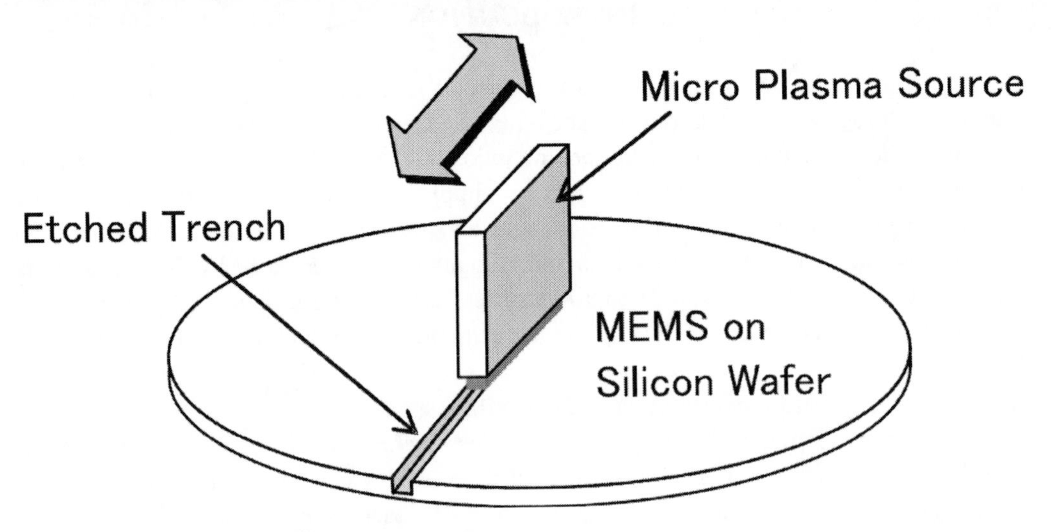

Figure 1. Dicing using a line-shaped microplasma.

The important criteria for such a line-shaped etching are high speed, uniformity of process width at the processing direction, uniformity of etching speed, and thin line width. Improvement of etching performance is attempted by investigating microplasma sources with various structures and process conditions.

In this chapter, the structures of line-shaped microplasma sources are described, followed by the thinning of the line width of the plasma and improvement of the efficiency of fluorine radical (F*) generation. Furthermore, experimental result of micro-width line etching of Si substrate and influence of substrate temperature on the micro-width line etching of the Si substrate is presented. Finally gas flow numerical simulation is used to discuss the result of etching.

2. Structure Of Line Shaped Microplasma Etching Instrument

In this work, three different microplasma etching systems are used; type-A in which rectangular copper electrode is covered with ceramics so that an electrode is protected against plasma exposure, and types-B and -C in which the end of an aluminium electrode is knife-edged and exposed into a plasma.

Figure 2 shows the schematic diagram of the type-A microplasma source. The microplasma source consists of four thin ceramic plates with grooves of gas passages, and a copper cathode electrode. The two outer gas feeds are interconnected each other via through-holes in the ceramic plates. He gas or a gas mixture of He and reaction gas is introduced into the inner gas feeds, and exhausted from the outer gas feeds, or N_2 or reaction gas is supplied into the outer gas feeds. An RF voltage of 13.56 MHz is applied to the electrode to generate a plasma between the electrode and the substrate. The inner gas outlet measures 100 μm x 30 mm.

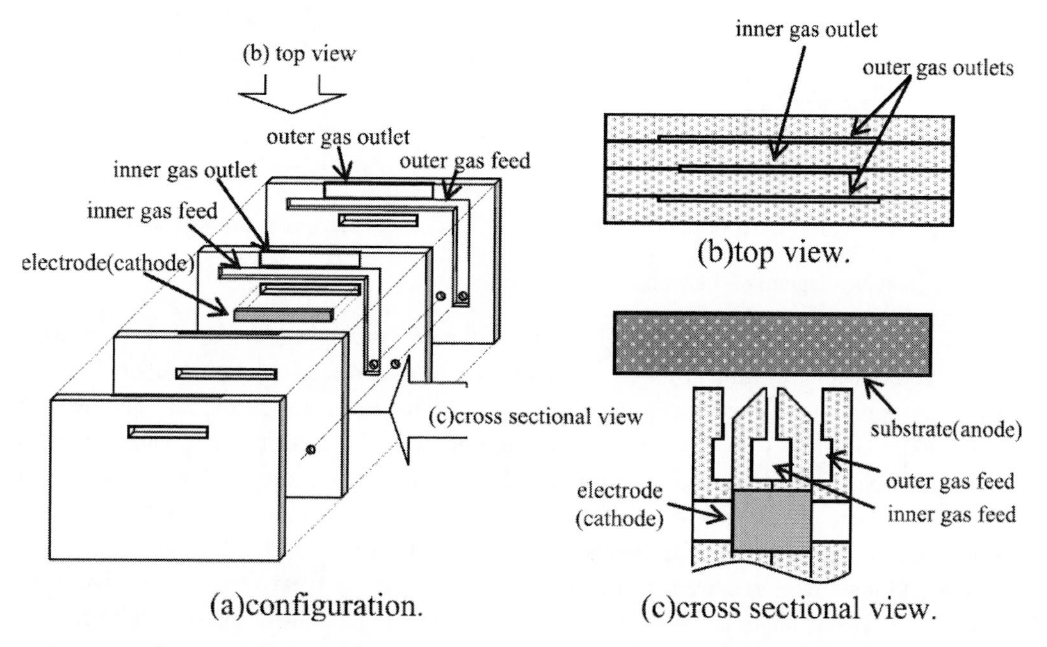

Figure 2. Schematic diagram of the type-A microplasma source.

Figure 3 shows the schematic diagram of the type-B microplasma source. It consists of four thin ceramic plates with grooves for gas feeds, and an aluminium cathode electrode. The end of the aluminium cathode is knife-edged so that electric field is concentrated at the electrode end. The two outer gas feeds are interconnected via through-holes in the ceramic plates. The two inner gas feeds are also interconnected in the same manner. Inert gases and reaction gases are supplied into the inner and outer gas feeds, respectively. An RF (13.56 MHz) voltage is applied to the electrode to generate a plasma between the electrode and a substrate. The inner gas outlet (at two positions) measures 50 μm x 30 mm, and the distance between the two outer gas outlets is 3 mm.

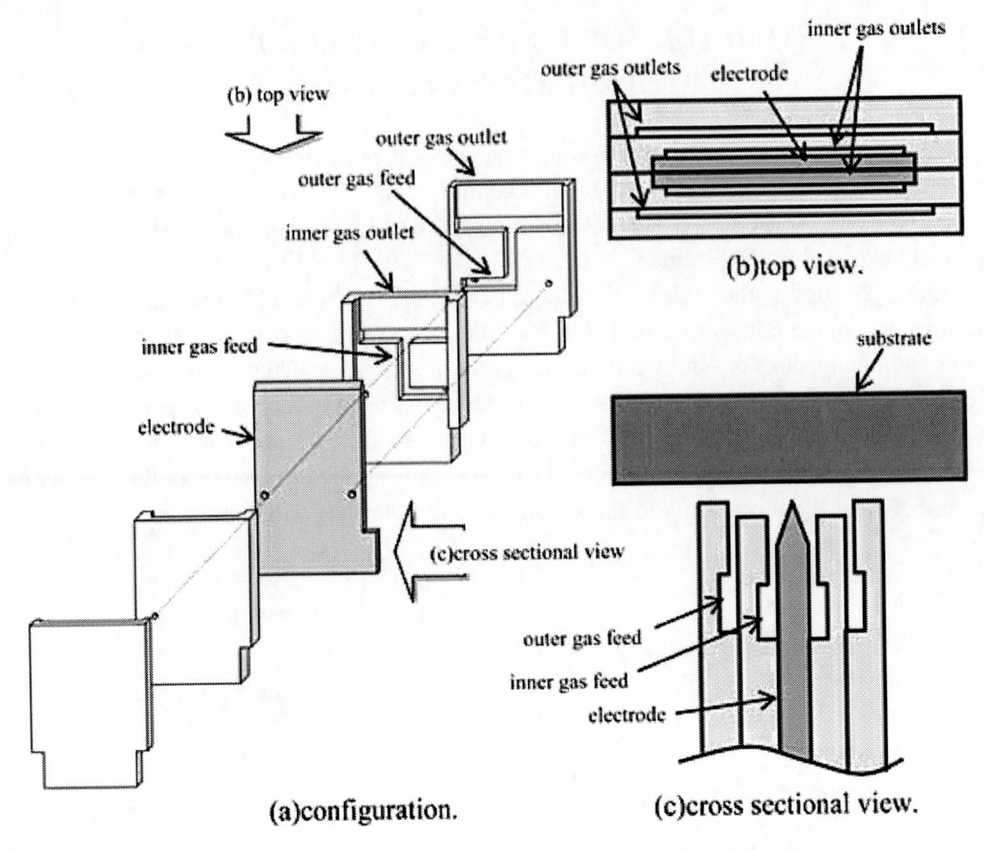

Figure 3. Schematic diagram of the type-B microplasma source.

Figure 4 shows the schematic diagram of the type-C microplasma source. It consists of two thin ceramic plates with grooves for outer gas feeds, two thin ceramic plates without grooves, an aluminium anode electrode with a groove for inner gas feeds, and another thin ceramic plate with fine slits. The end of the aluminium cathode electrode is knife-edged so that electric field is concentrated at the electrode end. The two outer gas feeds and the two inner gas feeds are interconnected via through-holes in the ceramic plates and the electrode. The thin ceramic plate with the fine slits overtops the electrode and the four ceramic plates. The inert and reaction gases are supplied into the inner and outer gas feeds, respectively. An RF (13.56 MHz) voltage is applied to the electrode to generate a plasma between the electrode and a substrate. The dimension of the inner gas outlet is 130 μm x 30 mm, and the distance between the two outer gas outlets is 650 μm [7].

The schematic diagram of the etching apparatus equipped with such a microplasma source is shown in Figure 5. In order to suppress the RF noise and avoid diffusion of toxic gas, a Si substrate, which is a part of the microplasma source and acts as a specimen to be exposed by a plasma, is placed in a shield cage. The shield cage consists of a double layer of an acrylic plate and a metallic mesh so that inside the cage can be observed from outside. The RF power is supplied to the plasma source or the substrate via an RF matching box connecting to an RF power supply. A several tens pF ceramic condenser is connected to the load (plasma) in parallel at the output side of a commercially available L-type RF matching box. Gases are fed to the plasma source finely controlled by mass flow controllers.

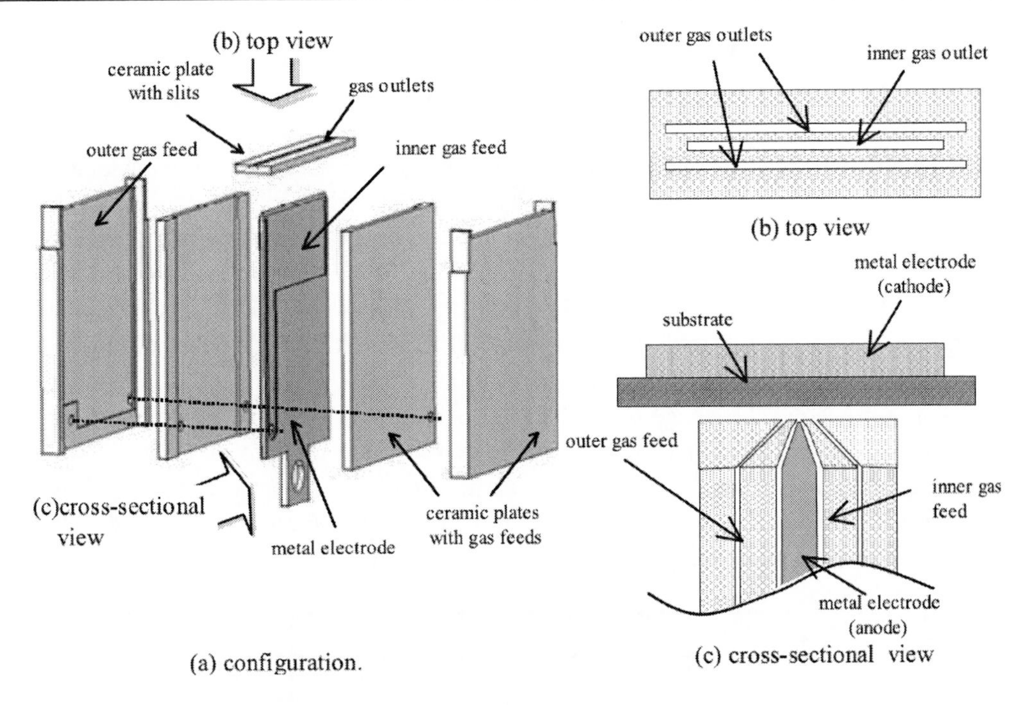

Figure 4. Schematic diagram of the type-C microplasma source.

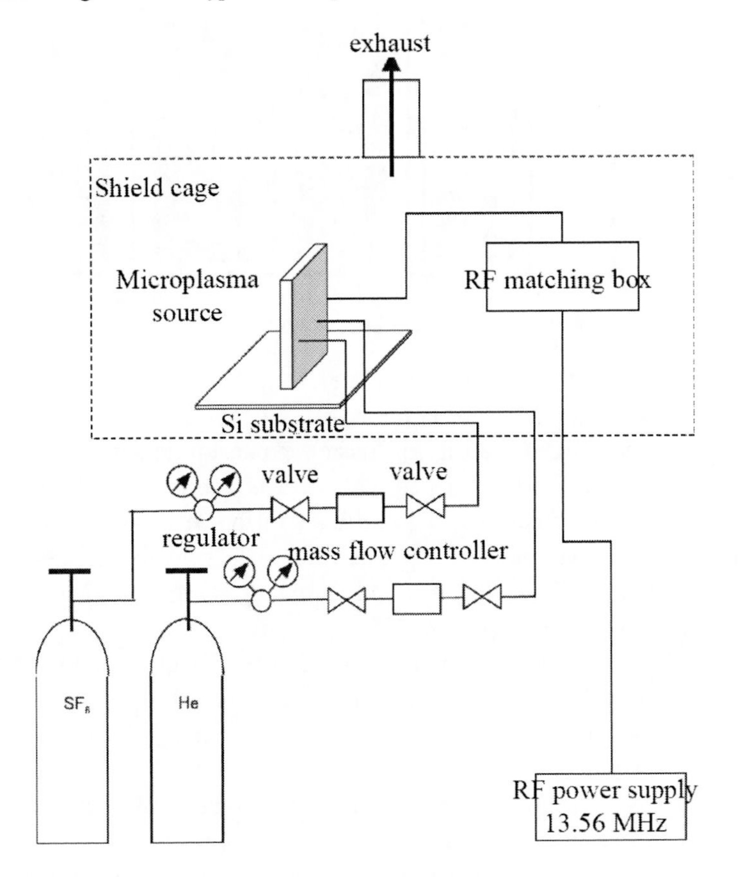

Figure 5. Schematic diagram of the microplasma etching system.

3. Narrowing the Line-Width of a Plasma and Improving the Efficiency of F* Generation

The line-width of a generated plasma must be thin so as to decrease the line-width of the etching process. The optical emission of a plasma is observed with different ways of gas supply. In the experiment a glass substrate on which an ITO film is deposited is used as an anode. Figure 6 shows the images of optical emission of the type-A microplasma source under the following conditions: gas pressure = 100 kPa, PF power = 100 W, distance between the plasma source and a substrate (hereafter referred to as the "gap") = 0.3 mm. Figure 6 a) shows the image of the plasma observed when 1 slm He gas was supplied into the inner gas feed. Figure 6 b) shows the image observed when 1 slm He was supplied into the inner gas feed and exhausted from the outer gas feeds. Figure 6 c) shows the image observed when 1 slm He and 400/600 sccm CF_4/N_2 were supplied into the inner and outer gas feeds, respectively. Figure 6 d) shows the image observed when 1 slm/16 sccm He/CF_4 and 1 slm N_2 were supplied into the inner and outer gas feeds, respectively.

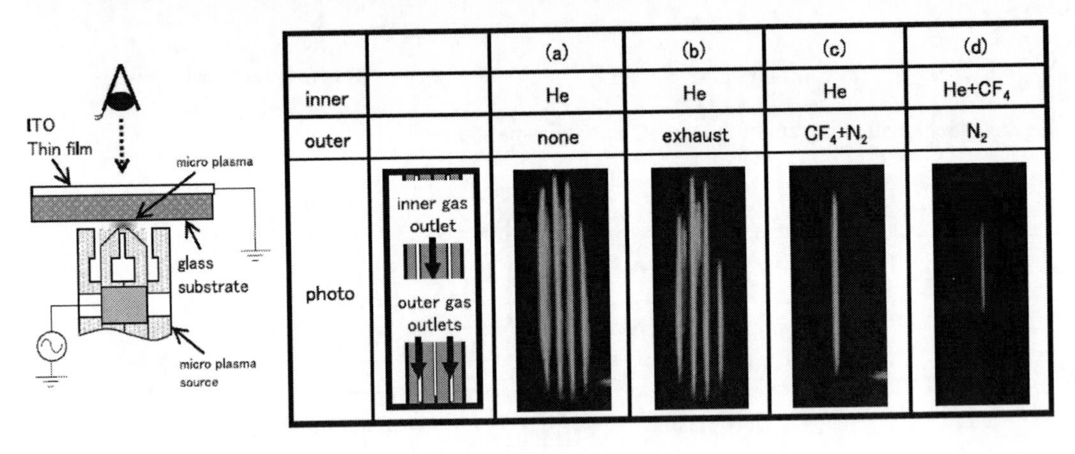

Figure 6. Optical emission from the type-A microplasma source.

Plasma expands outside four ceramic plates as shown in Figure 6 a), while plasma expansion is somewhat smaller in Figure 6 b). This is because the area of high He concentration is narrower in b) than in a). Plasma expansion is suppressed significantly in Figure 6 c), resulting in a fine line-shaped plasma. It is due to the supply of CF_4 + N_2 gas mixture into the outer gas feeds which is difficult to ionize at atmospheric pressure. The length of fine line-shaped plasma is shorter when He + CF_4 gas mixture was supplied into the inner gas feed as shown in Figure 6 d) than in Figure 6 c). Plasma was extinguished when the CF_4 flow rate increased to 20 sccm or more.

It is concluded that in order to generate a line-shaped plasma it is efficient to supply He and reaction gases into the inner and outer gas feeds, respectively. The optical emission intensity of F* at the wavelength of 738.3 nm is measured which is generated by the dissociation of the reaction gas CF_4 in a plasma. Figure 7 shows the optical emission intensity of F* when CF_4 gas is supplied into the inner and outer gas feeds, respectively. The measurement was carried out under the following discharge conditions: pressure = 100 kPa, RF power = 100 W, total gas flow rate in the inner or outer gas feed = 1 slm, gap = 0.3 mm. When CF_4 gas was supplied into the outer gas feeds, the optical emission intensity increased with the CF_4 concentration, eventually exceeding the maximum emission intensity obtained when CF_4 gas was supplied into the inner gas feed. This result indicates that the developed microplasma source provides a higher etching rate when CF_4 gas is supplied to the outer gas feeds than the inner gas feed. Considering these results, at the following parts, He and reaction gases were supplied into the inner and outer gas feeds, respectively, when a fine line-shaped etching was attempted so that a fine line-shaped plasma was generated and a significant amount of etchant was efficiently produced.

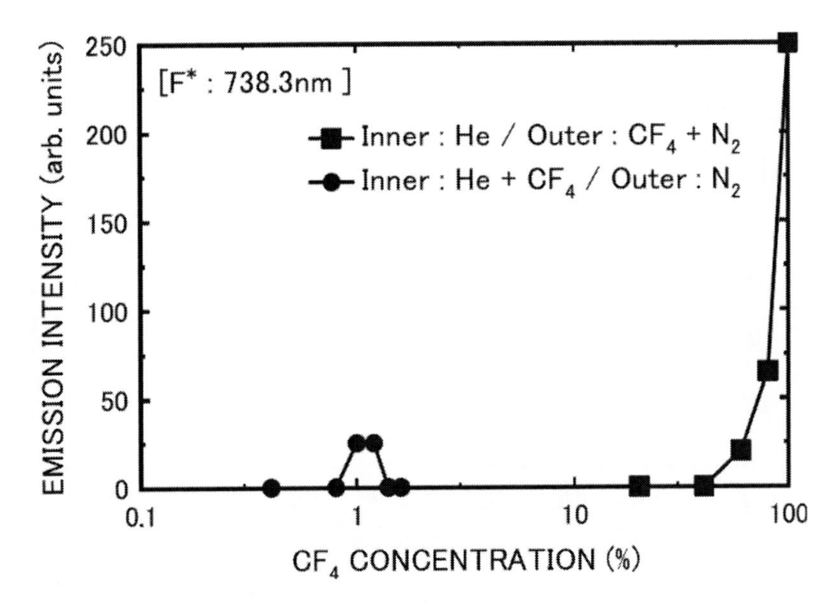

Figure 7. Optical emission intensity of F*.

4. Si Fine Line-Shaped Etching Process

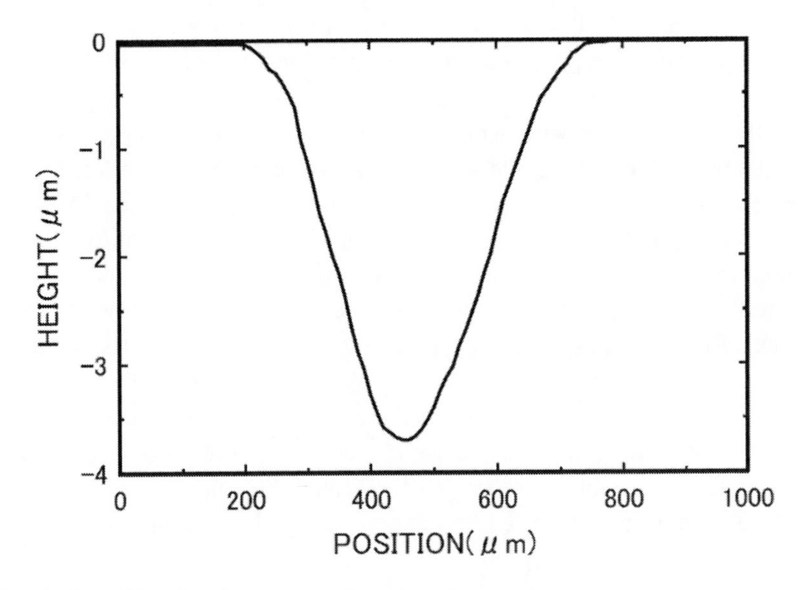

Figure 8. Si etched profile using the type-A microplasma source.

A Si substrate surface was etched using the type-A microplasma source. Figure 8 shows the etched profile of the Si substrate. The etching was carried out under the following conditions: pressure 100 kPa, RF power = 100 W, He flow rate in the inner gas feed = 1 slm, SF_6 gas flow rate in the outer gas feed = 500 sccm, gap = 0.3 mm, etching time = 15 s. The resulting etching rate was 14.7 μm/min.

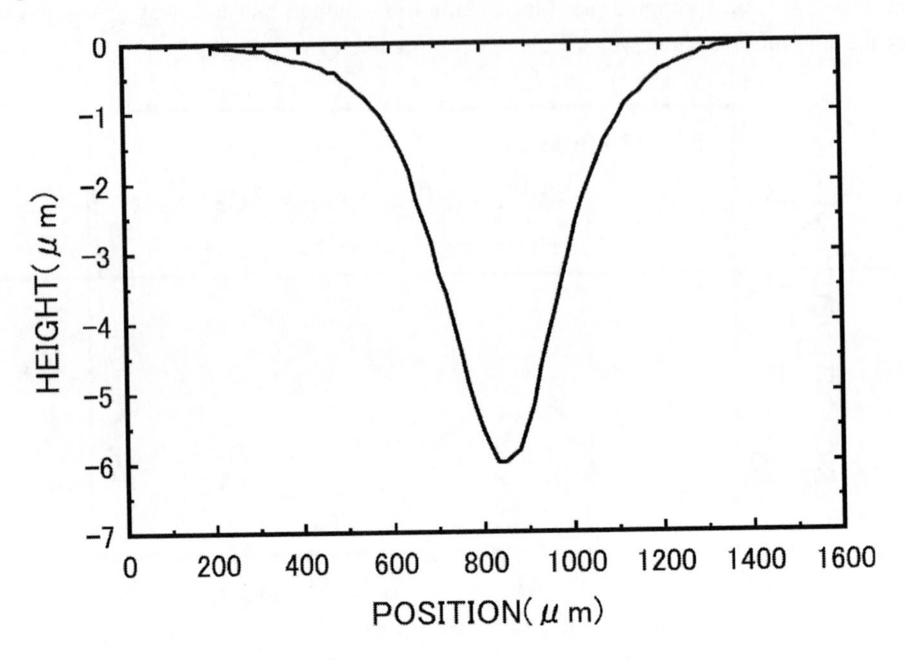

Figure 9. Si etched profile using the type-B microplasma source.

A Si substrate surface was also etched using the type-B microplasma. Figure 9 shows the etched profile of the Si substrate using the type-B microplasma source. The etching was carried out under the following conditions which are similar to the conditions above: pressure 100 kPa, RF power = 100 W, He flow rate in the inner gas feed = 1 slm, SF_6 gas flow rate in the outer gas feed = 500 sccm, gap = 0.3 mm, etching time = 5 s. The etching rate was 71.8 μm/min, which is distinctively higher than that obtained using the type-A microplasma source. Since the distance between the electrode and a substrate with the type-B microplasma source is substantially shorter than that with the type-A microplasma source, the electric field is concentrated at the electrode end, resulting in generation of a higher density plasma.

Under the assumption of the maximum etching depth to be D, the line width at depth 0.2D is defined to be top width W1, and at depth 0.8D is defined to be bottom width W2 as shown in Figure 10. Figure 11 shows the etching rate, the top width W1 and the bottom width W2 with various SF_6 flow rates using the type-B microplasma source. Etching was carried out under the following conditions: pressure = 100 kPa, He flow rate in the inner gas feed = 1 slm, gap = 0.3 mm, etching time = 30 s. SF_6 was supplied into the outer gas feeds. The bottom width W2 was almost unchanged with various SF_6 flow rates. However, as the SF_6 flow rate increased, the etching rate tended to increase and the top width decreased. It is because as the SF_6 flow rate increased, the region of high He concentration was suppressed, and thus the RF power absorbed per unit volume increased.

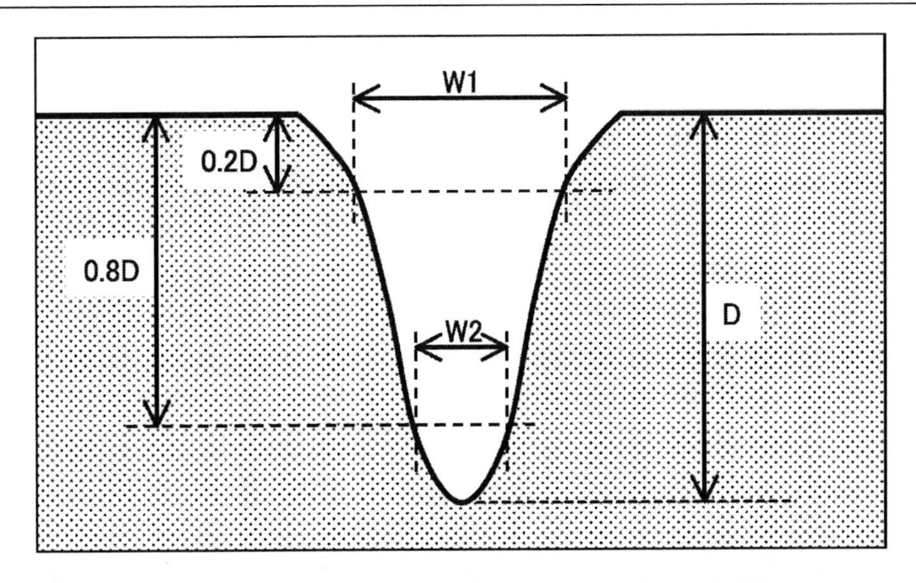

Figure 10. Definition of the top width W1 and the bottom width W2 of an etched line.

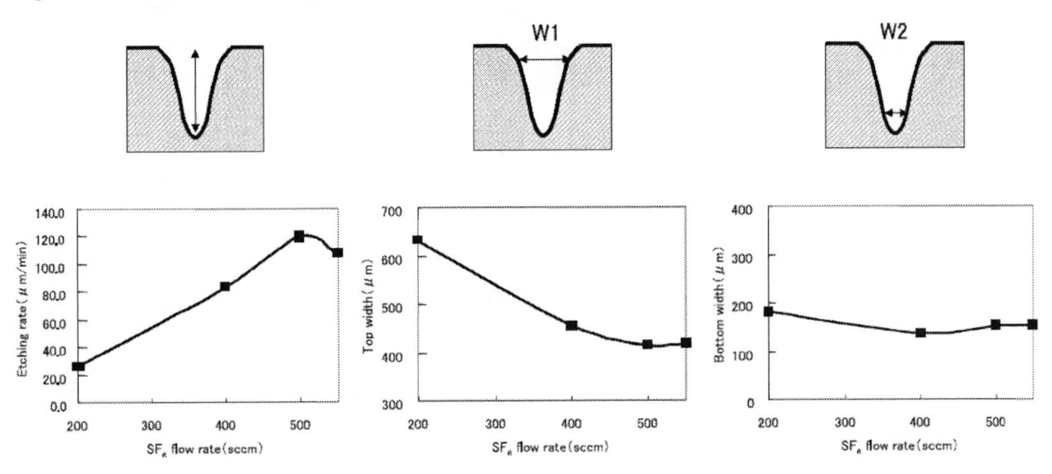

Figure 11. The etching rate, the top and bottom widths W1 and W2 at various SF_6 flow rates using the type-C microplasma source.

The etching properties using the type-C microplasma sources were also measured. Figure 12 shows the cross-section of an etched profile of a Si substrate using the type-C microplasma source. Etching was carried out under the following conditions: pressure = 100 kPa, RF power = 80 W, He flow rate in the inner gas feed = 1 slm, SF_6 flow rate in the outer gas feed = 215 sccm, gap = 0.1 mm, etching time = 30 s. The etching rate was 113 μm/min.

Figure 13 shows the etching rate, the top width W1, and the bottom width W2 along the line direction. The averaged etching rate was 109 μm/min and the uniformity was ±12.9 %. The top width W1 was 234 μm ±8.2 %, and the bottom width was 114 μm ±6.5 %. These results indicate that using the type-C microplasma source, the etching rate can be increased and the process-line-width can be drastically decreased compared with the type-B microplasma source. It is because the type-C microplasma source has only one inner gas outlet, and the distance between the outer gas outlets was decreased from 3 mm to 650 μm, contributing to decrease the thickness of the plasma line-width. The comparison of the type-B and type-C microplasma sources using the gas flow simulation is discussed later.

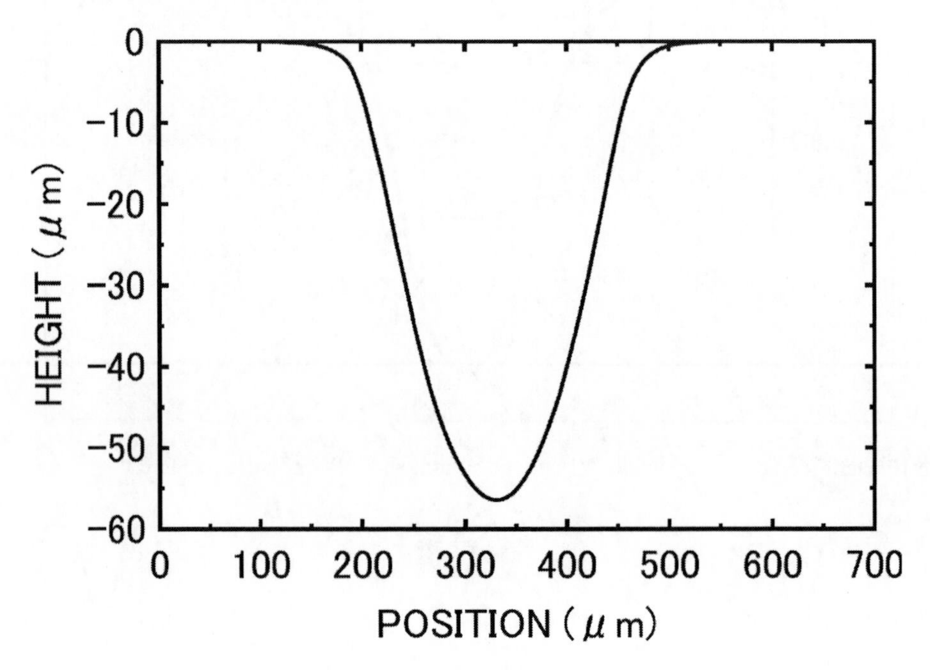

Figure 12. Si etched profile using the type-C microplasma source.

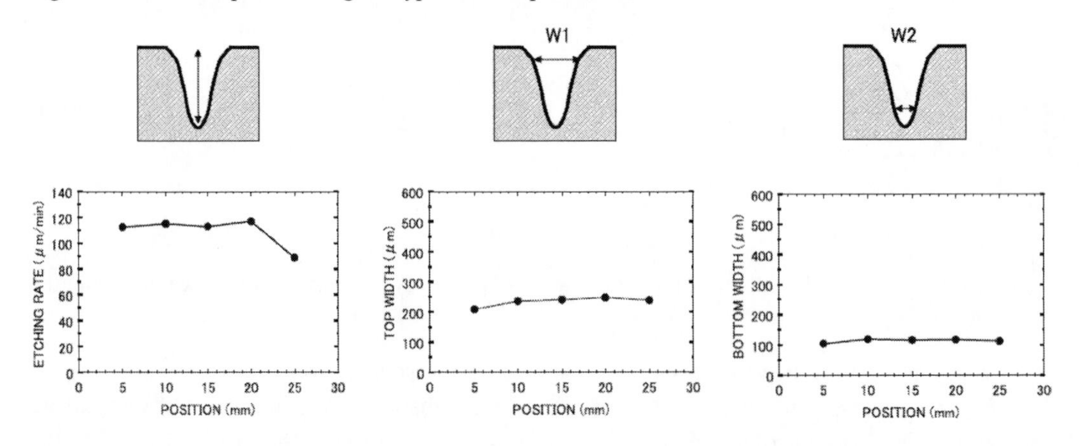

Figure 13. The etching rate, the top and bottom widths W1 and W2 at various positions along the line direction using the type-C microplasma source.

5. Influence of Substrate Heating

In order to increase the etching rate, enhancing etching reactions by substrate heating is considered to be effective. The etching properties with various substrate temperatures were measured. Figure 14 shows the etching rate, the top width W1, and the bottom width W2 with various substrate temperatures. Etching was carried out using the type-B microplasma source under the following conditions: pressure = 100 kPa, RF power = 100 W, He flow rate in the inner gas feed = 1 slm, SF_6 flow rate in the outer gas feed = 400 sccm, gap = 0.3 mm, etching time = 30 s.

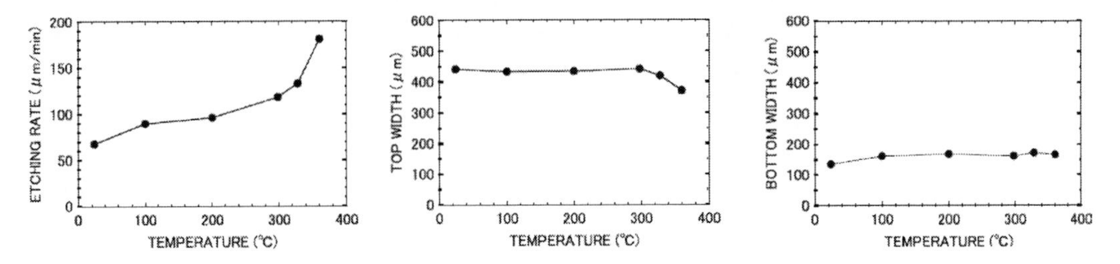

Figure 14. The etching rate, the top and bottom widths W1 and W2 at various substrate temperatures using the type-B microplasma source.

The etching rate and the top width W1 at room temperature (24 °C) are 67.8 μm/min and 441 μm, respectively. With increasing the substrate temperature from room temperature to 300 °C, the etching rate slightly increases while the top width W1 remains almost unchanged. However, the etching rate markedly increases and the top width W1 decreases at more than 300 °C. Meanwhile, the bottom width W2 remains almost constant even when the temperature was higher than 300 °C. The etching rate, the top width W1, and the bottom width W2 at 360 °C were 181 μm/min, 371 μm, and 166 μm, respectively.

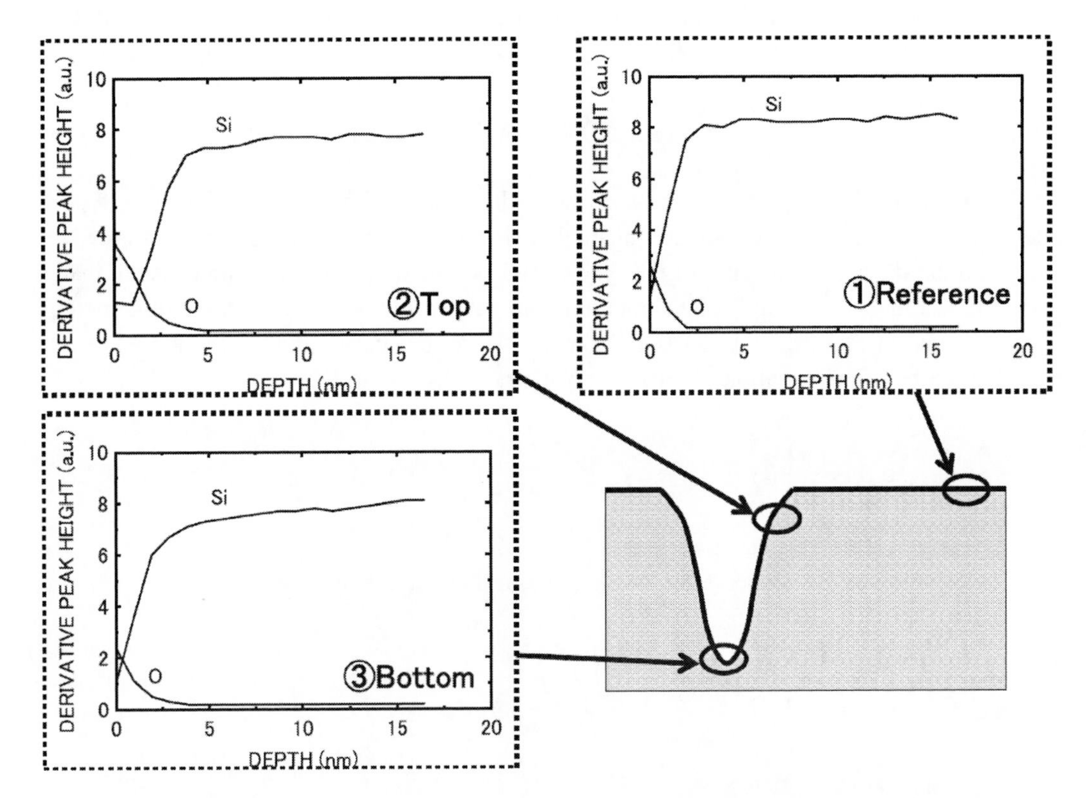

Figure 15. Depth profiles of the Si substrate etched by the type-B microplasma source at 360 °C, measured at the reference site with no plasma exposure ①, at the top shoulder of the etched line②, and at the bottom of the etched line ③.

The reason for the increase in the etching rate at higher temperatures was thought to be the enhancement of the reaction of F* at the Si substrate surface and the vaporization of etching by-products from the substrate surface. In order to investigate the reason why the top width W1 decreased at more than 300 ºC, Auger electron spectroscopic (AES) analysis was performed. Figure 15 shows the AES depth profiles of Si and O when the substrate temperature was 360 ºC. The horizontal axis of each Figure indicates the thickness converted into SiO_2 thickness. ① is measured at the reference site with no plasma exposure. ② and ③ are the sites of the top shoulder and the bottom of the etched line, respectively. It is clearly seen that a larger number of oxygen atoms was detected on the line shoulder ② than that on the reference site ① and the bottom of the etched line ③. It is therefore indicated that a silicon oxide film which is thicker than natural oxide is selectively formed on the line shoulder ②. As a result, the etching rate at the line shoulder ② decreased, decreasing the top width W1. It is suggested that oxygen needed for the oxidation at the line shoulder ② would come from the ambient air. Since the numbers of oxygen atoms at the bottom and reference sites were almost the same, the oxide at the bottom of the etched line was thought to be a natural oxide film formed after the etching experiment.

The depth profile of Si and O at the top shoulder was also measured when the substrate temperature was 200 ºC. In this case, it was observed that the oxide film formed at the line shoulder was not particularly thick. Namely, the formation of the thick oxide film at the line shoulder is specifically observed only when the substrate temperature is no less than 300 ºC.

6. Gas Flow Simulation

In order to investigate the reason why the etched line-width using the type-C microplasma source is much smaller than that using the type-B microplasma source, the concentration of the gas mixture of the type-B and type-C microplasma sources is compared using the gas flow simulation. The simulation was carried out with the finite volume method using the analytical cord Star-CD (CD Adapco). Here, the basic equation in the analytical cord is described. The equation of the conservation of mass is as follows;

$$\frac{1}{\sqrt{g}}\frac{\partial}{\partial t}\left(\sqrt{g}\rho\right)+\frac{\partial}{\partial x_j}\left(\rho\tilde{u}_j\right)=s_m \tag{1}$$

The equation of the conservation of momentum (Navier-Stokes equation) is

$$\frac{1}{\sqrt{g}}\frac{\partial}{\partial t}\left(\sqrt{g}\rho u_j\right)+\frac{\partial}{\partial x_j}\left(\rho\tilde{u}_j u_i-\tau_{ij}\right)=-\frac{\partial p}{\partial x_j}+s_i \tag{2}$$

x_i: rectangular coordinate system (i=1,2,3),
t: time
u_i: the absolute vale of the x_i direction component of the velocity of the fluid
$\tilde{u}_j = u_j - u_{cj}$: the relative speed of the fluid to the local (moving) axis with the speed of u_{cj}

p: pressure
ρ: density
τ_{ij}: a component of stress tensor
s_m: mass source
s_i: a component of momentum source
$\sqrt{g}$: determinant of metric tensor

The k-ε model was used for the simulation of turbulent flow. Three dimensional structural model was constructed for the simulation.

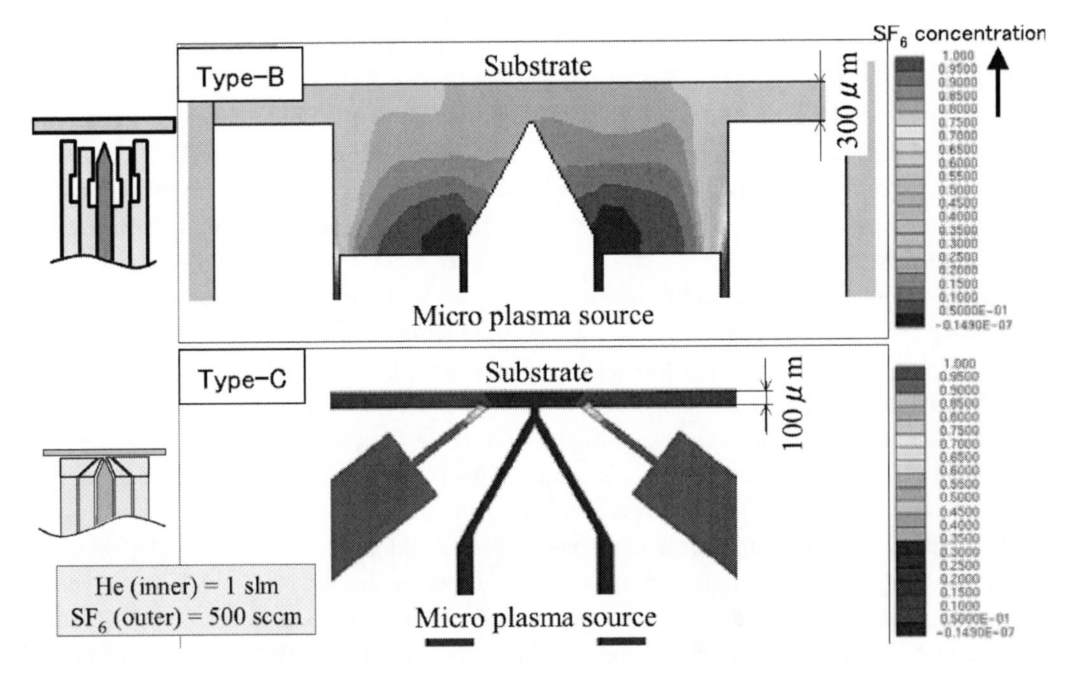

Figure 16. The gas simulation results for the type-B and type-C microplasma sources.

Figure 16 shows the results of the gas flow simulation for the type-B and type-C microplasma sources. In both cases, the flow rates of He (inner gas feed) and SF_6 (outer gas feed) were 1 slm and 500 sccm, respectively. In the case of the type-B microplasma source, the region where He concentration exceeds 95 % is around the inner gas outlet only, the region of high He concentration extends broadly near the substrate surface, and the change of the He concentration of the surface direction is moderate. On the other hand, using the type-C microplasma source, even near the substrate surface the region where the He concentration exceeds 95 % exists, and the change of the He concentration of the surface direction is significantly steep. It is indicated that these differences resulted in the different region of plasma generation, and subsequently different etching line width.

SUMMARY

Micro-scale mask-less Si line-etching was presented in this chapter.

The three different atmospheric pressure sources were compared; the type-A in which the copper electrode is protected against plasma exposure, the type-B in which the aluminium electrode is exposed in a plasma and type-C with the shorter distance of the outer gas outlets using the ceramic plate with the slits. Study of the inner and outer gas feeds indicated that extension of a microplasma can be significantly suppressed and that the thinnest line-plasma can be generated by supplying He and reaction gases in the inner and outer gas feeds, respectively. When the reaction gas was supplied into the outer gas feeds, the optical emission intensity of F* increased as the increase of its flow rate, and is higher than that when the reaction gas was mixed with He and supplied into the inner gas feed.

Fine pattern etching of a Si substrate was attempted using the microplasma sources. The type-B microplasma source demonstrated the higher etching rate than the type-A microplasma source. Using the type-C microplasma source, the etching rate further increased up to 113 μm/min. The etching rate and the etched line width were generally uniform toward the line direction.

The dependence of the substrate temperature on the etching properties was investigated. The etching rate drastically increased at the substrate temperature over 300 ºC, and reached 181 μm/min at 360 ºC, while the top width W1 decreased. On the other hand, the bottom width W2 remained almost unchanged even at the temperature over 300 ºC. AES analysis indicated that an oxide film thicker than the natural oxide film was selectively formed at the shoulder part of the etched ditch. The existence of this thick oxide film caused the decrease in the etching rate, resulting in the decrease in the top width W1.

The type-C microplasma source demonstrated much thinner process line-width than the type-B microplasma source. The gas flow simulation indicated that when the type-B microplasma source was used, the region with high He concentration extends broadly around the substrate surface, and the change of the He concentration of the surface direction is moderate. On the other hand, using the type-C microplasma source, the change of the He concentration of the surface direction is significant. It is indicated that these differences resulted in the different region of plasma generation, and subsequently different etching line width.

REFERENCES

[1] Inomata, K.; Ha, H.K.; Li, B.J.; Koinuma, H. *Shingaku-Gihou* (Japanese). 1994, SDM94-116, 35-40.

[2] Sankaran, R.M.; Giapis, K.P. *Appl. Phys. Lett.* 2001, *79(5)* 593-595.

[3] Ichiki, T.; Taura, R.; Horiike, Y.; *J. Appl. Phys.* 2004, *95* 35-39.

[4] Ideno, T.; Kanou, H.; Kondoh, M.; Hori, M. *66th Jpn. Soc. Appl. Phys. Abstract.* 2005 9p-ZG-15.

[5] Okumura, T.; Saitoh, M. In *Proc. Plasma Mater. Sci. 153 Committee 62nd research meeting* 2003 21-28.

[6] Okumura, T.; Saitoh, M.; Yashiro, Y.; Kimura, T. *Jpn. J. Appl. Phys.* 2003 *42(6B)* 3995-3999.

[7] Okumura, T.; Saitoh, M.; Matsuda, I. *Jpn. J. Appl. Phys.* 2004 *43(6B)* 3959-3963.

In: Generation and Applications of Atmospheric Pressure Plasmas ISBN: 978-1-61209-717-6
Editors: M. Kogoma, M. Kusano and Y. Kusano

Chapter 14

SiO_2 ETCHING

Masaru Hori
Department of Electrical Engineering and Computer Science,
Nagoya University, Nagoya, Japan

INTRODUCTION

High speed etching of Si and SiO_2 films is one of the core technologies in the manufacturing processes of ultra-large scale integrated circuits (ULSI), micro electro-mechanical systems (MEMS) and liquid crystal devices (LCD). In the ULSI processing, manufacturing methods based on low pressure low temperature plasma processing are currently used. It takes substantial time for μm order processing. Therefore, high-speed damage-freed processing is demanded. Furthermore, majority of the treatment in the manufacturing processes are performed in vacuum or reactive gas atmosphere. Consequently the manufacturing systems tend to be complicated and costly with low yield. In addition vacuum processing is disadvantageous in that it needs certain time for pumping out.

Recently low temperature processing at atmospheric pressure is proposed as a tool to overcome these disadvantages, and its intensive researches and development are subsequently carried out. The use of an atmospheric pressure plasma enables us to simplify vacuum system, and is significantly advantageous for its low cost and short processing time. It is possible to generate a very high density plasma so that high speed processing is expected.

However, an atmospheric pressure plasma generally tends to have high gas temperature and to become an arc discharge easily. In such an arc discharge, gas and electrons are in an equilibrium state, and gas temperature may be more than thousands K, at which condition control of plasma chemistry is very difficult. Low temperature plasma chemistry is commonly utilized in a low pressure non-equilibrium plasma. In order to achieve such low temperature plasma chemistry at atmospheric pressure, generation of a non-equilibrium plasma (low temperature plasma) at atmospheric pressure is necessary. Here a non-thermal plasma is a plasma in which gas temperature is lower than electron temperature. It enables control of chemical reactions (reaction control of gas phase radicals) which is impossible with the thermo-chemical method. An atmospheric pressure non-equilibrium plasma can be generated by two different methods as shown in Figure 1.

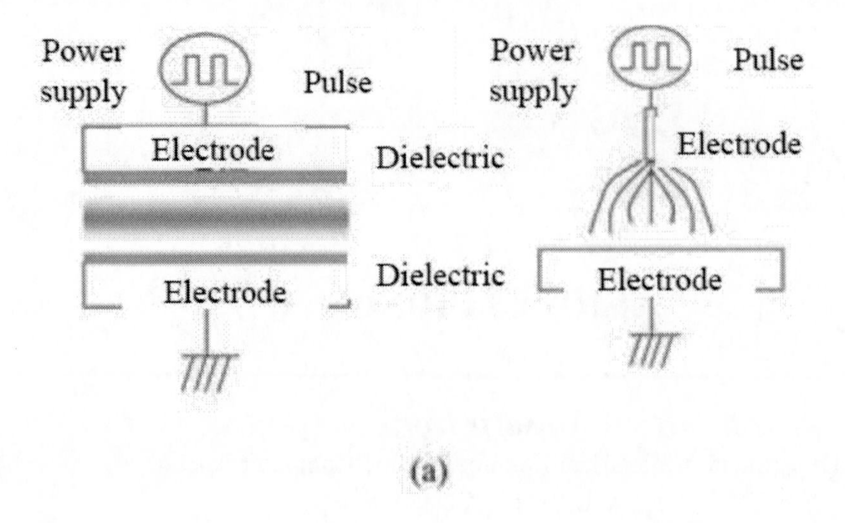

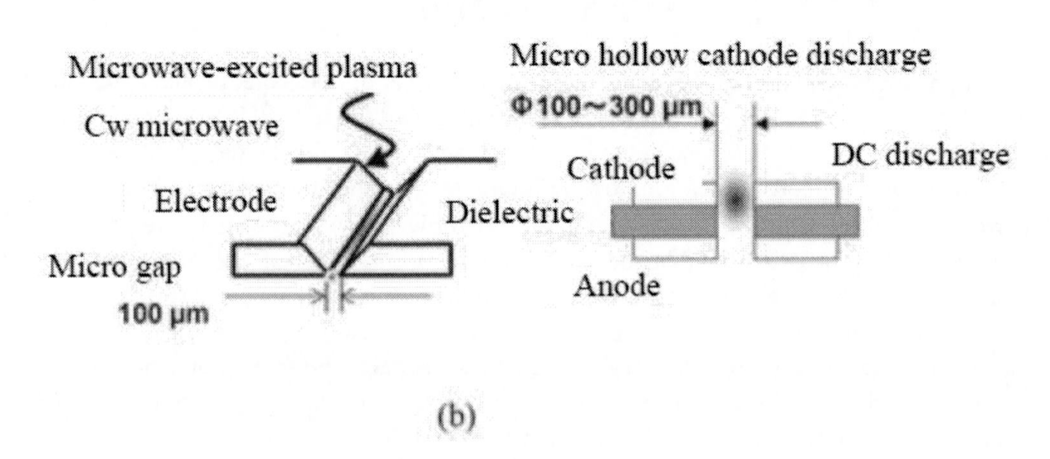

Figure 1. Generation methods of atmospheric pressure non-equilibrium plasmas (dielectric barrier discharge (DBD), streamer discharge, (b) micro hollow cathode discharge, micro gap discharge.

Figure 1 (a) shows the methods to extinguish a discharge just after its ignition using pulsed excitation and to use dielectrics with electrodes so that arc transition is avoided. Using these methods, a non-equilibrium plasma can be generated in a short time at a wide range of gas pressures [1,2]. Atmospheric pressure non-equilibrium plasma can also be generated by limiting a discharge volume in a microscopic space (e.g. micro-gap, micro-hollow structure) as shown in Figure 1 (b). Gas temperature increases due to multiple collisions of neutral species with electrons in a discharge. However, by limiting the size of the discharge volume in a microscopic space, neutral species can diffuse onto cooled walls before its temperature increases. Consequently increase in gas temperature is suppressed and non-equilibrium plasma can be generated. Its examples include a micro-hollow cathode discharge and a discharge in a micro-gap [3,4].

The author noticed a continuously generated atmospheric pressure non-equilibrium plasma in a micro-gap excited by microwave (electron temperature: approximately 1 eV, electron density: 10^{15} cm^{-3}), considering to make the best use of the advantages of an atmospheric pressure non-equilibrium plasma [4]. The plasma density of such a plasma is about four orders of magnitude greater than that of a normal low pressure non-equilibrium high density plasma. It enables ultrahigh density radicals and ions to be injected onto a substrate surface. In addition, since it is also a non-equilibrium plasma, the low temperature plasma chemistry which has been used for conventional low pressure plasma processing can also be used for design of processing. Furthermore a liquid solution can be insufflated into a plasma since it is at atmospheric pressure. Utilizing these advantages, ultrahigh speed processing of silicon oxide films is performed, based on the proposal of a novel plasma processing technology using new physical chemistry which is a result from fusion of properties of the dry and wet processing [5,6].

In order to clarify etching mechanisms of silicon oxide films using an atmospheric pressure non-equilibrium plasma, high speed processing of SiO_2 films is studied using a commonly used atmospheric pressure non-equilibrium pulsed plasma which is thought to be advantageous for large area treatment [7]. Its gas phase diagnostics is also carried out.

1. Ultrahigh Speed Processing of SiO_2 Films Using a Microwave-Excited Atmospheric Pressure Non-Equilibrium Plasma

1.1. Introduction

Recently micro-electro-mechanical systems (MEMS) and bio-nanotechnology applications demand high speed etching of over several-μm-thick SiO_2 films. Such etching is commonly performed using a low pressure high density plasma. However, a maximum etching rate is only about 1 μm/min in a fluoro-carbon gas, and thus high speed etching process is strongly demanded. In order to achieve the high speed etching, developing a new process is necessary which can be a breakthrough over the conventional plasmas. In the present study, in order to apply the microwave-excited atmospheric pressure non-equilibrium plasma to the material processing, a special micro-gap structured slip is developed for the introduction of microwaves, and a reactor is constructed, installing this plasma source. High speed etching is studied with using the microwave-excited atmospheric pressure non-equilibrium plasma, and ultrahigh speed etching of 14 μm/min was successfully demonstrated with gas mixtures of He, NF_3 and H_2O. On the other hand, Si was hardly etched and the selectivity of over 200 is obtained. Namely the microwave-excited atmospheric pressure non-equilibrium plasma enables construction of the high selectivity of SiO_2 over Si. For the understanding of this etching mechanism, the image-intensifying charge-coupled device camera (ICCD camera) is used for gas phase diagnostics using a 2-dimensional optical emission spectroscopy and Fourier Transform Infrared (FT-IR) absorption spectroscopy.

1.2. Experimental Setup

Figure 2 shows schematic diagram of the microwave-excited atmospheric pressure non-equilibrium plasma system. 2.45 GHz continuous wave (CW) microwave is introduced from the upper part of the apparatus through a quartz window. A stable slit-shaped plasma is generated by concentrating an electric field at the knifed-edged slit electrode (gap width: 0.2 mm). The slit width was 60 mm. Gas was fed from the upper part of the slit and flown to the bottom part of the slit. The process gas was a mixture of He and NF_3. In addition, H_2O was introduced into the chamber by a bubbler using He as a carrier gas. H_2O flow rate was estimated by a dew point hydrometer. These gases were introduced via mass flow controllers. The pressure in the chamber was kept at atmospheric pressure (1 atm) by pumping out the same flow rate at the downstream. A substrate holder is facilitated with a cooling system, and its temperature was controlled. A SiO_2 substrate (Borophosphosilicate glass (BPSG)) was placed at 5 mm downstream from the electrode, and was etched.

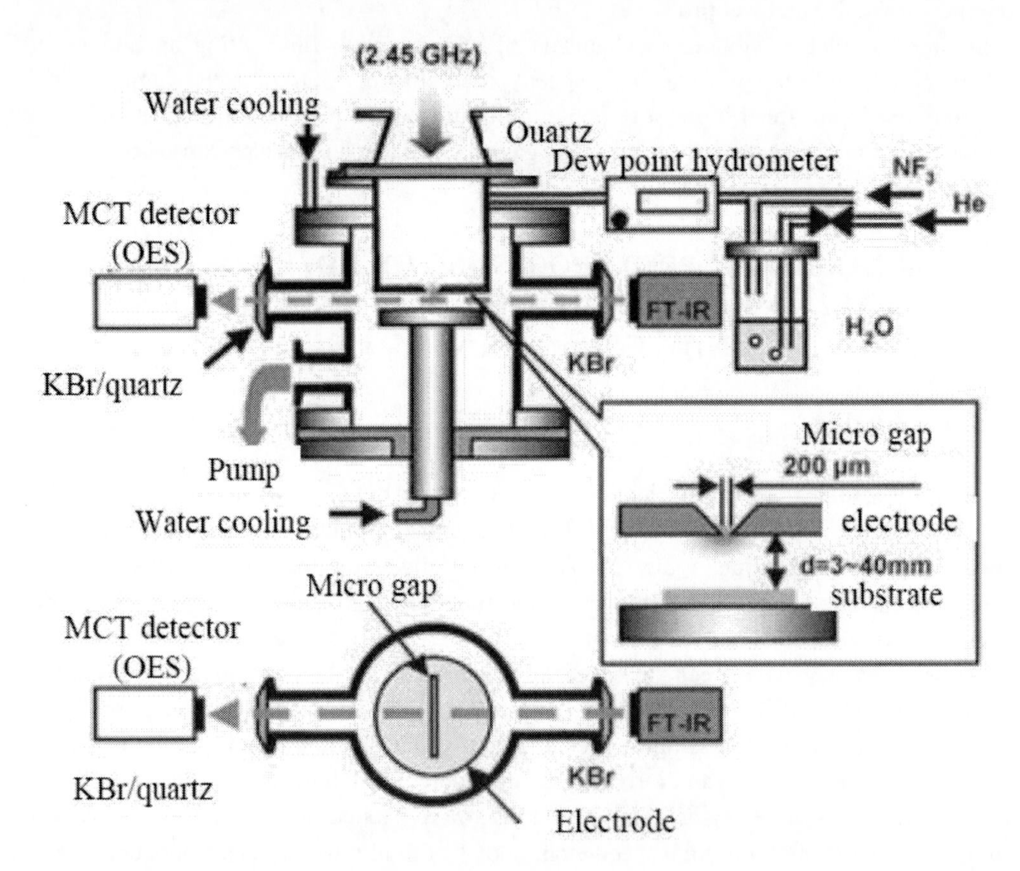

Figure 2. Schematic diagram of the microwave-excited atmospheric pressure none-equilibrium plasma etching apparatus.

KBr windows were disposed in opposition with each other attached at the viewports at the sides of the chamber, and FT-IR absorption spectroscopy was used to characterize the gas species and these densities between the slit electrode and the sample holder. Optical emission spectroscopy (OES) and special distribution measurement of optical emission of He, NF_3,

H_2O plasma were performed via the quartz viewport using the ICCD camera. From these measurements, radicals and special behaviour of photo-emitting species were estimated.

SEM was used to observe etched morphology of an SiO_2 substrate on which resist was patterned.

1.3. Results and Discussion

When the microwave power was 500 W, the pressure was 1 atm, and the flow rates of He, NF_3 and H_2O were 16 L/min, 60 sccm, and 196 sccm, respectively, a stable glow discharge was generated at atmospheric pressure, extending approximately 1 mm long in the direction of the gas blow, and 30 mm long in the direction of the slit. A high density plasma of approximately 2 x 10^{14} cm^{-3} was successfully generated, which was estimated by the Stark broadening of the H_β optical emission spectrum.

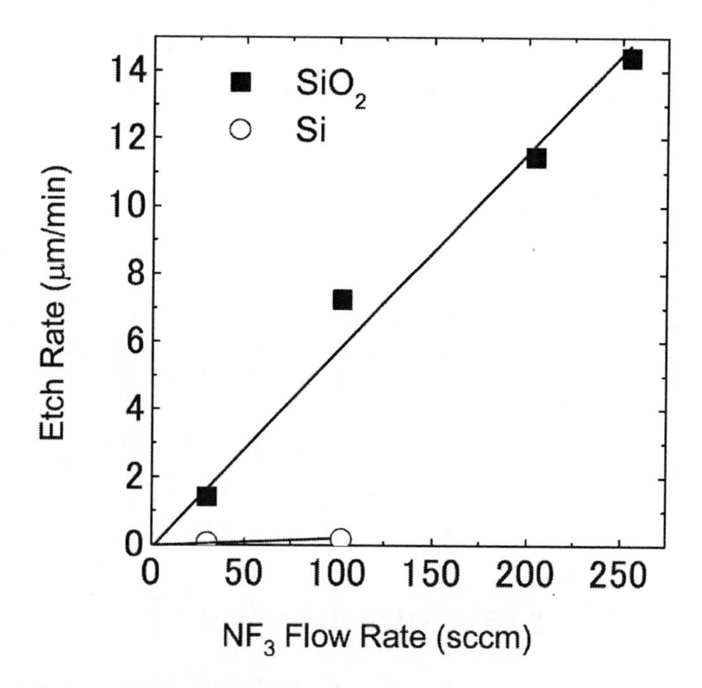

Figure 3. SiO_2 etch rate at various NF_3 flow rates.

Figure 3 shows the etching rates of SiO_2 (BPSG) and Si at various NF_3 flow rates. The etching conditions were as follows; gas pressure was 1 atm, microwave power was 500 W, substrate temperature was 18 °C. He mixed with NF_3 and H_2O were fed in to the plasma. H_2O was introduced by bubbling it with He as a carrier gas. The flow rate of H_2O was controlled by the flow rate of the carrier gas. The flow rate of H_2O was approximately 392 sccm, estimated by the dew point hydrometer. As the flow rate of NF_3 increased, the etching rate of SiO_2 increased. The high rate etching of approximately 14 μm/min was achieved at the NF_3 flow rate of 250 sccm, while the SiO_2 etching rate using a conventional low pressure high density plasma is approximately 1 μm/min. It can therefore be said that a markedly high speed etching is achieved using the microwave-excited non-equilibrium atmospheric pressure plasma. In addition, Si was seldom etched using this etching process, and the selectivity over Si was over 200. This indicates that this etching process provides the property of the wet etching. An etching model is suggested to be the formation of a liquid HxFy layer at the substrate surface by supplying H and F from H_2O and NF_3, respectively, and subsequent high speed etching by NF_2^-.

In order to investigate the etching mechanisms, gas species and their densities were measured using FT-IR in a 5-mm gap between the slit electrode and the substrate. Figure 4 shows the absorption spectrum of the He, NF_3, H_2O plasma. The spectrum was obtained by subtracting the spectrum without a discharge from that with a discharge. The downward peak appeared at around 907 cm^{-1} [8] indicates the dissociation of NF_3. The degree of NF_3 molecular dissociation was estimated to be approximately 60 %. The positive absorption bands between 3800 and 4200 cm^{-1} observed in Figure 4 indicates the production of HF. This result indicates the production of HF by the chemical reaction of H_2O and NF_3 in the plasma.

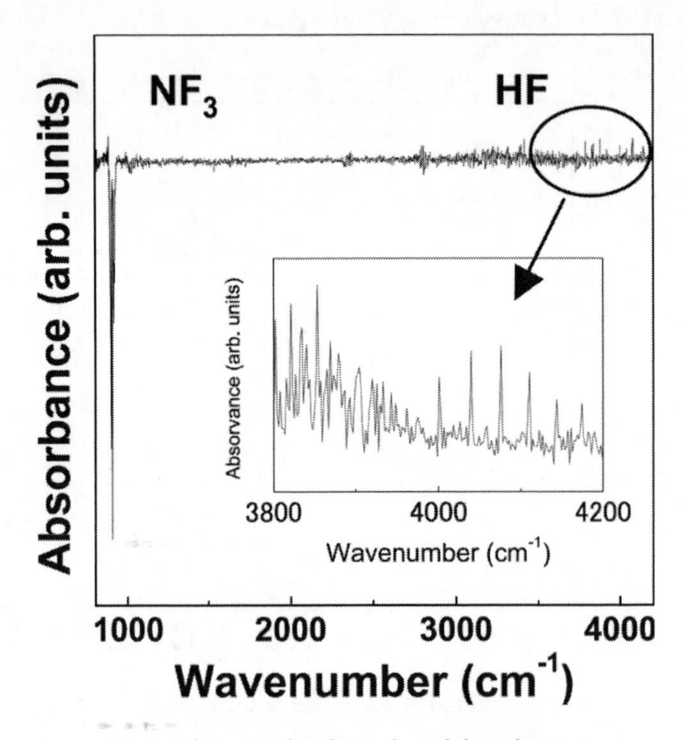

Figure 4. FT-IR spectrum of the gas between the electrode and the substrate.

Figure 5 shows the SiO_2 etch rate at different HF densities in the 5-mm gap between the substrate and the electrode. In the calculation of the HF density, it was assumed that gas temperature was 300 K, and that the absorption line was a Voigt profile. The absorption coefficient of HF at 4142.8 cm^{-1} corresponding to the vibrational and rotational energy level ($v=1 \leftarrow 0$) was used for the calculation. It is seen in Figure 5 that the SiO_2 etching rate increased linearly with the increase of the HF density. It is indicated here that HF contributes to the SiO_2 etching rate in the 5-mm gap between the substrate and the electrode.

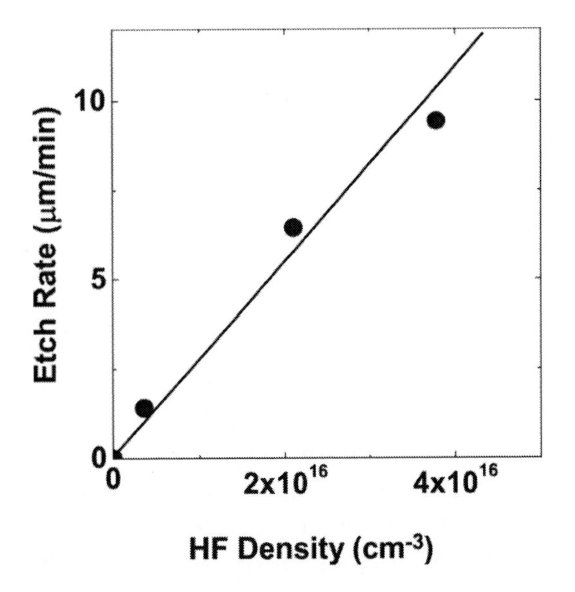

Figure 5. SiO_2 etch rate at various HF density.

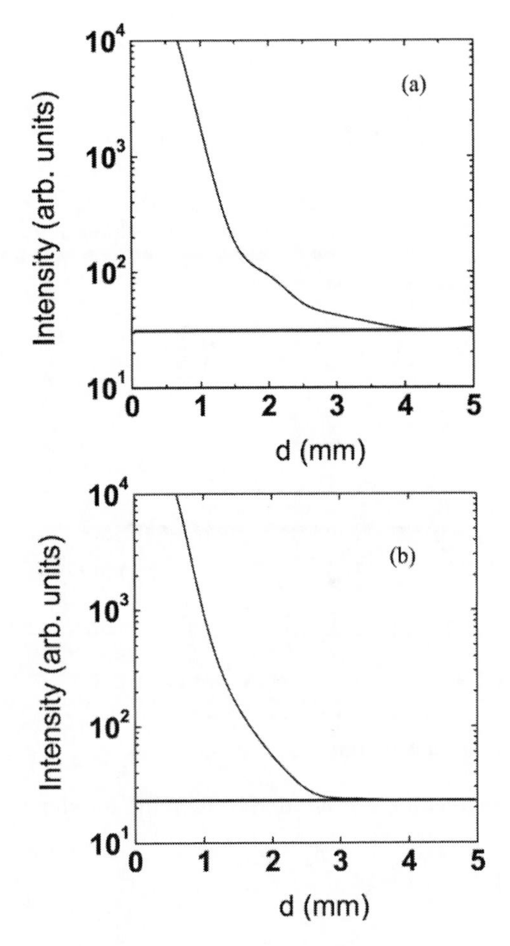

Figure 6. F radical emission intensity at various distances between the electrode and the substrate; He, NF_3 plasma without addition of H_2O (a), and He, NF_3, H_2O plasma (b).

For the understanding of the generation process of HF, the photoemission intensity of F radical at 704 nm at the distance *d* from the electrode with/without addition of H_2O was measured using the ICCD camera as shown in Figure 6. Without H_2O, the photoemission of the F radical was confirmed at the distance of 2.5 mm as shown in Figure 6 (a). On the other hand, with addition of H_2O the photoemission was extinguished when d = 2.5 mm. It is indicated that before they reach the distance of 2.5 mm, the F radicals collide and react with particles (H_2O, OH, H, O) which are induced by H_2O addition, and subsequently HF is generated. Namely such reactions generated HF.

The reaction model of ultrahigh-rate, high-selectivity etching reactions in this etching process is considered as shown in Figure 7. It illustrates two reaction schemes of He, NF_3, H_2O and He, NF_3. Optical emission spectroscopy identified the emission from N, F, H, O, OH, and N_2 corresponding to the first scheme, and that from N, F, and N_2 to the second scheme, and confirmed the existence of these radicals. It is thought that etching mainly by those reactive species proceeds near the electrode. Near the electrode at the distance of 5 mm, high rate etching of not only SiO_2 but also Si was seen, and subsequently the selectivity of SiO_2 over Si was as low as 5.1 and 1.2, respectively. It namely indicates that Si etching is mainly dominated by the F radical.

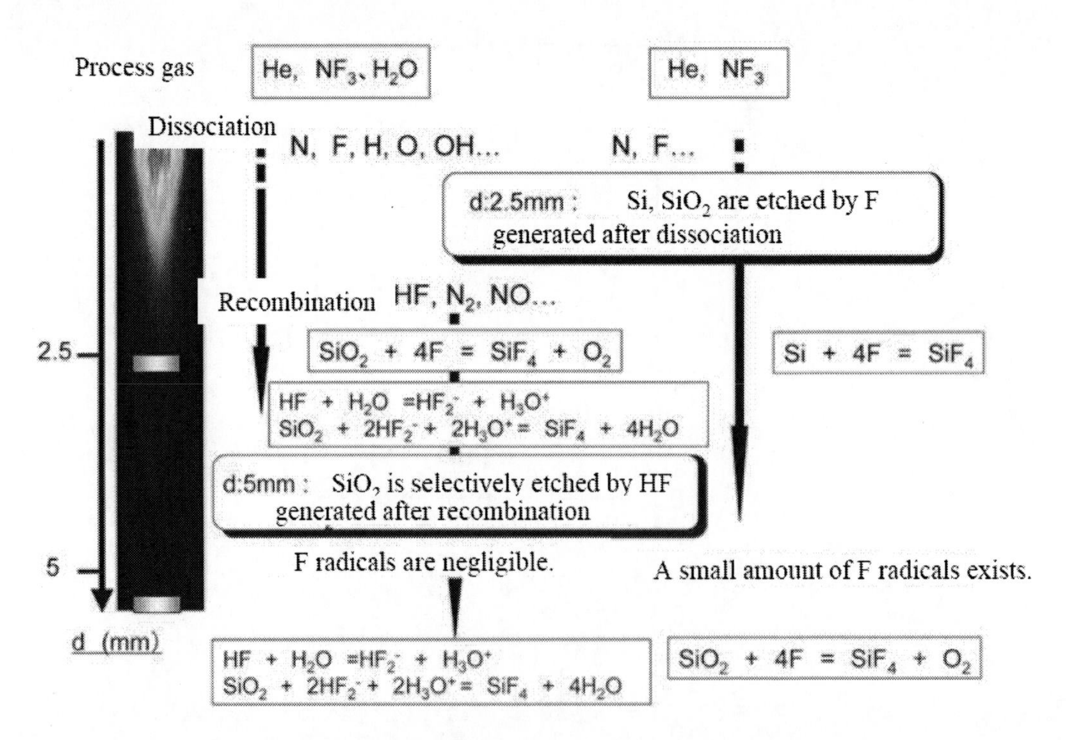

Figure 7. Reaction models of the etching process.

It is considered that the etching of SiO_2 and Si with the F radical proceed as;

$$SiO_2 + 4F \rightarrow SiF_4 + O_2$$

$$Si + 4F \rightarrow SiF_4.$$

On the other hand, with addition of H_2O, at the distance of 5 mm from the electrode, the number of the F radical decreases by the collision with other particles, and subsequently HF is generated as was already described. Since the selectivity of SiO_2 over Si becomes over 200, the F radials, which are used to etch Si, seldom reach the substrate. It is therefore indicated that etching by HF mainly proceeds. It is reported that with the existence of H_2O, the etching reaction of SiO_2 with HF proceeds with the following reactions;

$$HF + H_2O \rightarrow HF_2^- + H_3O^+$$
$$SiO_2 + 2HF_2^- + 2H_3O^+ \rightarrow SiF_4 + 4H_2O.$$

Without addition of H_2O, the etching rate was significantly low, and the selectivity was as low as only 2. Namely, the F radicals are thought to reach the substrate, although they are only a small amount. The number of the F radicals rapidly decreases with the addition of H_2O, and the reaction generates HF. Consequently density control of F and HF is important for high rate etching with high selectivity. Practically, density control of F and HF is achieved by optimizing the flow rates of NF_3 and H_2O, and the distance between the substrate and the electrode. In an atmospheric pressure plasma, generated reactive species will react in a gas-phase and be inactivated, as the mean free path is approximately 1 μm. Since the process highly depends on the plasma photoemission distribution and the distance from the substrate, it is necessary for a fine processing to control the distance between the substrate and the generated plasma.

Furthermore, the He, NF_3, H_2O plasma was applied for the pattern etching of resist-masked SiO_2. As a result, SiO_2 was etched down to the bottom part of the resist, and isotropic radical etching occurred. A plasma in He, NF_3, H_2O and CF_4 gas mixture was also used for etching. In this case, Si was almost vertically etched. This result indicates the possibility of anisotropic processing of SiO_2 using the microwave-excited atmospheric pressure non-equilibrium plasma. It is suggested that deposition of protective layer at the side wall during etching would cause the vertical etching.

1.4. Summary

In order to apply the microwave-excited atmospheric pressure non-equilibrium plasma for material processing, a special micro-gap structured slit was developed, and the reactor involving this plasma source was constructed. The high speed processing of SiO_2 (BPSG) films was investigated using the microwave-excited atmospheric pressure non-equilibrium plasma. The ultrahigh speed etching of SiO_2 at 14 μm/min was successfully demonstrated with the gas mixture of He, NF_3 and H_2O. Under this condition Si was seldom etched, and the selectivity was over 200. Spatial distribution measurement of the optical emission using the ICCD camera and gas phase diagnostics using the FT-IR measurement were performed for the investigation of the etching mechanism. It was confirmed that adding H_2O into the He, NF_3 plasma induced the reaction between F radicals from NF_3 and H from H_2O and subsequent production of HF. The relation between the HF density and the etching rate clarified that the HF molecules contribute the etching. In order to achieve the ultrahigh speed etching with high selectivity, it is necessary to control the densities of F radicals and HF particles by optimizing the flow rates of NF_3 and H_2O, and the distance between the substrate and the electrode.

2. High Speed Etching of SiO_2 Film Using Atmospheric Pressure Non-Equilibrium Pulsed Plasma and Gas Phase Diagnostics

2.1. Introduction

This section presents high rate processing of SiO_2 film in a fluorocarbon gas with H_2O, using the atmospheric pressure non-equilibrium pulsed plasma which is known to be advantageous for large area treatment. In order to investigate the mechanism of the etching process, optical emission measurement using the ICCD camera, which was also used in the previous section, and gas diagnostics of the exhaust using Ion attachment mass spectrometer (IAMS) etc. were used. The IAMS is a novel mass spectrometer which ionizes gas atoms and molecules by attaching ions. It is useful for the measurement of gas molecules with high mass numbers, since such a measurement is difficult using conventional mass spectrometers with electron sources. Therefore this method is useful for the measurement of gases and activated species generated in an atmospheric pressure plasma.

2.2. Experimental Setup

Figure 8 (a) shows a schematic diagram of the atmospheric pressure non-equilibrium pulsed plasma setup (dielectric barrier discharge (DBD)). The gap of the two parallel plate electrode covered with dielectrics is fixed at 1 mm. A plasma is generated by applying several tens kHz pulsed voltage (variable in several μs order). A gas mixture of Ar, CF_4 and H_2O (or O_2) was fed from the upper electrode. For mixing H_2O, the process gas was fed through an H_2O vessel for bubbling. The flow rate of H_2O was measured by the dew point hydrometer as well. At first the chamber was filled with N_2 gas at atmospheric pressure, and then the process gas was introduced. The etching was carried out at 5 mm below the electrode in the region of the remote plasma. SiO_2 was used as a substrate. Synthesized quartz glass was attached at the side of the electrodes for the optical emission spectroscopic (OES) measurement between the electrodes. Figure 8 (b) illustrates the schematic diagram of the plasma unit and the IAMS setup. A part of the exhaust gas from the plasma was introduced for the mass analysis. The ion attachment is the method to ionize a molecule (M) by attaching it with alkali metal ion such as Li^+ ion ($M + Li^+ \rightarrow MLi^+$). This method enables measurement of molecules with large mass numbers. These large molecules are inevitably dissociated if conventional electron beam irradiating mass spectroscopy is used. The IAMS can be a very useful method for the characterization of atmospheric pressure plasmas since it can directly monitor its exhaust gas. In the IAMS a Li^+ ion is attached to a specimen gas molecule in the first chamber (ionization chamber), passing the second and third chamber which are differentially pumped out individually, and mass analysis is carried out by quadrupole mass spectroscopy (Q-MS) in the forth chamber. By directly introducing calibration gases such as C_4F_8 into the IAMS chamber, calibration of the absolute values can be employed by the comparison of the intensity of the mass spectra. In the present work the IAMS (CANON ANELVA Technics; L-240G-IA) was used.

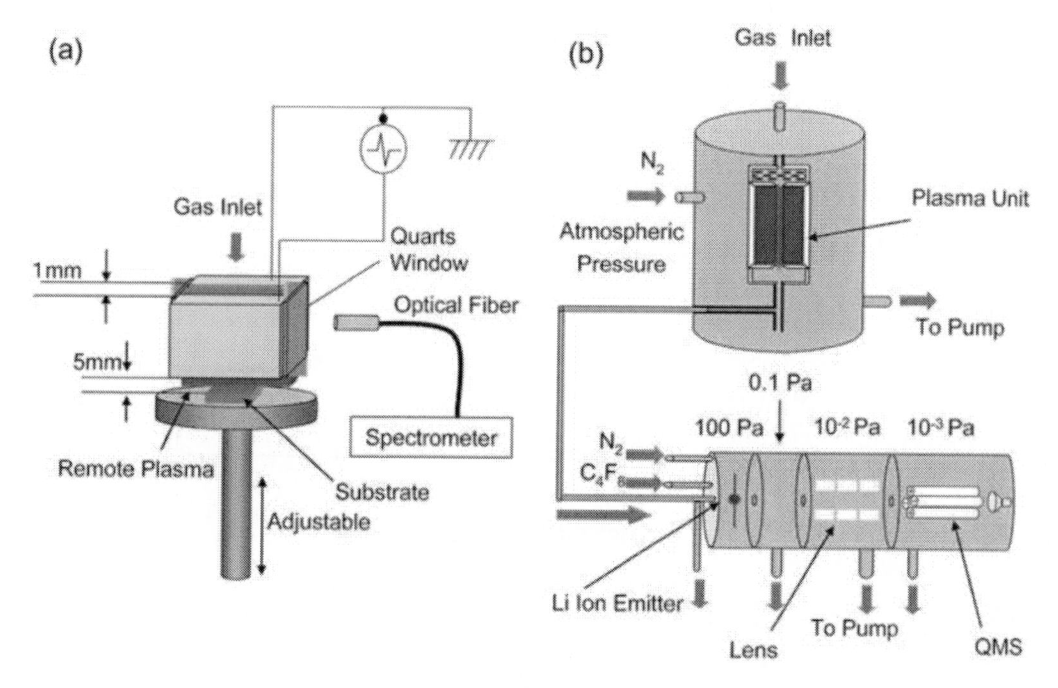

Figure 8. Schematic diagrams of the atmospheric pressure non-equilibrium pulsed plasma etching apparatus (a), and the system when the IAMS is used (b).

2.3. Results and Discussion

Figure 9 shows the etching rate of SiO_2 at the place where the exhaust gas is erupted from the atmospheric pressure plasma driven by the frequency between 5 and 20 kHz with the $CF_4/O_2/Ar$ flow rates of 60/0/240 sccm or 55/5/240 sccm. The SiO_2 etching rate monotonically increased with the increase of the frequency. Since the etching rate increased with the addition of O_2, it is suggested that increase of certain activated species by the O_2 addition would contribute to the etching.

Figure 10 shows 2-dimensional distribution images of optical emission from CF_3 of the plasma generated with the above mentioned condition. It is seen that the optical emission intensity of CF_3 between the electrodes decreased along the downstream. Similar experiment for the optical emission from Ar showed homogeneous photoemission, and thus such an inhomogeneous emission is confirmed to be specifically corresponding to CF_3. Under the assumption of the uniform electron density between the electrodes due to the homogeneous photoemission of Ar, it is suggested that the density of CF_3 would decrease along toward the downstream. This result indicates that CF_3 further dissociates at the downstream region.

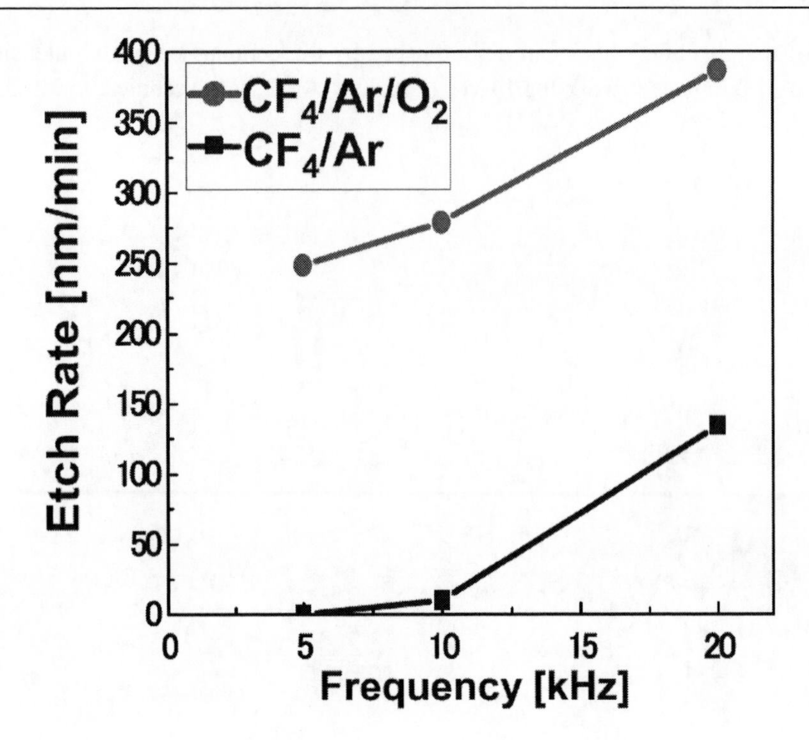

Figure 9. SiO_2 etching rate with/without addition of O_2 in the process gas.

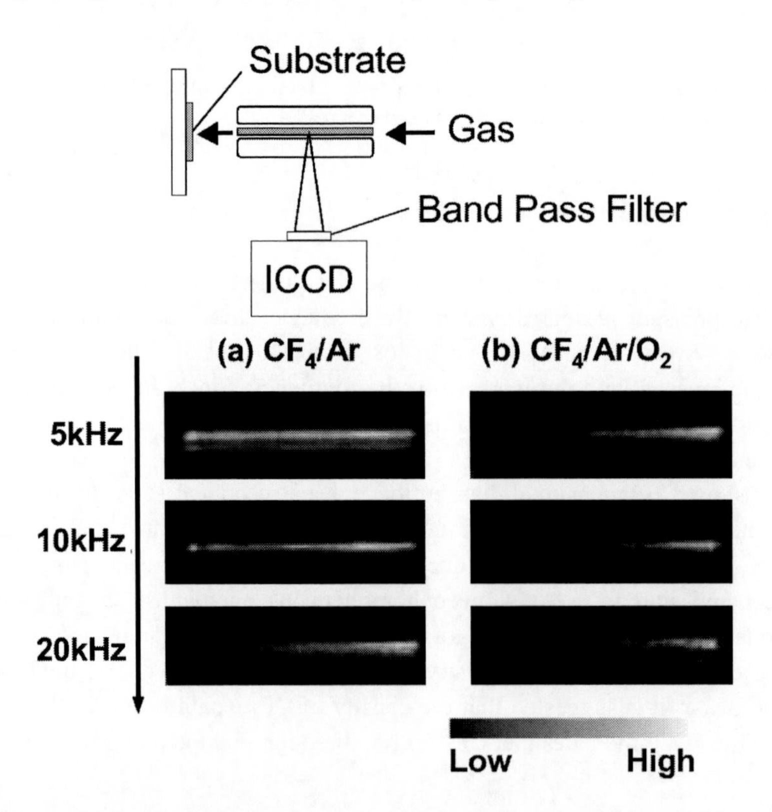

Figure 10. Two-dimensional distribution of CF_3 photoemission between the electrodes in CF_4/Ar (a), and $CF_4/O_2/Ar$ (b) gas mixtures.

Figure 11 shows the etching rate when H_2O was added by bubbling. The flow rates of the process gases $CF_4/H_2O/Ar$ were 60/5.8/240 sccm. The etching rate when O_2 was added with the $CF_4/O_2/Ar$ flow rate of 55/5/240 sccm is also shown in Figure 11. High rate etching of SiO_2 at over 8 mm/min was obtained at the driving frequency of 20 kHz. In order to investigate the reason why the etch rate changed drastically, the exhaust gases were characterized from the plasmas without H_2O ($CF_4/O_2/Ar$ = 60/0/240 sccm) and with H_2O ($CF_4/H_2O/Ar$ = 60/5.8/240 sccm), respectively. The mass spectra of the exhaust gases from the plasmas without and with adding O_2 are shown in Figure 12 (a) and (b), respectively. When the gas mixture without adding H_2O was fed into the plasma, polymerized particles such as C_3F_6 and C_3HF_7 were significantly generated. On the other hand, when H_2O was added, generation of smaller mass-molecules was pronounced. Particles with high mass numbers are the final molecules of higher order radicals generated in a gas phase. The deposition of these particles is thought to be significant [15, 16]. On the other hand, particles with small mass numbers are generated by multiple polymerization of H and O. The CF_2 radical is thought to be a precursor of the higher order radicals [17-19]. The overlap of the spectra prevented some species from identification.

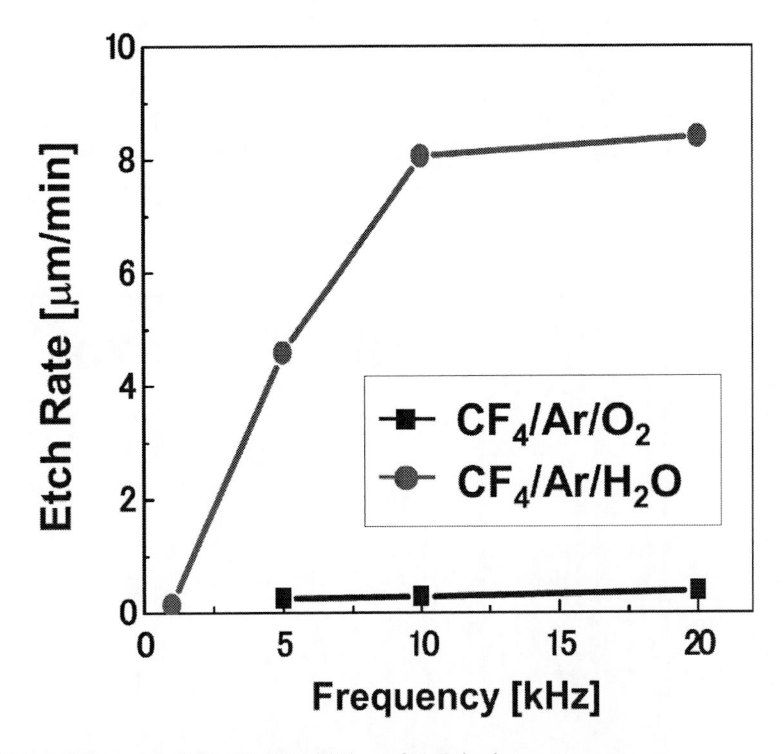

Figure 11. SiO_2 etching rate with/without addition of H_2O in the process gas.

However, densities of HF and COF_2 were measured at different driving frequencies as shown in Figure 13. As was discussed in the previous section, HF is the important particle for the high rate etching of SiO_2 using an atmospheric pressure plasma. Since COF_2 is a stable gas molecule, its existence suppresses production of the higher order radicals [20]. Therefore COF_2 can be used as an indicator for estimating the amount of the higher order radicals generated. Figure 13 shows the frequency dependence of HF and COF_2 in the exhaust gases of the plasmas of the systems without adding H_2O ($CF_4/O_2/Ar$ = 60/0/240 sccm) (a), adding

O_2 ($CF_4/O_2/Ar$ = 60/5/240 sccm) (b), and adding H_2O ($CF_4/H_2O/Ar$ = 60/5.8/240 sccm) (c). In each case, these densities increased with the increase of the driving frequency. Higher densities of HF and COF_2 were generated with the addition of O_2 or H_2O than without the addition of H_2O. The result indicates that the addition of H_2O suppressed the generation of the higher order radicals and subsequent deposition of carbon coatings on to the substrate surface, and thus the etching by HF is progressed.

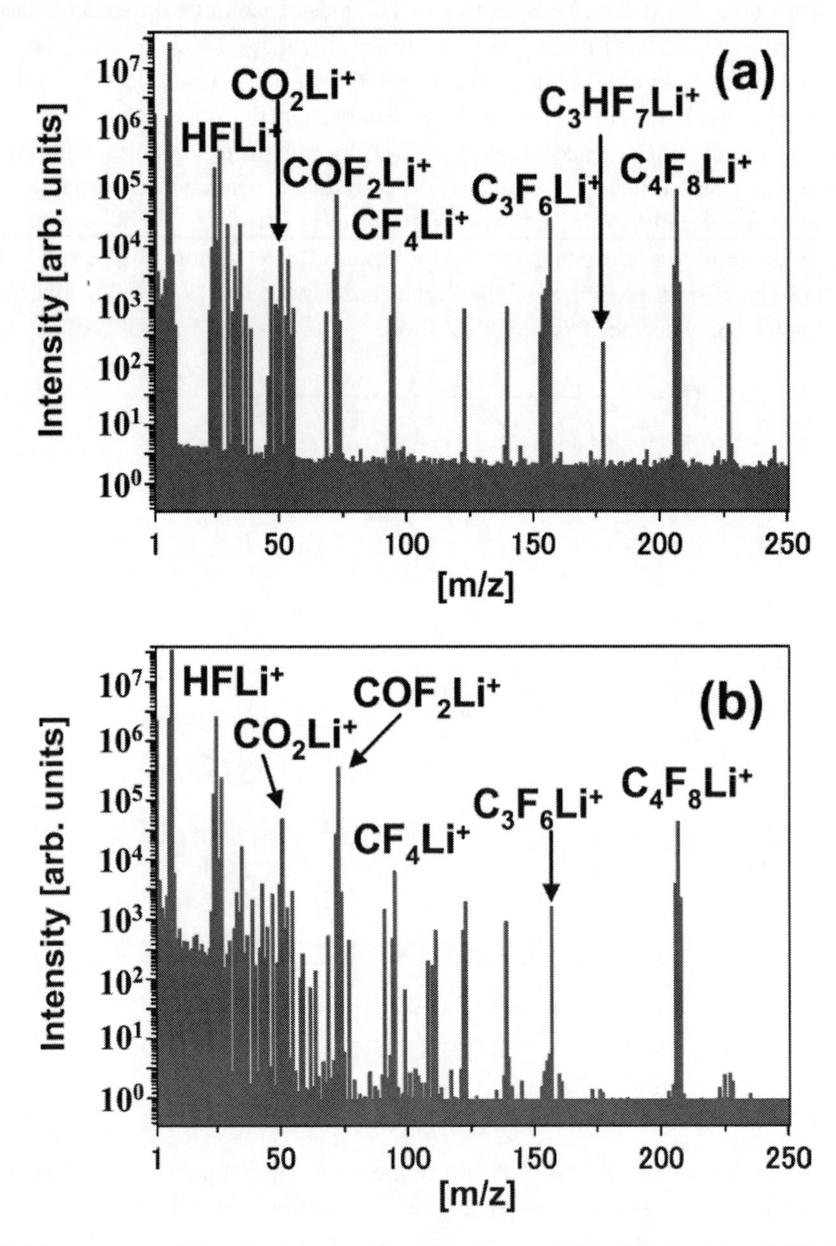

Figure 12. Mass spectra of the exhaust gases from the plasmas in the CF_4/Ar (a) and $CF_4/H_2O/Ar$ (b) gas mixtures.

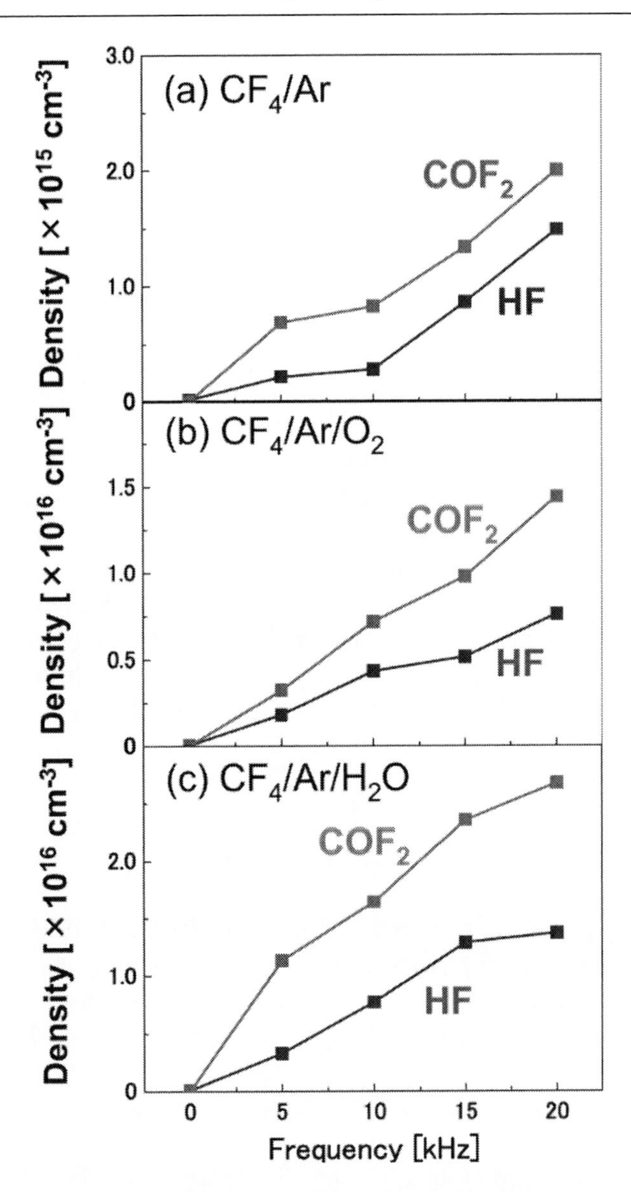

Figure 13. Densities of HF and COF_2 in the exhaust gas from the plasma in CF_4/Ar (a), $CF_4/O_2/Ar$ (b), and $CF_4/H_2O/Ar$ (c) gas mixtures.

2.4. Summary

The etching process apparatus was constructed for the application of SiO_2 etching using the atmospheric pressure non-equilibrium pulsed plasma source, which is applicable to large area treatment. The high rate etching of silicon oxide film using the atmospheric pressure non-equilibrium pulsed plasma was investigated. The products dissociated from CF_3 are generated more in the $CF_4/O_2/Ar$ gas mixture than the gas mixture without O_2. It is thought that the etching proceeded by the F radicals. The high rate etching of SiO_2 at over 8 μm/min was achieved with the gas mixture of $CF_4/H_2O/Ar$. The IAMS was used for the diagnostics of the exhaust gas from the plasma for the understanding of the etching mechanism. The result indicates that adding H_2O increased the density of HF, and polymerization of the higher order radicals on to the substrate was suppressed. If a high-rate uniform etching technique is established for a large area processing, its applications to various processing are thought to be possible.

CLOSING REMARK

The microwave-excited atmospheric pressure non-equilibrium plasma and the atmospheric pressure non-equilibrium pulsed plasma (DBD) are significantly advantageous for decreasing the cost and process time. Using these plasmas the high rate etching of SiO_2 was successfully demonstrated by adding H_2O into the fluorocarbon- or NF_3-containing process gas. In order to investigate the etching mechanism of the atmospheric pressure non-equilibrium pulsed plasma, 2-dimensional optical emission spectroscopy using the ICCD camera, gas phase diagnostics using absorption FT-IR, mass spectroscopy using the IAMS were employed. The results indicate that the etching process of silicon oxide coating using the atmospheric pressure non-equilibrium plasmas is highly influenced by the recombination of F radicals, generation and suppression of higher order molecules, and the generation of HF.

It is expected that high rate etching process using atmospheric pressure non-equilibrium plasmas will be applied in various areas furthermore.

REFERENCES

[1] Okazaki, S.; Kogoma, M.; Uehara, M.; Kimura, Y. *J. Phys. D* 1993, *26*, 889-892.

[2] Massines, F.; Rabehi, A.; Decomps, R.; Gadri, R.B.; Segur, P.; Mayoux, C. *J. Appl. Phys.* 1998, *83*, 2950-2957.

[3] Schoenbach, K.H.; El-Habachi, A.; Shi, W.; Ciocca, M. *Plasma Sources Sci. Technol.* 1997, *6*, 468-477.

[4] Kono, A.; Sugiyammma, T.; Goto, T.; Furuhashi, H.; Uccida, Y. *Jpn. J. Appl. Phys.* 2001, *40,* L238-L241.

[5] Yamakawa, K.; Hori, M.; Goto. T.; Den, T.; Katagiri, T.; Kano, H. *Appl. Phys Lett.* 2004, *85*, 549-551.

[6] Yamakawa, K.; Hori, M.; Goto. T.; Den, T.; Katagiri, T.; Kano, H. *J. Appl. Phys.* 2005, *95*, 013301-1-13301-6.

[7] Iwasaki, M.; Ito, M.; Uehara, T.; Nakamura, M.; Hori, M. *J. Appl. Phys.* 2006, *100*, 093304-1-093304-5.

[8] Ogawa, H.; Arai, M.; Yanagisawa, M.; Ichiki, T.; Horiike, Y. *Jpn. J. Appl. Phys.* 2002, *41,* 5349-5358.

[9] Saito, M.; Kataoka, Y.; Homma, T.; Nagatomo, T. *J. Electrochem. Soc.* 2000, *147,* 4630-4632.

[10] Nakamura, M.; Hino, K.; Sasaki, T.; Shiokawa, Y.; Fujii, T. *Vac. Sci. Technol. A.* 2001, *19,* 1105-1110.

[11] Fujii, T.; Arulmozhiraja, S.; Nakamura, M.; Shiokawa, Y. *Anal. Chem.,* 2001, *73,* 2937-2940.

[12] Fujii, T.; Nakamura, M. *J. Appl. Phys.* 2001, *90*, 2180-2184.

[13] Nakamura, M.; Hirano, Y; Taneda, Y.; Shiokawa, Y. *J. Vac. Soc. Jpn.*, 2005, *48*, 619-624.

[14] Nakamura, M.; Hirano, Y.; Shiokawa, Y.; Takayagi, M.; Nakata, M.; *J. Vac. Sci. Technol. A.* 2006, *24,* 385-389.

[15] Teii, K.; Hori,; M.; Goto, T.; Ishii, N. *J. Appl. Phys.* 2000, *87*, 7185-7190.

[16] Teii, K.; Hori, M; Ito, M.; Goto, T.; Ishii, N. *J. Vac. Sci. Technol. A.* 2000, *18,* 1-9.

[17] Cunge, G.; Booth, J. P. *J. Appl. Phys.* 1999, *85*, 3952-3959.

[18] Booth, J. P. *Plasma Sources Sci. Technol.* 1999, *8*, 249-257.

[19] Sasaki, K.; Takizawa. K.; Takada, N.; Kadota, K.; *Thin. Solid. Films* 2000, *374*, 249-255.

[20] Plumb, I. C.; Ryan, K. R.; *Plasma Chem. Plasma Proc.* 1986, *6*, 205-230.

In: Generation and Applications of Atmospheric Pressure Plasmas ISBN: 978-1-61209-717-6
Editors: M. Kogoma, M. Kusano and Y. Kusano

Chapter 15

ULTRA PRECISION MACHINING USING PLASMA CHEMICAL VAPORIZATION MACHINING (CVM)

Kazuya Yamamura, Yasuhisa Sano and Yuzo Mori
Graduate School of Engineering, Osaka University, Osaka, Japan

INTRODUCTION

Machining is a processing that creates a surface with a shape or properties as designed for the purpose of realizing aimed characteristics. Human being has kept developing itself by various production activities under the concept of realizing convenient and comfortable life. It is not an exaggeration to say that such development has been succeeded by the development of machining technologies. For example, it is now possible to acquire any information instantaneously and communicate with others using communication equipment such as PCs and mobile phones. This is in fact owing to appearance of semiconductor devices which require sub-μm level ultra precision fabrication technologies. In such a way, since the dawn of the history human being has always been seeking high-quality and high-precision fabrication technologies. At present in the 21st century, ultra precision optical devices such as X-ray mirrors for the synchrotron radiation and reflection mirrors for the extreme ultra violet lithography (EUVL) are required. SOI wafers for high speed low energy consumption LSI also require precision of less than 1nm of the figure errors, thickness distribution, and surface roughness [1]. It is close to the atomic size. However, as shown in Figure 1, when conventional machining such as lapping and polishing is applied, though the removal rate is high, it often suffers influence of vibration and thermal deformation because of its contact removal mechanism. In addition, it is unavoidable to introduce a significant amount of defects at the subsurface layer, considering the principle of machining utilizing plastic deformation and brittle fracture. In order to process the surface without introducing such defects, we must rely on chemical methods. The machining phenomenon of currently used low pressure dry etching is chemical process. However, its main purpose is fine patterning using a resist. It does not show high removal efficiency comparable to mechanical machining.

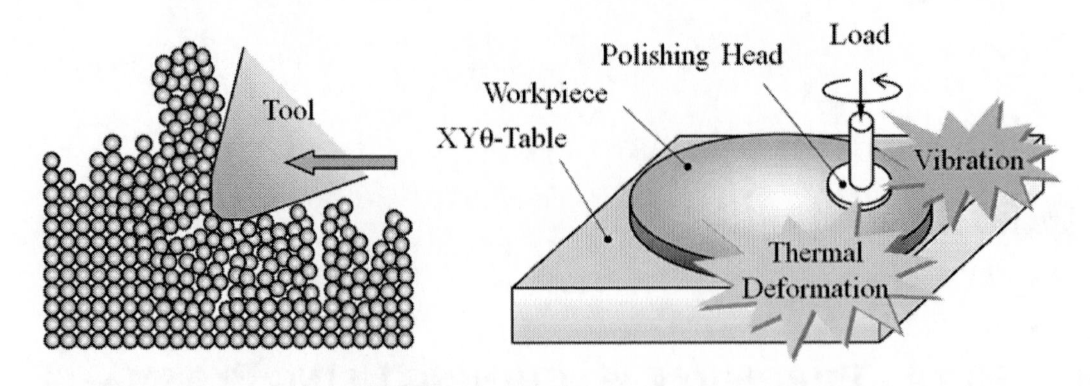

Figure 1. Principle of conventional machining and its problems.

Under these backgrounds, we developed (1998) novel figuring technique which is comparable both to mechanical removal techniques for the high spatial controllability and removal rate using an atmospheric pressure plasma and with chemical reactions. It was named plasma chemical vaporization machining (plasma CVM) [2]. This figuring technique is a method with which high density reactive species are generated in a spatially localized radio-frequency plasma, and reacted with the atoms at the processed material surface to generate volatile reaction products that are subsequently removed [3,4]. Since this method is an atomic-scale technique and does not require a contact of a machining tool, geometrically excellent figured surfaces can be obtained without influence of external disturbance. In addition, as the removal mechanism is purely chemical reactions, it can be expected that the intrinsic properties of the bulk materials will remain unchanged and that substantially excellent processed surfaces can be created in terms of crystallography.

Figure 2 shows the surface state density in a band gap existing at silicon surfaces after various removal processes evaluated by the surface photo-voltage spectroscopy (SPVS) [4]. The compared removal techniques are the mechanical polishing using colloidal silica, argon ion sputtering, and chemical etching using mixture of nitric acid and hydrofluoric acid as shown in Table 1. Single crystal silicon (p-type, CZ, $\rho = 10\ \Omega$ cm) was used as a specimen. It is shown that even polishing, which is thought to be the mildest machining method and is comparable to the final polishing of a general commercially available wafer, forms numerous defect levels. The broad peak located 0.17 eV below the bottom of the conduction band (1.1 eV) is a defect level called the A peak, and is reported to be complex defects of oxygen and vacancies [5,6]. On the other hand at the processed surface by the plasma CVM, the defect level density is at least two orders lower than that after polishing and is almost as same as the chemically etched surface by the purely chemical processing. It is also lower than that of the surface after Ar ion sputter etching. It can be concluded that in the atmospheric pressure plasma the kinetic energy of ions is very low due to short mean free path and high collision frequency, and the damage of a substrate by the ion bombardment is drastically decreased.

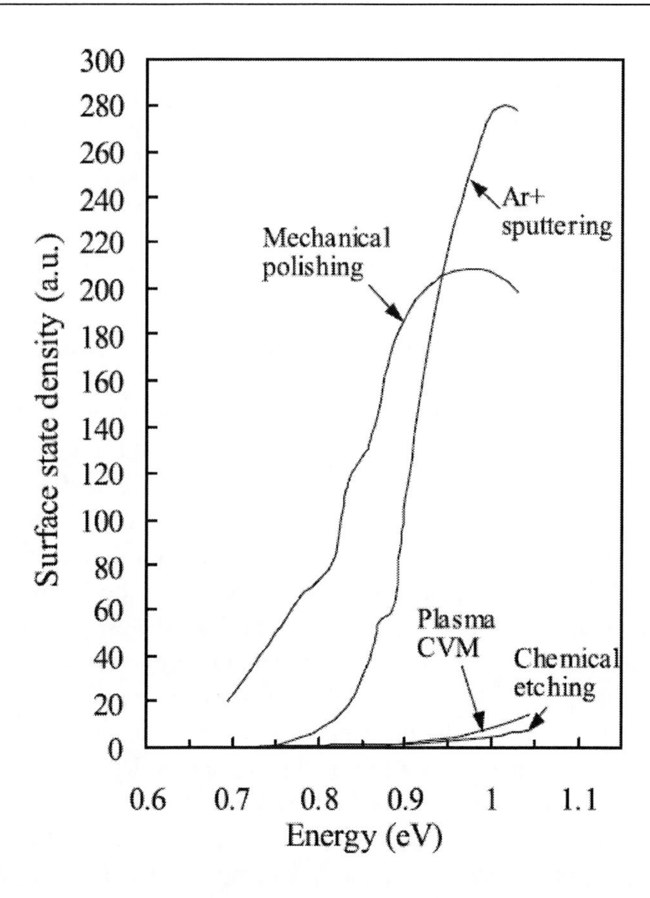

Figure 2. Surface state densities of variously processed Si (100) surface measured by SPVS.

Table 1. Process parameters.

Plasma CVM	Process gas SF_6 : 1% (He balance) RF power : 100 W (f = 150 MHz)
Mechanical polishing	Ultrafine particle: 0.1 μm, SiO_2 Polishing pressure: 150 gf/cm^2
Ar^+ sputtering	Accelerating voltage: 1 kV Ion current density: 5 μA/cm^2
Chemical etching	HF : HNO_3 : H_2O = 1 : 6 : 8

At atmospheric pressure, the mean free paths of gas molecules are as low as approximately 0.1 μm [7]. Consequently as a plasma is generated locally only around electrodes, optimal shape of electrodes enable maskless cutting and figuring by numerically controlled scanning. It can therefore be said that plasma CVM can potentially be an alternative of the conventional machining in terms of removal rate and spatial resolution.

This chapter presents numerically controlled plasma CVM using atmospheric pressure plasma applied to nm-level-precision-machining of a surface.

1. ULTRAPRECISION FABRICATION USING PLASMA CVM

1.1. Concept of Figuring by Numerically Controlled Scanning

A localized plasma can be generated in a plasma CVM. Therefore by changing the shape of the electrodes for the generation of a plasma, various machining such as cutting, planarization, and figuring can be realized as shown in Figures 3 and 4 [4]. Among them, figuring by numerically controlled scanning is performed by generating a localized plasma using a pipe-type or circular-type electrode as shown in Figure 5, following the procedure shown in Figure 6 [8]. In this process, a shape of the surface to be figured is first precisely measured in a constant temperature room, and deviation from the objective shape is obtained. Based on the principle that the amount of removal volume proportionally increases with the dwelling time of a plasma on the workpiece, the adequate scanning speed data of the worktable is calculated by deconvolution simulation in order to minimize the deviation at each removal point. This data is forwarded to the NC controller, and figure correction is performed by controlling the scanning speed of the worktable. Finally, the figured shape is measured again, and the above mentioned process is repeated until the figure error becomes below the permitted value and the aimed form accuracy is achieved. Figure 7 shows the developed numerically controlled figuring equipments.

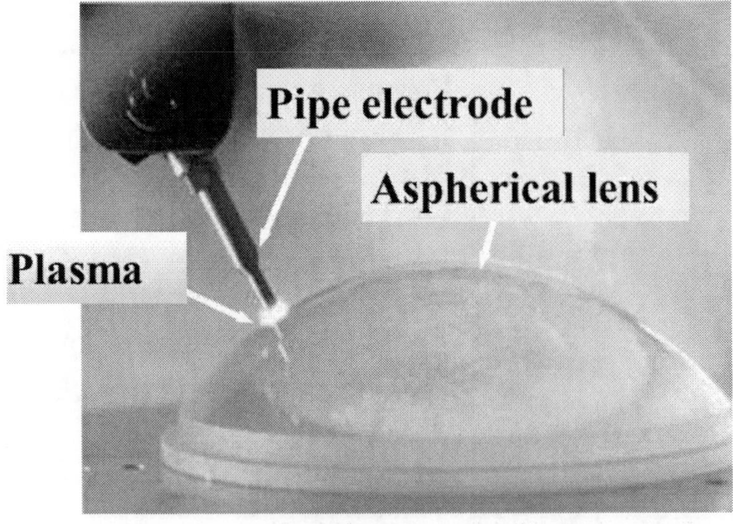

H. Takino et. al. Nikon Co., Ltd.

Figure 3. Figure correction of aspherical lens using pipe electrode.

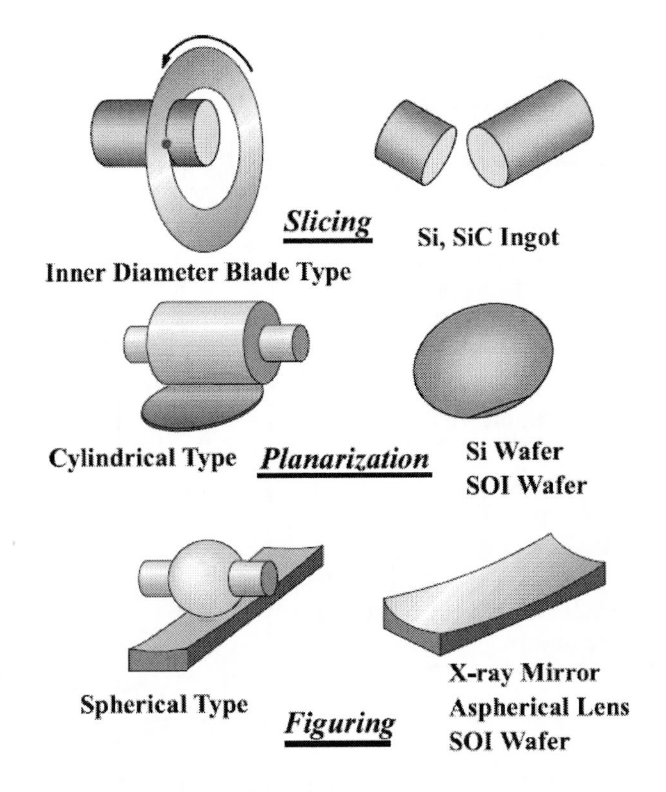

Figure 4. Application examples of rotary electrode.

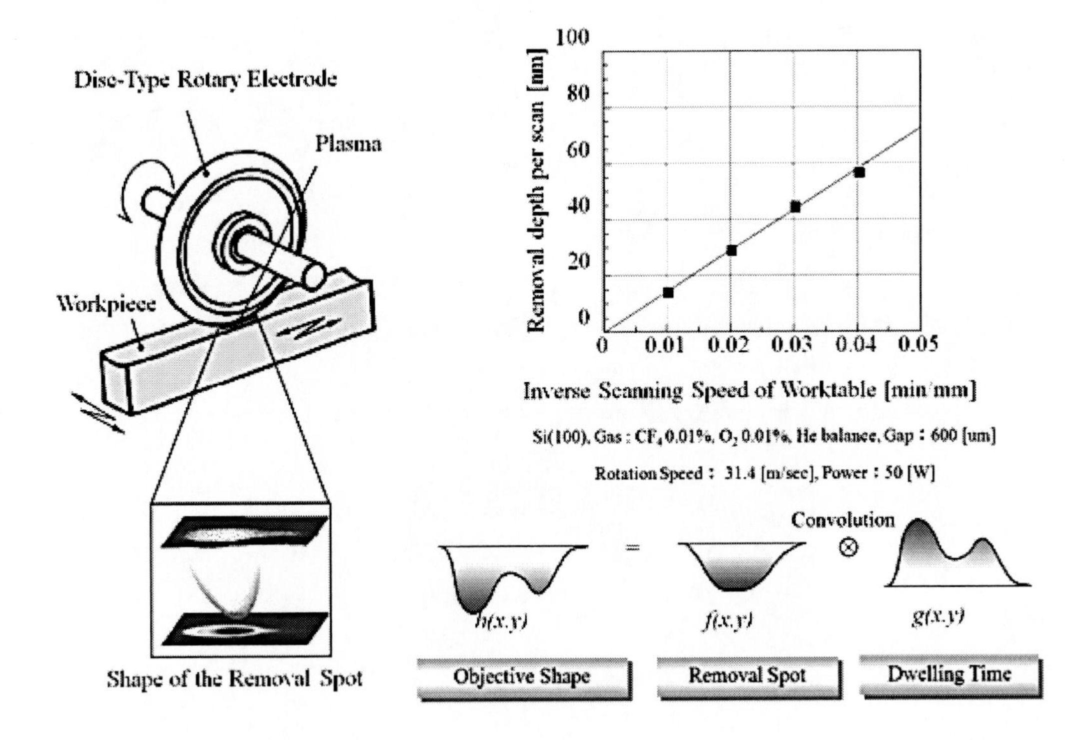

Figure 5. Principle of numerically controlled figuring in plasma CVM.

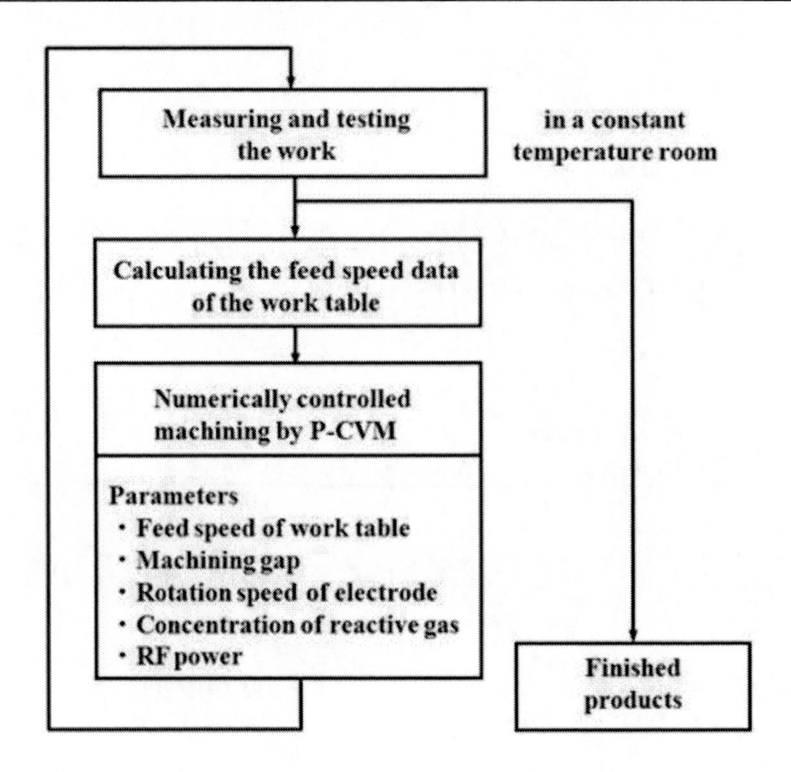

Figure 6. Procedures of numerically controlled figuring.

(a) Hydrostatic Bearing Type

(b) Linear Guide & Ball Screw Type

(c) Open-Air Type

Figure 7. Photographs of NC-PCVM apparatuses.

1.2. Fabrication of Hard X-Ray Reflection Mirror

In a beam line of synchrotron-radiation such as SPring-8, a large size flat mirror is used for eliminating high order harmonics, utilizing photon energy dependence of the total reflection and critical angle. Here finishing of this mirror was performed using the numerically controlled plasma CVM (NC-PCVM). The size of the fabricated mirror is 400 mm long, 50 mm wide and 30 mm thick. It is a single crystal silicon, and the surface orientation of the mirror is (100). Figure 7 (a) shows the machining equipment used for this figuring, which is suitable for large substrate. The region of the shape evaluation was 320 mm x 40 mm. It was measured by using a laser interferometer (GPI-XPHR, Zygo).

Figure 8 shows the result of the figure correction. The flatness of the surface finished by lapping and polishing before NC-PCVM was 158 nm (p-v). After figure correction by NC-PCVM, the flatness of 22.5 nm (p-v) was obtained [8].

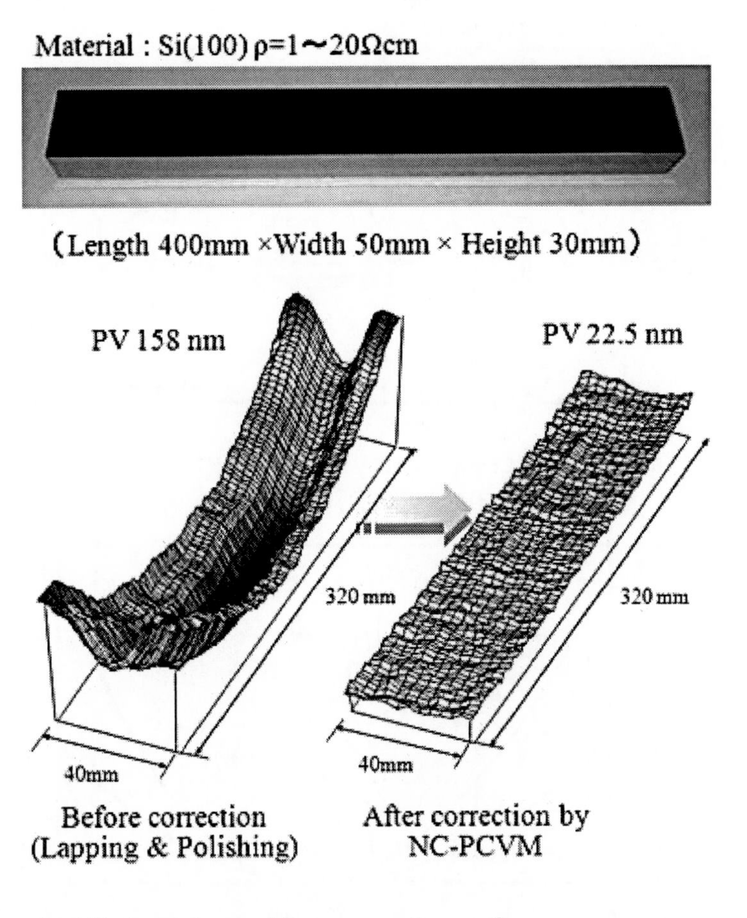

Figure 8. Photograph of Si plane mirror and figure correction result.

Fabrication of a focusing mirror of hard X-ray (E = 15 keV, λ = 0.8 Å) in the 1 km beam line of SPring-8 is presented here. The focusing system is Kirkpatrick-Baez (K-B) type and its shape is elliptical as shown in Figure 9. The mirror is made of a single crystal silicon, and the surface orientation of the mirror is (111). Its size is 100 mm long, 50 mm wide and 10 mm thick. For the fabrication of the mirror, first using a 200 mm diameter rotary electrode (the diameter of the removal footprint is approximately 20 mm), the figure error was reduced down to approximately 50 nm (p-v). Subsequently minor correction was carried out using the 1-mm-outer-diameter 0.5-mm-inner-diameter pipe-type electrode (the diameter of the removal footprint is approximately 2 mm). Figure 10 (a) shows the figure error of the fabricated elliptical mirror. It is seen that the figure error of less than 3 nm (p-v) over the effective length of 80 mm. Figure 10 (b) shows the power spectral densities (PSDs) of the figure error in the steps of before correction after the correction using the rotary electrode and the finishing using the pipe electrode. The figure correction of the spatial wavelength was approximately 10 mm and 1 mm after correction with the rotary and pipe electrodes, respectively. The lowest correctable spatial wavelength corresponds to approximately half of the size of the removal footprint. Figure 11 shows the focused beam profiles using the fabricated elliptical mirrors. The focal widths of 200 nm and 120 nm were obtained for the mirrors with the focal lengths of 300 mm and 150 mm, respectively. This indicates that diffraction-limited-focusing is realized [9].

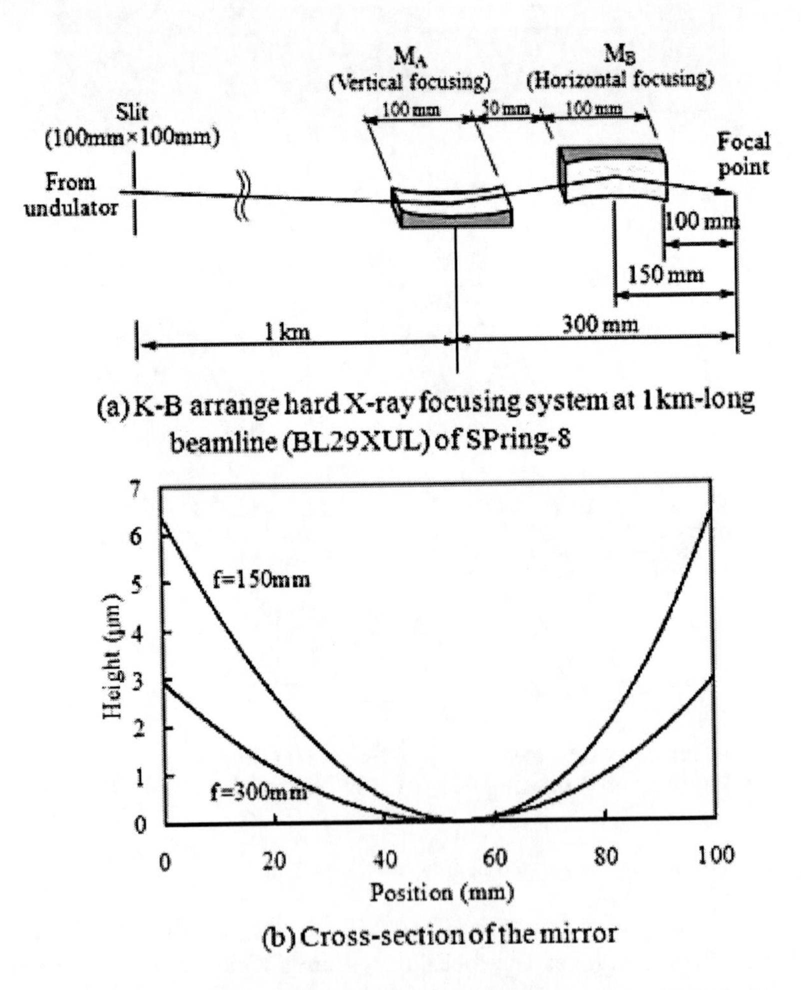

Figure 9. K-B arranged X-ray focusing optical system and shapes of plano-elliptical mirrors.

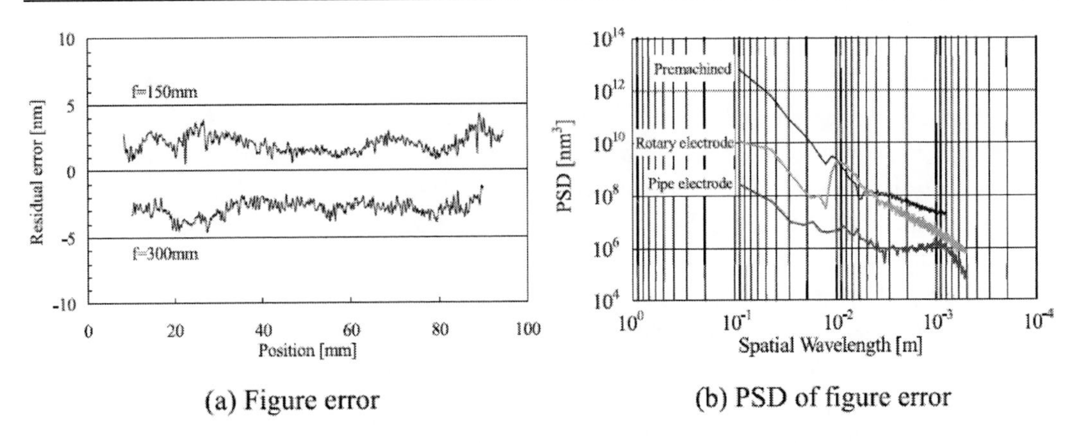

Figure 10. Figure errors and PSDs of fabricated plano-elliptical mirrors.

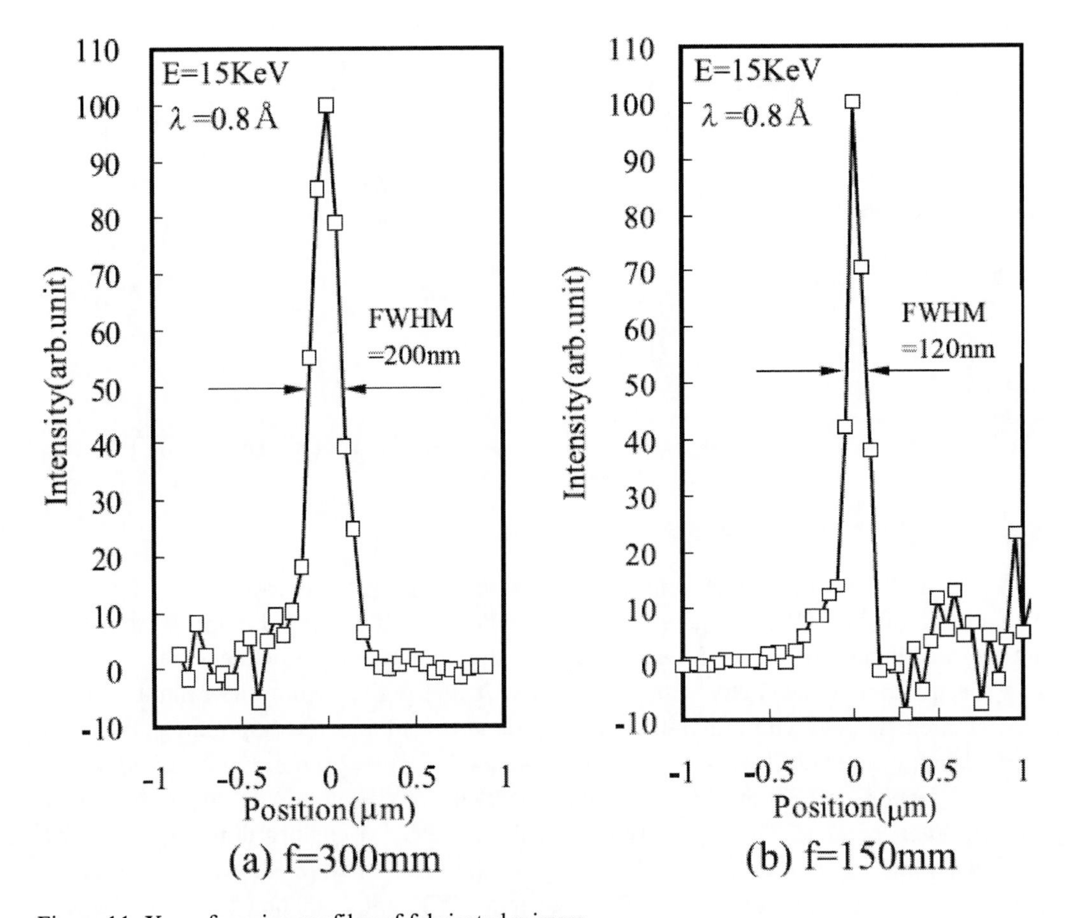

Figure 11. X-ray focusing profiles of fabricated mirrors.

1.3. Uniformity and Reduction of SOI Wafer Thickness

A silicon-on-Insulator (SOI) wafer is expected to be a substrate for high-speed and low-power-consumption semiconductor integrated circuit. It consists of a buried oxide (BOX) layer on the silicon substrate covered with a thin single crystal silicon (SOI) layer. A metal oxide semiconductor field effect transistor (MOSFET) is constructed on this SOI layer as shown in Figure 12 (b). Such a structure enables reduction of a parasitic capacitance. As the device can also run with a small amount of charges, the high-speed low-power-consumption device can be possible, and high performance MPU and low-power-consumption LSI for watches are already applied. The International Technology Roadmap for Semiconductor (ITRS) anticipates that ultra thin SOI wafers with the initial thickness of the SOI layer of 15-nm level will be required in about 2010 with miniaturizing transistors. In order to suppress the variability of operating properties of the transistors on the wafers, the high uniformity of the SOI layer is required as well. Since it is very difficult to fabricate such an ultra thin SOI wafer using conventional techniques, uniformity and reduction of SOI thickness is investigated by using the NC-PCVM [10]. A device was also fabricated on the thinned SOI wafer, and its operating properties were evaluated [11].

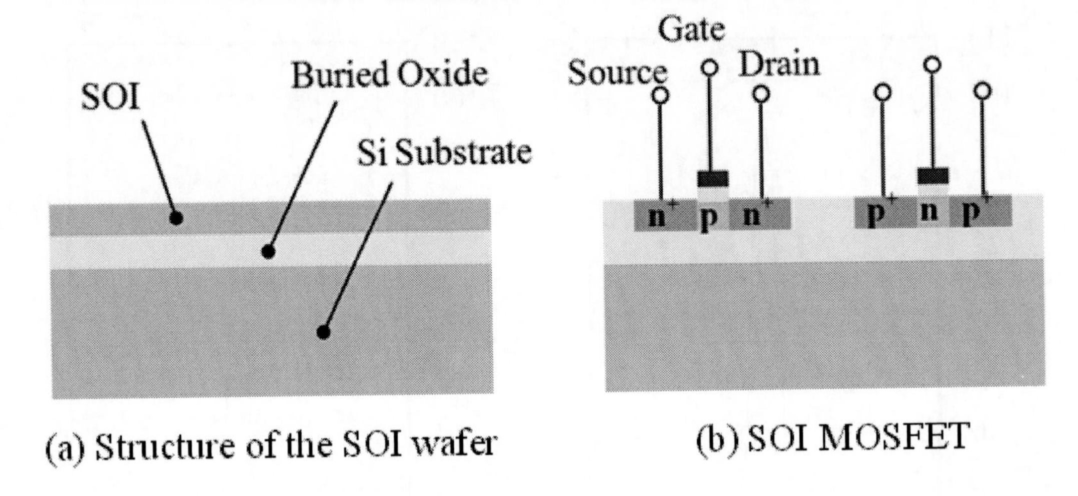

Figure 12. SOI wafer and SOI MOSFET.

A 6 inch UNIBOND® wafer with a 200 nm thick SOI layer was used as the SOI wafer. The thinning of the SOI layer down to 10 nm order was attempted. Figure 13 shows the thickness distribution of the SOI before and after the thinning. The thickness was measured using spectroscopic elipsometry with 1 mm pitch. At the region within the central 120 mm diameter circle the SOI with initial thickness of 200 nm was thinned to approximately 13 nm. The deviation of the SOI thickness improved from ±4.2 nm to ±2.0 nm. Figure 14 shows the AA cross sections of Figure 13. It is confirmed that deviation of the SOI thickness is improved after thinning. The result shows that the NC-PCVM enables thinning of the SOI film down to the 10 nm order. AFM observation of thinned SOI wafer shows that the surface roughness was 1.45 nm (p-v), 0.12 nm R_a in the area of 500 nm x 500 nm. It is comparable to the surface roughness of a commercially available silicon wafer. It is therefore indicated that the surface roughness is not degraded after the NC-PCVM.

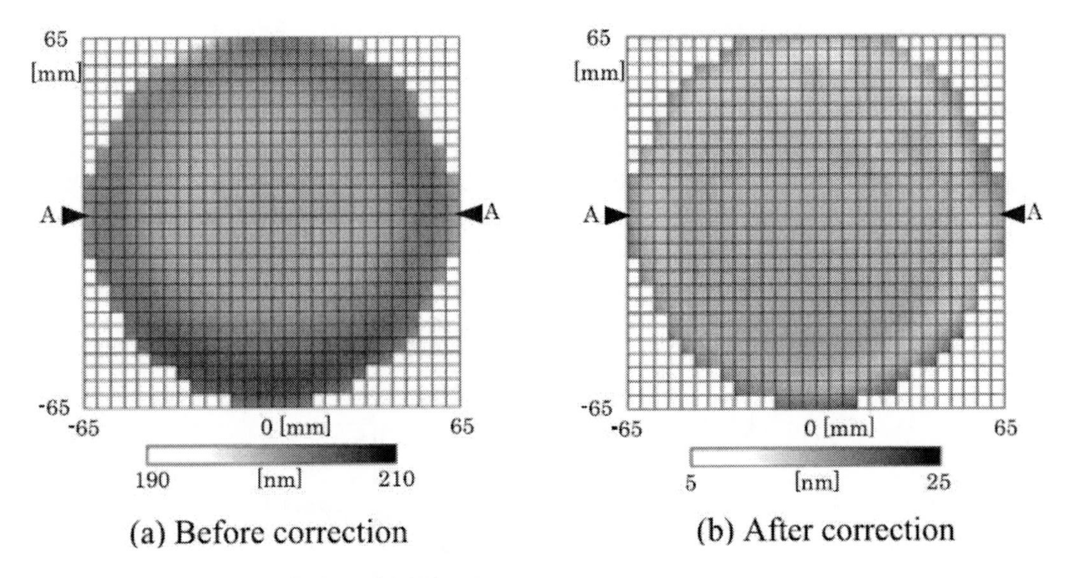

Figure 13. Thickness distribution of SOI layer.

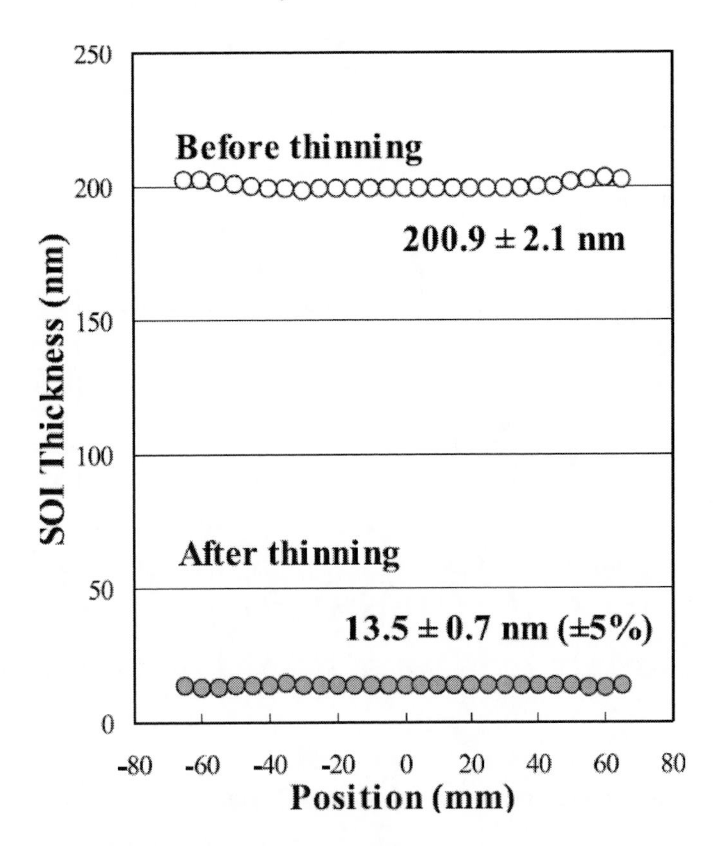

Figure 14. Thickness distribution at the AA cross section.

Devices such as MOSFET and resistances were fabricated on all over the 8 inch diameter SOI wafer thinned to approximately 60 nm and a reference SOI wafer using the NC-PCVM and that for reference, and then they were compared. Here, the reference SOI wafer was fabricated by thinning the SOI layer by thermal oxidation and etching of the oxide layer and adjusting the averaged thickness to be approximately 60 nm. Figure 15 shows an example of the drain current as a function of the gate voltage for the fabricated MOSFET. Good operating properties was demonstrated for both cases, showing no difference in the leak current and steepness of the slopes when the current significantly increases as the voltage increases (sub-threshold property: S = 65 mv/dec). The areal distribution of the resistivity of the resistances was measured, showing that the SOI wafer thinned by the PCVM has lower deviation of the resistivity. It is because the SOI wafer thinned by the PCVM shows better uniformity of the SOI layer thickness. In conclusion, it is shown that PCVM technique does not degrade the crystalinity and cleanliness of the SOI wafer and is applicable to a substrate for the semiconductor integrated circuit.

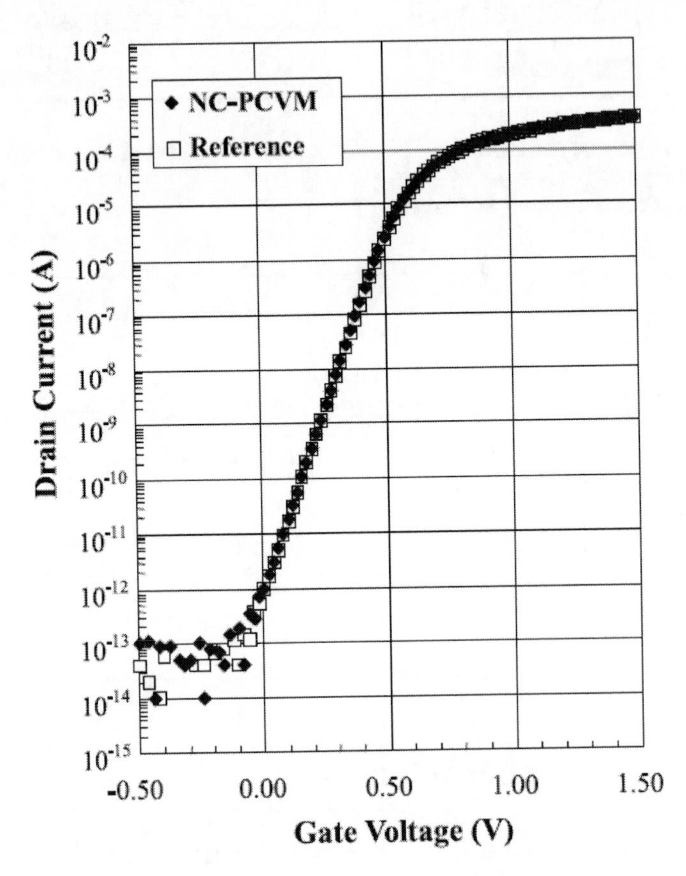

Figure 15. Gate voltage – drain current characteristics of fabricated nMOSFET (gate width/gate length = 10/0.35 μm, drain voltage = 0.1 V).

1.4. Uniformity of the Quartz Crystal Wafer Thickness

Quartz resonators play an important role in the development of optical communications technology. Higher frequency oscillation is required for high speed electronic communications [12, 13]. A quartz resonator consists of a quartz crystal plate sandwiched by space gap electrodes, or metallic coating electrodes deposited by evaporation. Depending on applications, various cutting methods are proposed. Among them the most frequently used method is the AT-cut showing desirable properties of frequency-temperature characteristics at around room temperature. When an AC voltage is applied to the electrodes attached to the quartz crystal plate, an elastic oscillation is induced just near the electrodes due to the piezo-electric converse effect. This is strongly reflected at the both sides of the quartz crystal plate, and thus elastic standing wave is subsequently induced, satisfying the boundary condition for the thickness *t*. In the case of the AT-cut, the transversal plane wave vertically directing to the quartz crystal plate surface is used as the principal oscillation. This oscillation mode is called “oscillation of thickness shear modes”. Under the assumption that the quartz crystal is an infinite plate so that the contour of the quartz crystal plate is much larger than the thickness *t*, the resonant frequency of oscillation of thickness shear modes *f* can be expressed as a function of *t* only as shown in Equation (1), indicating that the plate must be thin for higher frequency operation.

$$f \approx 1670 / t \tag{1}$$

Here, *f* and *t* are the resonant frequency [MHz] and the thickness of the quartz crystal wafer [μm], respectively.

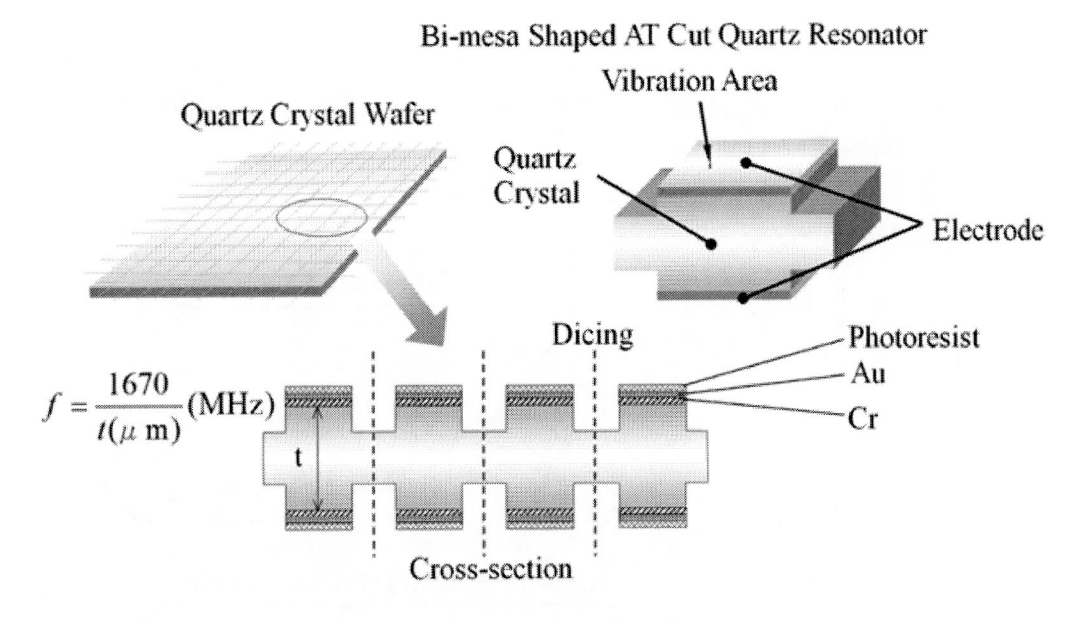

Figure 16. Fabrication process of mesa type quartz crystal resonator.

Figure 16 shows the manufacturing process of mesa type quartz crystal resonator which is assisted by photo-lithographic process. In order to improve the productivity, many resonators are fabricated at a large quartz crystal wafer, and then they are separated to the individual resonators by the dicing, as is the same as the semiconductor device manufacturing process. However, the conventional thinning process of quartz crystal wafers by both-side-mechanical-lapping can damage carrier or wafer itself, or cause thickness unevenness due to the unevenness of pressure during polishing. The uneven thickness causes variability of the resonant frequency of resonators at the quartz crystal wafer. In order to adjust the frequency, additional wet etching process is necessary so that the thickness of the quartz crystal resonators is adjusted. Up to now, in the case of the resonators whose frequency is more than 100 MHz, permitted thickness of the individual quartz crystal resonators after the wet etching is generally ± 10 nm. Therefore, it is expected that the productivity of the quartz resonator will be significantly improved by achieving the permitted thickness distribution at the stage of the quartz crystal wafer, and thus eliminating thickness unevenness of quartz crystal wafer is required. We propose a two step correction process consisting of 1-dimensional numerically controlled scanning using a cylindrical rotary electrode and 2-dimensional numerically controlled scanning using a pipe electrode, which can eliminate unevenness of thickness of the quartz crystal wafer and reduce the thickness down to below the limit by the mechanical machining [14]. This process combines coarse correction using the cylindrical rotary electrode and fine correction using the pipe electrode as shown in Figure 17. This process enables high productivity of the fabrication and high preciseness simultaneously. In case when the volume of the necessary removal is small, only the fine correction using the pipe electrode will be applied.

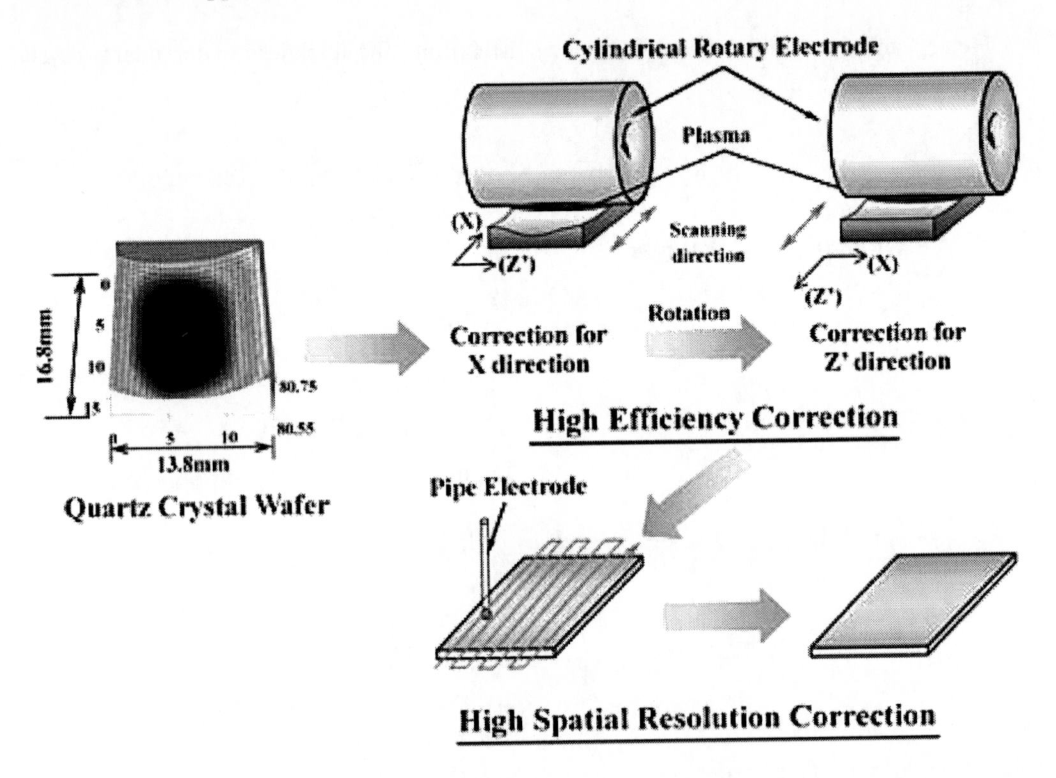

Figure 17. 2-step thickness correction process.

The result of unifying thickness distribution of the AT-cut quartz crystal wafer (25 mm x 20 mm x 80 μm^t) whose both sides were polished in advance is also presented [15]. The removal amount is set at approximately 250 nm for the coarse correction using the cylindrical rotary electrode, and at approximately 40 nm for the fine correction using the pipe electrode.

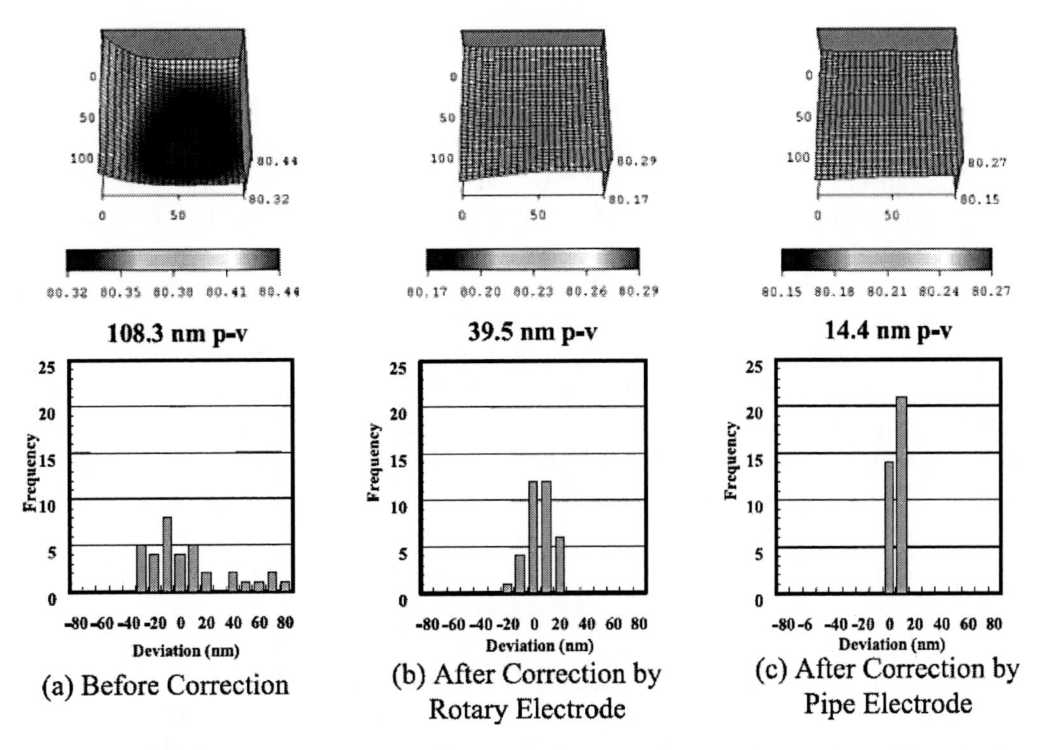

Figure 18. Improvement of thickness uniformity of quartz crystal wafer by numerically controlled plasma CVM.

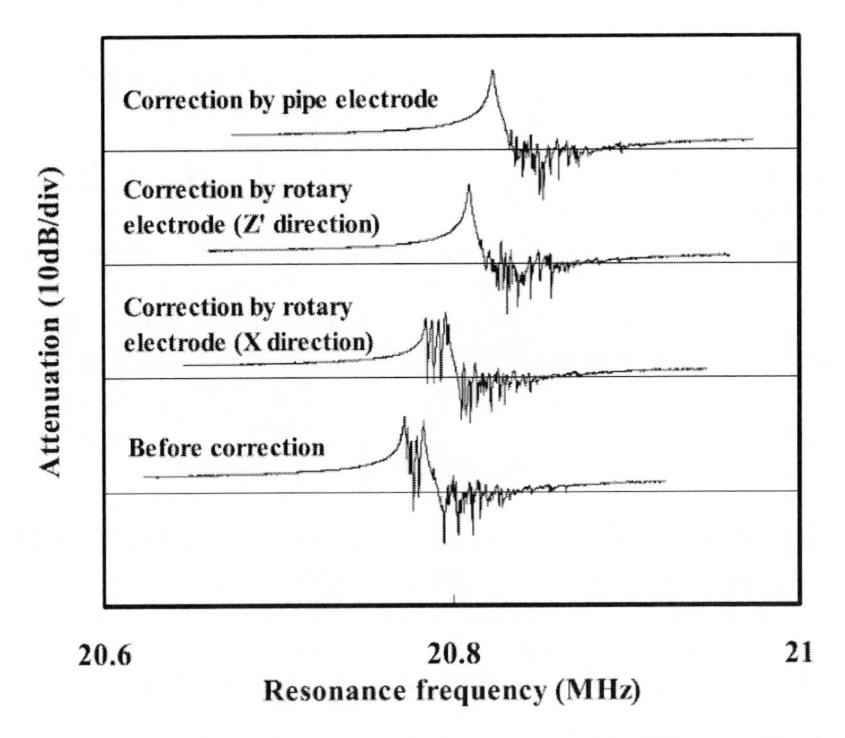

Figure 19. Improvement of resonant property by the improvement of the thickness uniformity.

Figure 18 shows the thickness distribution of the quartz crystal wafer before and after the correction using the cylindrical rotary electrode and subsequent pipe electrode for the evaluation area of 14.4 mm x 9.2 mm. The maximum value of thickness unevenness (p-v value) was 108.3 nm (frequency deviation 28.0 kHz) before correction. After the correction using the cylindrical rotary electrode and the pipe electrode, p-v values became 39.5 nm (frequency deviation 10.2 kHz) and 14.4 nm (frequency deviation 3.8 kHz), respectively. It is indicated that thickness evenness is significantly improved by the NC-PCVM correction. As the thickness correction proceeds, the intensities of spurious peaks decay, and finally desirable resonant properties are obtained as shown in Figure 19. This resonance curve is due to the oscillation of thickness shear modes excited by the 3 mm diameter probe electrode. It is thought that unwanted spurious peaks are suppressed after the thickness-evenness, namely improved parallelism in the region of facing a probe electrode. Furthermore, as generation of new spurious peaks is not seen, it is indicated that the subsurface damage layer [16] is not formed by the ion bombardment in a plasma.

REMARKS

When an ultra-precise optical device such as an X-ray mirror and aspherical mirror for stepper, and substrate for ultra high performance devices such as SOI wafers and quartz crystal wafers are fabricated, the current fabricating process based on the mechanical machining is approaching to a technical limit in terms of aiming further preciseness and higher qualities. The mechanical machining is advantageous in the high fabrication efficiency, but dislocation associated with the removal phenomena of deformation/fracture and crystal defects such as cracks can spatially occupy significantly comparing with the atomic level. In fact its size physically limits the decrease of the unit size of deformation and removal. Furthermore at the machined surface, a subsurface damage layer remains, consisting of the defects of dislocation and vacancies. Therefore in order to improve the preciseness of the fabrication process, development of fabrication process using chemical reactions, which is ideal as a fabrication phenomenon, is urgently required, instead of using such a mechanical machining process. The plasma CVM utilizing atmospheric pressure plasma is proposed as a completely new-concept chemical fabrication method which can replace the mechanical machining in order to meet the demand. This chapter has shown its examples.

REFERENCES

[1] Sweeney, D.W.; Hudyma, R.; Chapman, H.N.; Shafer, D. *Proc. SPIE* 1998, *3331*, 2-10.

[2] Mori, Y.; Yamauchi, K. *Jpn. Patent* 1996, No. 2521127.

[3] Mori, Y.; Yamamura, K.; Yamauchi, K.; Yoshii, K.; Kataoka, T.; Endo, K.; Inagaki, K.; Kakiuchi, H. *Nanotechnol.* 1993, *4*, 225-229.

[4] Mori, Y.; Yamauchi, K.; Yamamura, K.; Sano, Y. *Rev. Sci. Instrum.* 200, *71*, 4627-4632.

[5] Watkins, G.D.; Corbett, J.W. *Phys. Rev.* 1961, *121*, 1001-1014.

[6] Corbett, J.W.; Watkins, G.D. *Phys. Rev.* 1961, *121*, 1015-1022.

[7] Vincenti, W.G.; Kruger Jr., C.H. *Introduction to Physical Gas Dynamics*; John Wiley & Sons; New York, US, 1965.

[8] Mori, Y.; Yamamura, K.; Sano, Y. *Rev. Sci. Instrum.* 2000, *71*, 4620-4626.

[9] Yamamura, K.; Yamauchi, K.; Mimura, H.; Sano, Y.; Saito, A.; Endo, K.; Souvorov, A.; Yabashi, M.; Tamasaku, K.; Ishikawa, T.; Mori, Y. *Rev. Sci. Instrum.* 2003, *74*, 4549-4553.

[10] Mori, Y.; Yamamura, K.; Sano, Y. *Rev. Sci. Instrum.* 2004, *75*, 942-946.

[11] Mori, Y.; Sano, Y.; Yamamura, K.; Morita, S.; Ohshima, I.; Saito, Y.; Sugawa, N.; Ohmi, T. *Seimitsukougakkaishi (Jpn. Soc. Precision Eng.)*, 2003, *69(5)*, 721-715.

[12] In *Suishoudebaisu No Kaisetsu To Ouyou (Quartz crystal devices and their applications)*; Quartz Crystal Industry Association of Japan; Ed.; Quartz Crystal Industry Association of Japan: Tokyo, JP, 2002; pp 50.

[13] Sato, Y.; Hosokawa, Y.; Nishida, K.; Ogawa, M. *Proc. IEEJ Information Sys.* 2001, *2*, 215.

[14] Shibahara, M.; Yamamura, K.; Sano, Y.; Sugiyama, T.; Endo, K.; Mori, Y. *Rev. Sci. Instrum.* 2005, *76*, 096103 1-4.

[15] Shibahara, M.; Yamamura, K.; Sano, Y.; Sugiyama, T.; Yamamoto, Y.; Endo, K., Mori, Y. *Seimitsukougakkaishi (Jpn. Soc. Precision Eng.)*, 2006, *72(7)*, 934-938.

[16] Nagaura, Y.; Yokomizo, S. In *Proc IEEE Int. Freq. Control Symp.* 1999, *53(1)*, 425-428.

In: Generation and Applications of Atmospheric Pressure Plasmas ISBN: 978-1-61209-717-6
Editors: M. Kogoma, M. Kusano and Y. Kusano

Chapter 16

DESIGN OF PARTICLES USING ATMOSPHERIC PRESSURE PLASMA

Atsushi Takeda
ISI Limited, Saitama, Japan

INTRODUCTION

Fine ceramic particles were significantly developed in 1980s. This development enabled the mold casting process of a final product by mainly sintering. On the other hand, representative examples of particles utilizing finer level structure are powder catalysts, cosmetic materials, biomedical columns, and submicron particles which are dispersed in polymers or used as resin compounds of electronic devices.

In 1990s, development of submicron-size particles in a more strict sense is required, and many particle production processes were proposed. In particular it became difficult to produce submicron-sized particles using sintering and grinding methods which used to be the main process. On the other hand, the sol-gel method has become the major process for submicron-size particle production. However, without solving various negative effects due to the segregation of produced particles it is not still completely accomplished that who would design surface/interface for what purposes, how many amount, with what to be compounded, and dispersing all particles while achieving required quality conditions to be added to the final products.

At the same period, various developments of super fine particles such as metals, high purity oxides, ion doped oxides, and nitrides using inductively coupled thermal plasma (ICP thermal plasma) were initiated at private companies and the National Research Institute of Metals [1] in order to solve the most important problem of dispersion among these problems, and in order to add a hybrid type property to one particle. There were two types; one of them is production of particles by the reactions of metals or metal-organic compounds in a thermal plasma at several thousands K, the other is vacuum evaporation of oxides or metals, and spherical particles. Recently also outside of Japan, many companies and universities carry out various developments of particles targeting next generation nano-technology, and thus it can not be said that the research in this area in Japan is advanced in all respects [2]. In particular, collaborations of research and development in nano-bio, and nano-medical areas in the US and Europe are significant. Furthermore, the economical prospect in nano-technology in the US shows the estimate of 1 trillion dollar by 2010-2015, among which bio-industry dominates as high as 18 %.

These marketing trend and materials design are closely related with each other, and materials development contributing to protection of environment and resources is being required. In this respect, the following becomes the most important themes of development in this area; mass production and degree of freedom of materials design which are difficult to be accomplished by using the conventional plasma processing, research and development of the interface of particles for compositing, surface modification techniques, and mono separating dispersion (it is called here mono-dispersion) of materials and particles.

This chapter discusses the roles of atmospheric pressure plasmas and microwave chilled plasmas and prospective materials design as the particle synthesis and surface design systems which have potentials to overcome these themes.

Here 'powder' is used form the assembly of particles in order to clarify the terminology, while 'particle' is used when design of one particle is performed.

1. Early Stage Development of Functional Particles Utilizing Plasma

The attempt was initiated in the mid of 1990s for the above mentioned objectives. Its content is described below.

1.1. Synthesis of Multifunctional Thin Films for Bio-Compatible Powdered Pigments

Initial contact between the atmospheric pressure plasma developed by Okazaki, Kogoma *et al.* at the faculty of Science and Technology, Sophia University and functional particles was the anti- metal allergy colouring matter being developed by the author and Pias Corp. The first targets of the development were three kinds of iron oxides of red hematite (colcothar), yellow goethite (iron hydroxide) and black magnetite, and there was perfect dispersion in protection of elution of heavy metals such as Ni, Cr, and Co included in these particles and in cosmetics [3].

These oxides are apparently stable in the viewpoint of Mineralogy. However, they have crystal modification temperatures between 100 and 200 ^{o}C, and it is necessary to control the plasma environment at the temperature at least no more than 100 ^{o}C in order to deposit elution-protected coatings of heavy metals which are allergens and perfect dispersive coatings. Whenever this basic understanding of Mineralogy is neglected, it turned out that crystal modification inevitably occurs.

Furthermore, it was found that the CVD method can not synthesize perfect elution-protected coatings of previously mentioned metallic ions which are required by medical doctors. At that time, hematite euhedral mono-crystalline nano-particles were already successfully synthesized by the previously mentioned ICP thermal plasma. However, in terms of its cost and productivity, it was impossible to be used as the source material for cosmetics.

On the other hand, in order to form amorphous silica nano coating on the three kinds of iron oxides, TEOS was adsorbed at the first process, preliminary hydrolysis was acted at the coating, and a perfect anti-allergic iron oxide super dispersive particle was successfully produced by achieving oxidation of TEOS semi-dissociated coating by rotating it in atmospheric pressure plasma.

Furthermore synthesis of the similar silica coatings on six kinds of colouring matters was successfully realized using atmospheric pressure plasma, which was requested by medical doctors. The targeted organic colouring matters were organic pigments, organic dyes and aluminium lake pigments (the compounds which hold dyes in η- or γ- alumina). The acceptable plasma environment was at the ambient temperature of no more than 80 ^{o}C and the oxidative environment without cutting the main chain of colouring matter molecules. This condition was brought under control by the TEOS adsorption method in 1997.

1.2. Improvement of Magnetic Property of Magnetic Powders and Anti-Rust

The technique to coat each ferromagnetic particle with high resistance thin film in order to protect the ferromagnetic material against dielectric loss was developed by the author and Kogoma at Sophia University in the late 1990s. Before that, silica nano coatings have been exclusively developed. Utilizing this experience, synthesis of other oxide nano coatings was studied, and design and synthesis of

[1] zirconia,
[2] zirconia/silica two-layers,
[3] zirconica/silica/zirconia three layers,
[4] silica/titania/silica three layers

etc. were carried out and characteristics of each case were examined [4].

1.3. Formation of Barrier Coatings against Ions and UV Light on Phosphor Particles Used for Display Devices

In the development of flat panel display (FPD) devices, for the development of plasma TV, detailed study was performed for the synthesis of protective silica coatings as the most important objective for the degradation of PDP phosphor by UV and plasma ion bombardment, and oxidative modification of Eu^{2+} in the blue phosphor. The phosphor provided at that time is still used, which has been manufactured by a classical production method of particles which is called sintering/milling/screening. The diameters of these phosphor particles are approximately more than several tens times larger than the effective depth of photoemission from the individual particle by UV light. In order to overcome degradations of phosphor by photons and ion bombardment, amorphous silica coatings were synthesized by atmospheric pressure plasma. However, as mentioned previously, the nano-thin film effect was not sufficiently confirmed due to too large size of PDP particles. Recently, synthesis of nano-sized PDP phosphor is proposed, and its development is ongoing. The applicability of atmospheric pressure plasma for the synthesis of nano-phosphor can be highly expected. For the green phosphor, TEOS coating was synthesized on ZnO sub-micron particle surface, and immediate green photo-emission was observed in He/O_2 atmospheric pressure glow discharge. It is indicated that ZnO and SiO_2 inter-reacted, and that the main structure of $ZnSiO_3$ green phosphor was synthesized. As the current sizes are between 2 and 10 μm, the developed technique enables the size reduction of 1/20 – 1/100. Furthermore, development of preventing hydrolysis of dopants in blue PDP, and ion injection technique as well as improvement of UV emission efficiency is the largest theme.

2. Study, Synthesis, and the Necessity of Particles Using Atmospheric Pressure Plasma

2.1. Particle Design, Manufacturing Technique and Their Problems

Before developing powders utilizing atmospheric pressure plasma, the techniques such as the sol-gel method, RF plasma, DC arc discharge plasma, arc discharge in water, gas phase synthesis, and sintering and milling have been used. A representative example of mass-production is cement production. Its production of several hundred thousand ton a month became possible by converting the initial long-rotary kiln method to the suspension pre-heater method (SP/SF) in 1970s.

As the mid-size production, the sintering and milling method which enables production of no less than several tons a month is used. Though this method apparently looks old, it has been used for production of electronic materials like PDP phosphor etc. However, its drawback is that various transformations of particles, water absorption, segregation, fluctuation of granularity occur during micro-milling. Another production process is necessary for their control. It is very difficult for the existing milling technology to adjust the granularity in the range of sub-microns by the evaluation after dispersion.

There are many methods which belong to fairly small scale production. For example, oxygen combustion vapour phase synthesis oxidizes metal powder immediately by introducing it into the high temperature flame. Although the shape of this particle is substantially sphere, the granularity distribution is strongly influenced by the thermal space of the manufacturing system. A very thin and short furnace is necessary for manufacturing submicron or nano-level particles using this process, and the productivity will be significantly reduced.

Metal compounds undergo hydrolysis or chemically reacted in a liquid by neutralization with the sol-gel method, and submicron level particle can be produced. However, the most serious concern is significant segregation. For example, even if the size of first stage particles is between 10-30 nm, it is normal that they are coalescence more than 1 μm up to 10 μm. Furthermore, in order to dope ions and produce multilayered particles, it is necessary to ensure the resources of source materials, while they do not necessarily exist sufficiently. In addition, if inorganic particle is used as the first source material, its residual anion has a possibility of giving a fatal defect at the crystalline. In order to avoid this it is often the case that metal-organic compounds must be used.

2.2. Design of Particle Material Requiring Plasma Processing

What kinds of particles are synthesized only by plasma processing? This simple question is the most basic core issue that has been considered by ourselves who have developed silica and ion doped titanium oxides using a thermal plasma since late 1980s.

As was already described, several synthesis methods for submicron particles are already available. The production for these methods is typically inexpensive. These techniques often obstacle appropriate evaluation of plasma method in terms of the relation between particle manufacturing system using plasma and properties of the particles, and it is economically impossible that the realization of the particle manufacturing using plasma processing is established without considering this issue both now and in the future.

Therefore, in this chapter, the term of 'design' only indicates the design of particles which can be useful now and/or in the future, even if the production cost is expensive.

2.3. Two Flows of Materials Design

There are two methods of materials design with using particles in view of manufacturing final products. They are synthesis of nano-crystalline structures and formation of functional coatings on a surface.

Goal (manufacturing composite products) ← tolerance for influence of the change during molding by heat or polymerization ← solution of various segregations and particle growth ← choice between (1) and (2) ← matching to prescription of compounding and processing methods ← (1) development of new nano-structured particles/ (2) formation of surface nano-coatings

Here, (1) the new nano-structured particle means the particle that possesses optimum surface conditions and inner/outer structures of crystalline so that it can TANBUNSAN intrinsically without using special dispersion assistant or auxiliary particles and reach the goal with using one kind of particle. On the other hand, (2) surface nano-coating forms on each particle individually without accompanying second particle growth such as coalescence and aggregation. In addition, if necessary, more than one layer and more than one phase mineral phase or polymer in one layer is also involved.

3. Particle Design Model for Applying Atmospheric Pressure Plasma

It will certainly be an important theme in the future whether atmospheric pressure plasma would enable a novel identical crystalline model. Here, basic particle model is discussed in view of crystallography. If the design of such a model is an objective, what kinds of control and measurement of plasma would be required? This is the point which individual engineers working in the development and production must solve and devise by themselves. Figure 1 shows relation of atmospheric pressure plasma to the market needs of high functional added value for this present purpose based on a lot of literature.

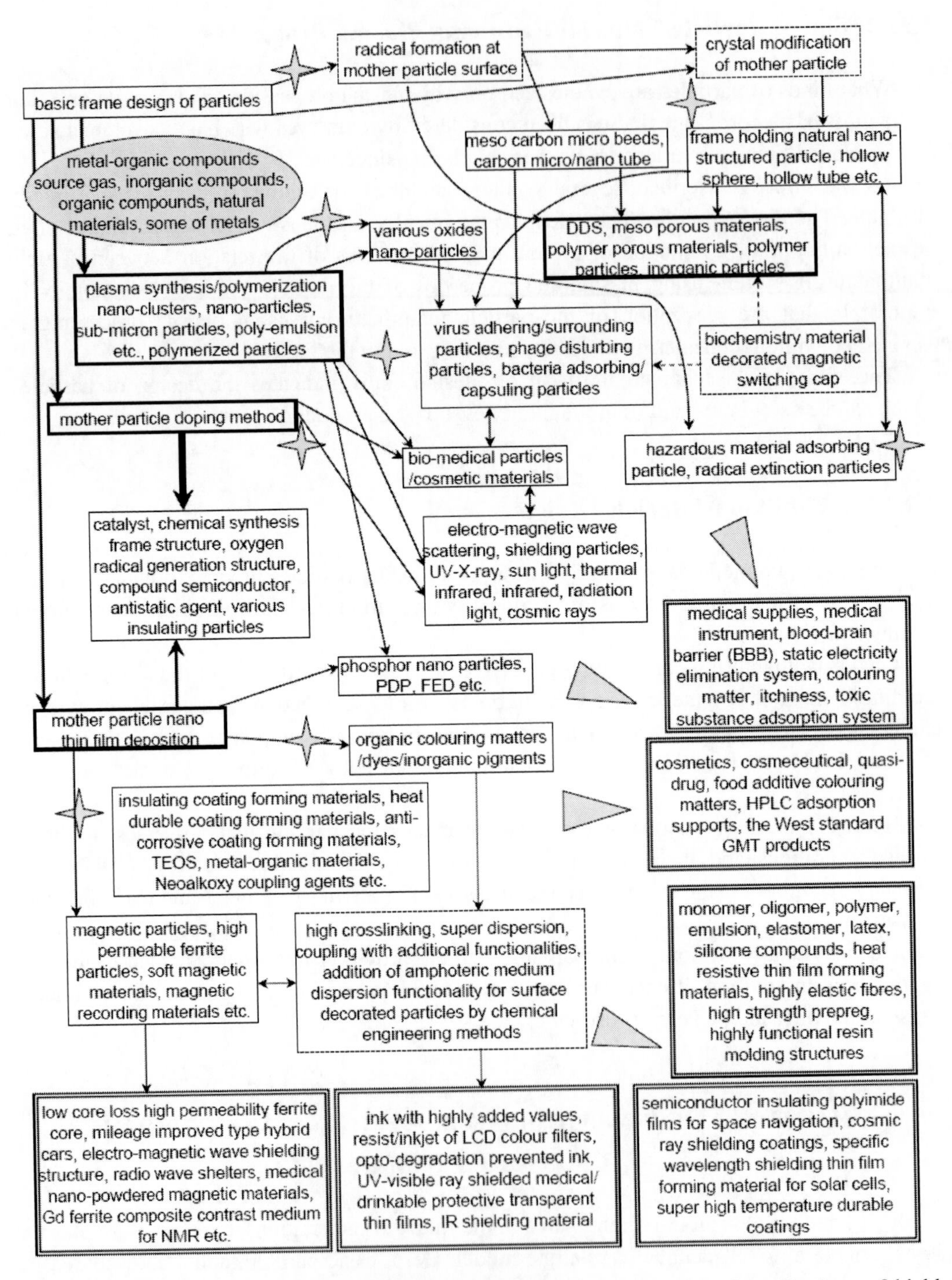

Figure 1. Examples of recent-next generation design targets and anticipated applications of highly functional particles utilizing atmospheric pressure plasma.

3.1. Construction of Crystalline Model Related to the Novel Nano-Structured Particle Design

The objectives of design for the novel nano-structured particles are;

A. synthesis of superstructure in a crystalline,
B. formation of functional nano-coating at a crystalline surface (single or more than one layer, each layer consists of 1 – N phases),
C. after filling a target material like medical material in a nano-structured space, a gap structure is formed, which can be taken away at the most outer surface of the particles when needed.

3.1.1. Typical Structure of Super-Lattice and New Materials Design Using Atmospheric Pressure Plasma

A representative super-lattice structure is a G-P zone which is seen in copper/aluminium alloys. Namely it is the structure regarded as segregation in which one element precipitation on the crystal plane of the other element at the inner structure of 2-element alloy.

As for the plasma synthesized oxide particles, crystalline analysis by means of high resolution electron microscopy (HREM) by Professor Kaito *et al.* in Ritsumeikan University [5, 6] clarified that this crystalline structure exists in the ion doped titanium oxide synthesized by our ICP RF thermal plasma between 1991 and 1993.

If super-lattice structure in a particle is developed by applying atmospheric pressure plasma and is involved in a production line, drastic improvement of productivity of functional particles and significant cost down will be established. Furthermore, social contribution which can not be anticipated so far can be performed. Such a technological innovation can not be realized only by generation, control of plasma.

Here, materials design is considered without trying to use plasma processing exclusively in all production process, combining other chemical and physical methods, producing mother particles inexpensively, and utilizing the crystalline structure and shape and properties of the particles.

The author already established synthesis of ion doped titanium oxide crystalline without using thermal plasma in 2004, and currently studies up-scaling of this method.

For the purpose of this particle design, atmospheric pressure plasmas or microwave chilled plasma operated at the temperature range at which the crystalline does not thermally transform are used. In addition, microwave plasma can be applied for large scale particle manufacturing equipment by up-scaling of plasma output power. Furthermore, as the important element of an epoch-making ion injection method, in the ion injection precursor production process prior to plasma reaction, treatment ranging from minimum 1 ton to maximum 400 ton a month can be achieved by the uniform surface treatment equipment enabling treatment of 5 – 100 kg a batch (20 minutes) by semi-dry process.

If up-scaling of plasma equipment is promptly achieved, and if gas recycling system is developed, it is expected that approximately 100 kg – tons can be treated in a month.

3.1.2. Details of Basic Properties of Ion Doped Oxide Particles Containing Super-Lattice Structure

The ion doped titanium oxide crystalline by thermal plasma in early 1990s was for the first time synthesized instantaneously in a gas phase by feeding metallic titanium particles as a source material and mixture of aluminium-atomized particles and iron particles flowing in argon gas, using a gas mixture of Ar/O_2, in RF thermal plasma at a temperature approximately at 4000 K. Its typical properties are described below;

[1] Mineral phase

The titanium oxide synthesized by ICP plasma shows minimum particle size of 15-20 nm, and central particle radius of 100 – 200 nm. The crystalline structure of titanium oxide is anatase-rutile mixture. The shape is generally preferable sphere. The Fe doped and Al doped types show between yellow and light flesh-colour, and between white and grayish white, respectively.

The difference of these colours depends on the concentration of the dopants. For example, in the case of the Fe doped type, concentrations of 0.05-0.1 weight %, 0.5-3 weight %, and the maximum saturated 5 weight % correspond to generally white, flesh-colour or light brown, and yellow brown, respectively.

[2] Crystalline lattice

Analysis of crystalline structure of XRPD data does not indicate camouflage of atoms by replacement of Ti by Fe until Fe is about to saturate. The strain of crystalline lattice is within measurement error. Furthermore, crystalline strain was seldom seen when Al is used as dopant instead of Fe. At this stage it is understood only that the limit of Al saturation is no less than 8 %.

[3] ζ potential which contributes dispersion

As for the dispersion of ion doped titanium oxides, the result of ζ potential measurement of dispersion of 5 % solid component indicates that equi-potential surface is approximately pH = 3.5. In particular for the aluminium doped type, ζ potential reaches between -50 and -75 mA when pH = 6, and the powder is completely dispersed in pure water without surfactant. Fe doped titanium oxide can similarly be self-dispersed. The reason of the self-dispersion is considered to be contribution of electrons supplied at the surfaces and $TiOH^-$ whose existence is identified using FTIR.

[4] Electrostatic self discharge property

The aluminium doped type shows the self discharge property. Its reasons is thought to be the existence of at least one significant Al doped side of the surfaces, and the tunneling effect of electrons at the surface. In addition, the hydroxyl group at the surface can emits electrons at the particle surface. Therefore absolutely novel crystalline structure is obtained utilizing titanium oxide crystalline as a basket, although the titanium oxide used for so-called condensers has same mother crystalline.

[5] UV shielding property

Figures 2 and 3 show the HREM image and the optical absorption spectrum of the aluminium doped titanium oxide particle, respectively. This spectrum contains not only the maximum absorption of the normal superfine titanium oxide particles at 300 ± 20 nm, but also other maximum absorptions between 340 – 345 nm and between 360 – 380 nm, and significant background absorption in the vacuum ultraviolet and visible ray ranges. These finest particles have a size distribution between 15 – 20 nm, and the central particle diameter is in the range between 100 – 200 nm. Furthermore, it was for the first time found that these powders have characteristics of shifting the maximum absorption band to 340-450 nm by pre-sintering.

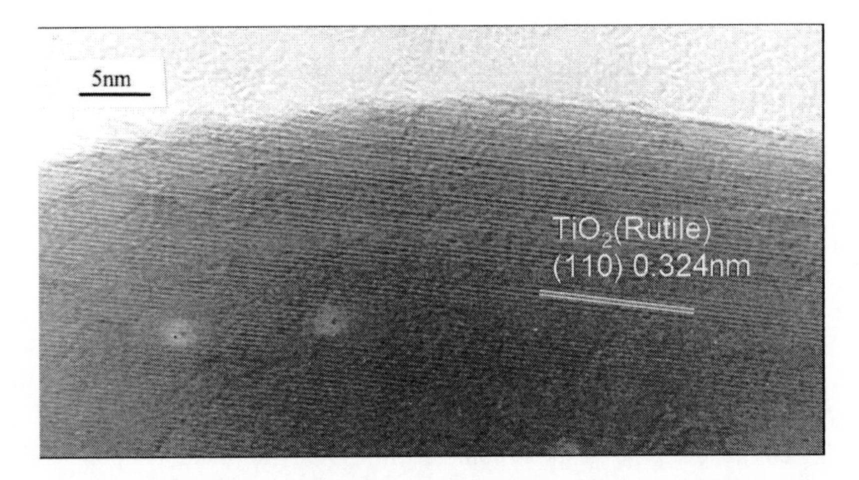

Figure 2. HREM image of aluminium doped titanium oxide particle.

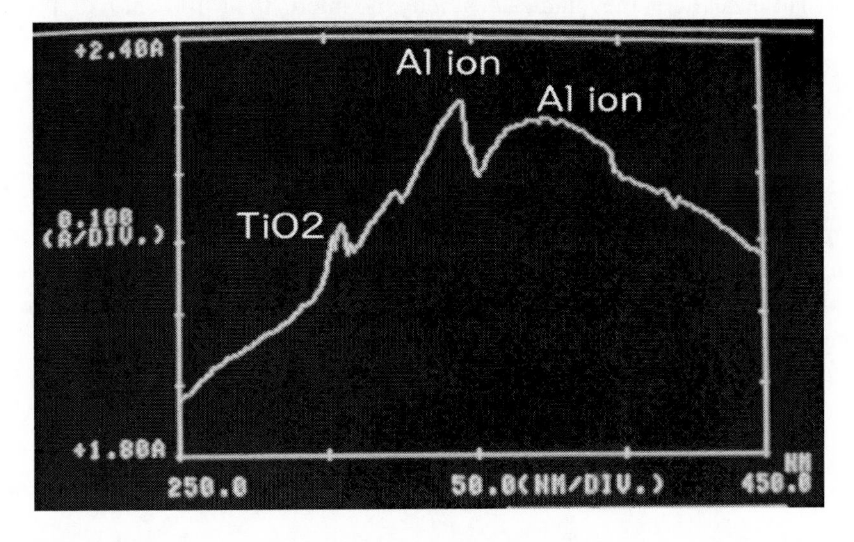

Figure 3. Photo-absorption property of aluminium doped titanium oxide.

[6] Self-protective function against photochemical reactions

Which radical reactions would occur with squalene ($C_{30}O_{50}$) was studied. Squalene is the main component of sebum with using plasma synthesized aluminium doped titanium oxide and normal titanium oxide crystalline, when ultraviolet is shut down. It is found that the only aluminium doped titanium oxide does not show photoemission by optical excitation. After irradiating ultraviolet at 254 nm, 305 nm and 365 nm, the other titanium oxides immediately emit photons between pink and orange, continuing for between several tens minutes and several hours. The above mentioned aluminium dope titanium oxide instantaneously emits slightly yellowish grey, immediately changed to a stable state of white grey. It is suggested that this is the photoemission by transparent squalene changing to light yellow by oxygen in Vascent state or radicals by anions contained in titanium oxides generated by the ultraviolet absorption of titanium oxide.

[7] Protection against oxygen drawing

The UV irradiated titanium oxide obtained in (6) was stored for half a year, and then obviously showed purplish colour tone, indicating oxygen vacancies. On the other hand, the aluminium doped titanium oxide did not show change of colour tone. It is therefore understood that the latter is significantly stable crystalline.

[8] Emersion property

Based on the background data described above, the aluminium doped titanium oxide was dispersed in pure water without blending surfactant, mixed with the oil phase at 80 ^{o}C by a high speed mixer, and perfectly dispersed skin-care cream was easily completed. This success resulted in putting in the market of the above mentioned aluminium doped titanium oxide compounding cosmetics at the cosmetic company in 1990. After that Fe doped titanium oxide was also offered for sale as skin-care cream and foundation. It was found that when only nano-level particles (here they indicate strictly no more than 100 nm) in the cosmetic manufacturing process, the rheological property of cosmetics is spoiled. Namely the sub-micron particles of plasma synthesized titanium oxide containing this crystalline structure show significantly high fluidity and dispersive property [6].

3.1.3. Crystalline Structure Analysis Using High Resolution Transmission Electron Microscopy (HREM) Etc. for the Particle Design by Atmospheric Pressure Plasma

The properties of ICP thermal plasma synthesized titanium oxide for the particle design by atmospheric pressure plasma are shown in Figures 4 and 5. The TEM images presented in this chapter were all taken by Professor Chihiro Kaito in Ritsumeikan University.

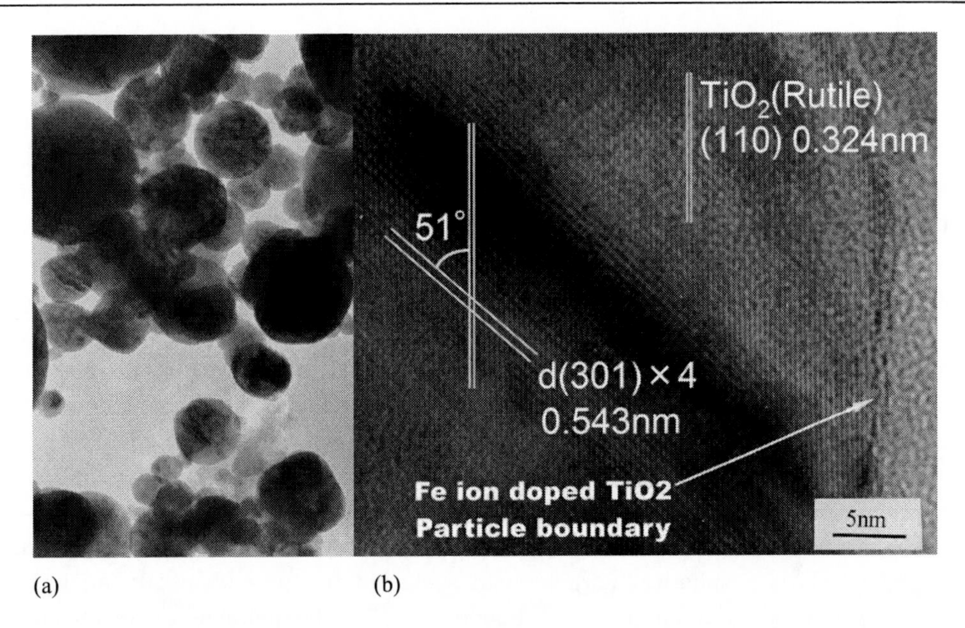

(a) (b)

Figure 4. Fe doped titanium oxide particle. HREM image (a), and crystalline analysis with HREM (b). G-P zone is indicated.

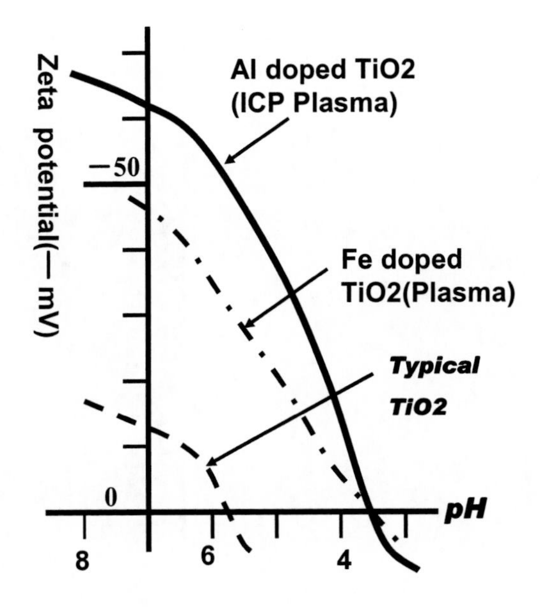

Figure 5. Zeta potential of plasma synthesized ion doped titanium oxide. Good dispersion when the potential is no less than 30 mV. General titanium oxide would not disperse at the mid range of the potential.

3.1.4. Particle Design by Using Atmospheric Pressure Plasma

Utilizing the results above, the design of crystalline which contains similar crystalline structure by ICP thermal plasma is performed by using atmospheric pressure plasma. Here, it is not appropriate to dope metals by vaporizing several tens μm metal particles in atmospheric pressure plasma because the boiling point of metals is no less than 1800 K. Therefore, the following two possible methods were attempted.

[1] Method with using metal-organic compounds as starting source material

This method is based on the technique of Ti-Al nitride gradient coating developed by Shimada *et al.* in Hokkaido University [7]. Here metal-organic compounds were used for the source of thin film materials for both Ti and Al, and thin films were deposited onto a substrate in RF plasma. When the mixed liquid of the metal-organic compounds was introduced into plasma reaction system, smooth processing was carried out by adding stabilizer in order to avoid curing during the process of introducing liquid drops by using capillary.

Oxidation was performed using the same equipment and same source materials mentioned above, which kind of particles would be synthesized was investigated. It is found that the use of thermal plasma is indispensable in order to reduce the residual carbon content in the particles. It is therefore indicated that even if particles are produced from similar source materials in atmospheric pressure plasma, it is difficult to synthesize particles with similar properties obtained by ICP thermal plasma. The particles obtained in the thermal plasma from the above mentioned source material has a size distribution within the strict region of the nano particles. It is a spherical particle with a size of no more than 50 nm, is highly dispersive, and forms aluminium doped titanium oxide. The titanium oxide obtained by evaporating metals using ICP plasma method shows broad size distribution and is disadvantageous only for UV shielding property. However, if metal-organic compound is used as the starting source material, it is advantageous in UV shielding property. Figure 6 shows the HREM image and electron diffraction pattern of this particle.

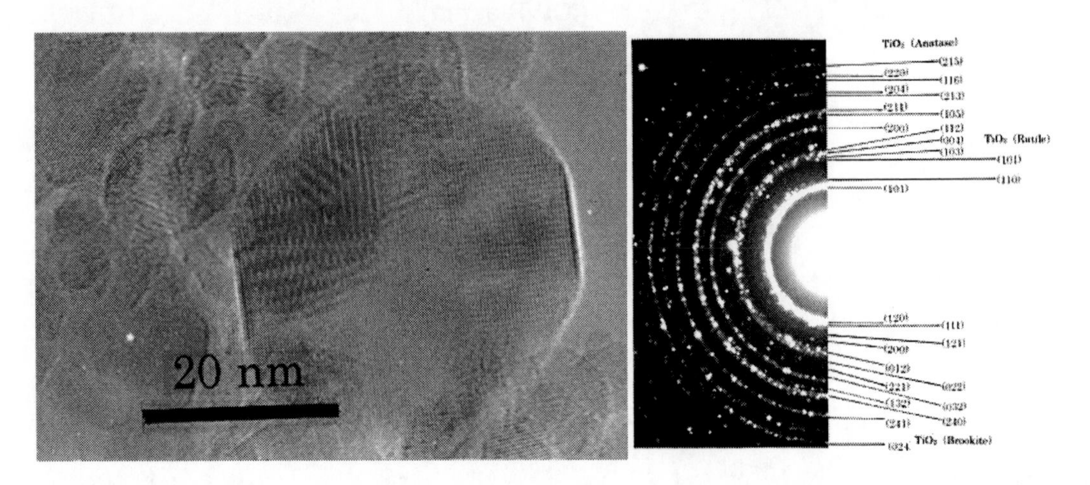

Figure 6. HREM image of aluminium doped titanium oxide (D_{50} = 30 nm) with using ICP thermal plasma. Mineral phases: Anatase, Rutile, Brookite.

[2] Method of doping to mother crystalline lattice space [8]

In order to perform doping, ions are injected among the crystalline lattices in the particle. In general, although inorganic materials apparently look dense materials, from the viewpoint of crystallography, it has a structure like a basket through which ions and atoms can easily pass.

As a similar structure, fullerene is well-known. On the other hand, natural clay mineral (e.g. allophane), silicate mineral, and sulfides exist, and in many cases impurities are contained in them. From the structural interpretation, some ion doping models can be considered; ion exchanging by camouflage of atoms, fractional crystallization, or introduction of atoms into the space which unit of crystalline forms.

Utilizing similar crystalline structure and atmospheric pressure plasma at relatively low temperature state or downstream of microwave plasma, it is thought that injection of ions and atoms into mother crystalline are possible with simultaneous cutting of molecular chains of metal-organic compounds. For example, titanium oxide crystalline can be observed by using the TiO_6 unit as shown in Figures 7-9.

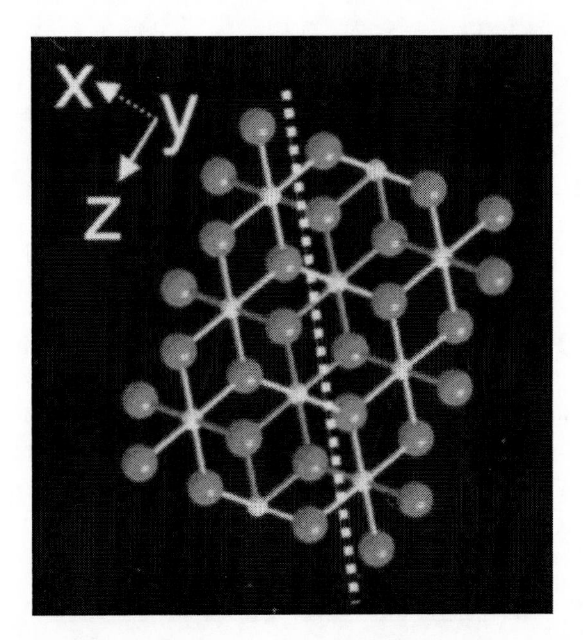

Figure 7. TiO_6 atomic bonding.

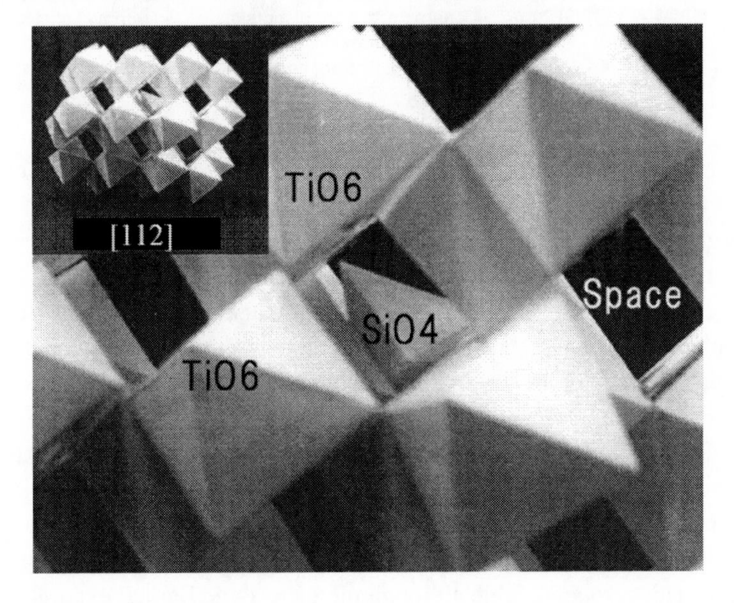

Figure 8. SiO_4 free path in TiO_6 space.

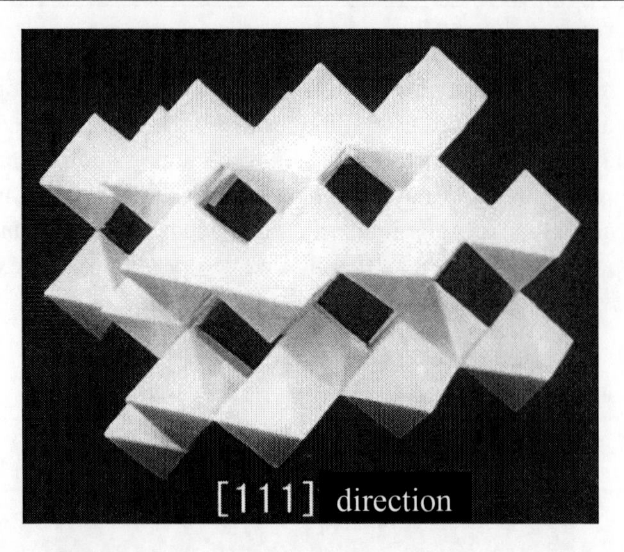

Figure 9. Crystal space of anatase.

3.2. Synthesis and Process of New Ion or Atom Doped Particles

The methods of atom- or ion-injection in the crystalline space are summarized below. The application of this method can revolve conventional ion injection technique.

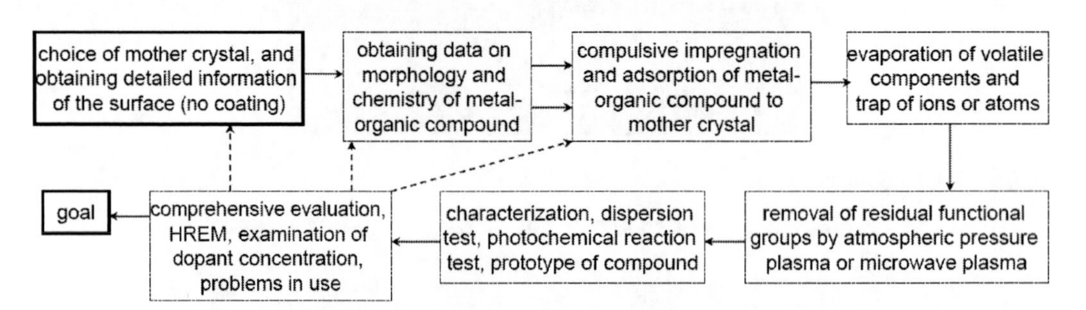

Here, the choice of metal-organic compound is very important. For example, when acetylacetonate is used, a practically almost impossible operation is required if its solubility is low, even though high concentration atom injection theory exists. In addition, acidic material sometimes reacts with mother crystalline. Furthermore, the valence of metals in the organic material may be different from objecting valence. Mother crystalline often strictly forbidden with water, and thus the choice of a solvent must be carefully performed. For example, engineering alcohol can not be used to treat nitrides, corrosive magnetic materials, and phosphor etc.

Next, an adequate amount of dopants solved in solvent is introduced into a crystalline space of mother particles using vacuum equipment etc. The size of the molecules must be small enough to accept impurities into the space in the mother crystalline. Therefore if necessary, molecular chain must be cut. This process requires various energy sources. The chemical hydrolysis is also applicable although there are some limited examples available.

After this process they are dried. After that the molecular chains are completely cut in a plasma, and then atoms or ions can be filled in the mother particle crystalline. However, during this process, the metal-organic compounds are sometimes simply coated at the surfaces in spite of being introduced into the crystalline. Therefore characterization of the properties of the product must be performed.

In these respects, this method requires high level characterization systems like HREM are necessary in order to perform highly precise doping. In addition, it is necessary to be mastery of the operation of them and knowledge of the characterization.

3.3. Case Study (Fe Doped Titanium Oxide: Rayleigh Scattering Applicable Type, Visible Ray Oriented Scattering Type)

A case study was performed as follows. The microwave plasma system in Sophia University was used.

[1] Mother particles: absolutely no surface treated titanium oxides which is for the electronic materials of 50 nm primary particle (Ishihara Sangyo)
[2] Metal-organic compounds: Fe acetyl-acetate (IAA) (commercially available)
[3] Solvent: anhydrous ethanol.

The following processes were carried out using the materials (1)-(3) above,

Process 1: No more than 4 wt.% IAA is solved in ethanol,

Process 2: The required amount of IAA and ethanol in order to introduce 1 % Fe into 1 kg TiO_2 is

$$X= \{(1000g \times 0.01) /(Fe/IAA \text{ molecular weight})\}/0.04=\{10/(56/350)\}/0.04=1562.5 \text{ g}$$

Subsequently IAA=62.5 g, EtOH=1562.5 g are obtained.

Process 3: Firstly IAA is completely resolved in the adequate amount of ethanol. Since the solubility of IAA is not high, IAA must be sufficiently reacted with ethanol, subsequently it is impregnated in vacuum by introducing 1 kg titanium oxide into the solution, ethanol is gently evaporated, and the precursor is obtained.

Process 4: IAA adsorbed titanium oxide particles are crushed down, and are exposed to a plasma introduced by the gas flow in a plasma chamber.

Process 5: If plasma dissociation in a gas flow is insufficient, organic molecules dissociated are recombined and strong smell is emitted, and it is difficult to eliminate functional groups. In the case that complete elimination of functional groups is necessary, the obtained particle must be cleaned using ethanol again.

Process 6: Ferric atom doped titanium oxide obtained by the processes above shows high dispersive property in both water and oil based solvents.

This particle has very smooth surface, and it was found that a transparent thin film could be grown with the titanium oxide with 50 nm mother particles. On the other hand, titanium oxide particles on average 200 nm showed multi-scattering particles with strong sense of transparency though semi-transparent. They showed an unprecedented broad shielding property in the UV-VIS range. These properties will be further investigated in the future.

However, the initial evaluation indicates that the property is at least comparable to those obtained by ICP thermal plasma.

3.4. Improvement of Design and Manufacturing of Nano-Particles, and Metal-Organic Compounds

Production of particles using atmospheric pressure plasma or microwave plasma which is systematically simple will be a big improvement in the future.

3.4.1. Productivity of Design/Manufacturing of Particles by ICP Thermal Plasma, Atmospheric Pressure Plasma/ Microwave Chilled Plasma

Table 1 shows representative parameters for particle design by plasma based on already obtained data. The properties of the metal-organic compounds etc. (common to synthesis of nano-coatings) which are used for particle design is shown in Table 2.

Table 1. Representative productivity and problems of manufacturing of particles by plasma processing

	ICP thermal plasma	Atmospheric pressure glow discharge plasma	Microwave chilled plasma
Input power [kW]	40-100	1-3	1-10
Treating amount of source material [g/h]	Bulk. 100	10-100	10-no less than 1000
Aimed yield of particles [%]	30-50	50-80	70-95
Problems	Difficulty in supplying source material	He gas recycling	Temperature control and gas recycling

Table 2. Representative metal-organic compounds and their applicability to plasma

	examples	Soluble solvents	Applicability to atmospheric pressure plasma	comments
Acetylacetonate	Fe, Ti, Zr etc.	EtOH etc.	Possible recombination of molecules	Strong smell
Lactate	Fe, Al, Ti etc.	Water etc.	applicable	Possible bivalence condition
Acetate	Ti, Zr etc.	Polar solvents, water etc.	applicable	
Alkoxide	Al, Sn etc.	IPA, non-polar solvents etc.		Can be dissociated during pre-treatment
Alcoholate	many	NA	NA	Explosive combustion in air
Carbonyl	Fe etc.	specified solvents	applicable	
neoalkoxide	Zr, Ti, Al etc.	specified solvents, soluble in water	Possible by molecular chain	Coupling agent

3.4.2. Problems of Metal-Organic Compounds

Many metal-organic compounds show different characteristics, and have specific problems. Careful preparation is necessary for design of particle materials in advance, as the following problems can be identified by a broad division.

[1] High viscosity materials

Aluminate, tin compounds, and some of zirconates correspond to this. These have high viscosity, and thus have to be diluted for use. The control of atmosphere is also necessary. If thick coating is synthesized in the pretreatment, it can not be dissociated by a plasma.

[2] Materials with intense hydrolysis

Special care must be paid for materials with intense hydrolysis like zirconates when they are handled in ambient air. In addition there is a possibility that the solvent chosen enhance hydrolysis, the suitability of the solvent must be checked before treatment, and it is necessary to complete precursors in a grow box in which environmental control is possible.

[3] Hazardous materials

Especially when hazardous elements are used, information of the hazardous material must be reconfirmed in advance. Since there are materials which can form crystals that will close respiratory system and result in fatal damage, such as V and Be. Therefore the administrator with scientific knowledge is indispensable.

3.4.3. When Mixture of More Than Two Compounds is Used

Even if each material is stable, when mixture of more than two stable materials is used, stability of the mixed phase is completely different. The Author utilizes mixture of dissimilar metal-organic compounds using the most stable metal-organic compound as a solvent. This method can be disadvantageous for doping. One example is the mixture of tetra-butoxy-zirconate with a silane compound. However, it is necessary to be ready for using hazardous materials which is described above, since probability of explosion by the chemical reactions is not negligible.

3.4.4. Use of Inorganic Compounds

If some materials which are generally obtainable like chlorides, sulfides, sulfates, nitrate, ammonium compounds are exposed in a plasma, they detach negative ions, and toxic gas is generated. In addition, gases generated at the downstream of plasma can recombine to form particles, and they can be concentrated at the particle surfaces. In particular, chlorine behaves in a tricky way, and so it can not be handled in some applications.

3.5. Summary of the Session and Economy

There are various available plasmas for compulsive injection of metal-organic compounds and promotion of structural defects in mother particles. In addition, not only plasmas but also electron beam, ultraviolet, and lasers can also be used. There is also an old technique of ion injection by injecting accelerated ions onto a target.

The atmospheric pressure plasma proposed in this chapter has a high potential of developing various application technologies. The next example is promising for synthesizing complex-type nano-structure based on marketing research.

3.5.1. Challenge for the Design of Plasma-Synthesized Particles for New Application Areas

Even for the synthesis of basic particles 'meso porous biomedical nano particles (particles with 2 – 50 nm continuous voids)' can be obtained by a proper choice of plasma environment and source materials. Not only the conventional DDS, but also silicas which can support many biochemical materials are useful for treating micro-scale cell mutation such as cancer treatment, treatment of incurable diseases, target treatment and marking treatment. In national and international medical industry, clinical trials and pilot tests were already initiated [10, 11]. In the biomedical areas in USA (e.g. National Cancer Institute (NCI)) and Europe, such areas are regarded as strategic national projects.

[1] Design of particles which promote extinction of cancer cells [12]

The use of particles for cancer treatment started in USA in 1950s is summarized in Table 3. For the design of such particles the use of plasma can be considered.

Among above mentioned examples, the usefulness of atmospheric pressure plasma was proved for nano-particles related to magnetic materials (pigment of cosmetic foundation etc.), meso-porous silica (application of plasma operation), improvement of hydrophilicity of polymer beads surfaces (particles which are difficult to be treated such as cross-linked polystyrene), and treatment of precursors for self cross-linking which will be described in section 4.

One promising application is plasma synthesis of gold nano-coatings. It is known that gold particles which were first spherical became arbitrary shapes by aging with ICP thermal plasma. On the other hand many researches are reported about carbon nano-tubes and fullerene. The detailed progresses in medical applications are reported outside Japan.

Theoretically expected particles are nano-particles like silica whose surfaces are charged, and can disperse individually in a living body. The particles on which various proteins or malignant-cancer-cell attacking peptide are attached are thought to be useful for early disappearance of cancer cells.

Table 3. Recent trend of nano-particles for cancer examination and treatment which potentially link to atmospheric pressure plasma techniques

	Shape and material of the particles	Destruction method of cancer cell	Device of combined use	Design of simplified structure
1 nanoshells	Sphere. core: SiO_2. outer layer: gold	Thermal destruction (occurrence of fatal fierce heat for cells)	Near infrared laser (transmittable several cm skin)	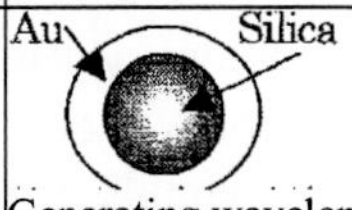 Au Silica Generating wavelength determined by the particle size.
2 Dendrimer	Spherical giant molecule where molecules are radially constructed. It can be grown either from the shell or the core.	DDS, acting as a photo-trapping antenna, and holding medicine. Difficult to prepare.	Infrared, ultrasound etc.	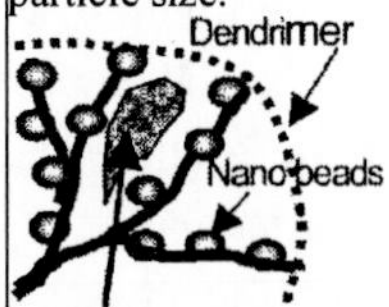 Dendrimer Nanobeads
3 Nano Magnetic Particles	Various shapes of nano-particles.	Destruction of cancer cell by magnetization heating.	Magnetizing device.	Anticancer drug etc. DDS
4 Porous Silica Particle	Silica particles with 2-50 nm continuous voidage.	Used as DDS.	Example of simultaneous use of ultrasound.	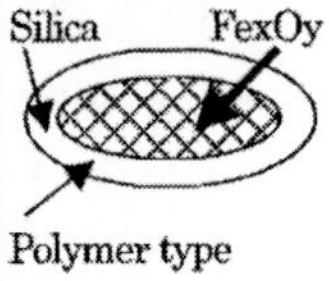 Silica FexOy
5 Hybrid Particles	Extensively studied mainly in the US and Europe. There is a type which directly react cancer cells. poly(lactic-co-glycolic acid) (polymer particle).	Various types exist. In general DDS, adsorption of malignant cancer cell, a type of holding photo-reactive excimer.	Ultraviolet etc. appropriately.	Polymer type
6 Cantilever (reference)	Nano-size "cantilever". an open sided small bar.	Detecting strain of the bar sue to the change of surface tension by bonding to cancer related molecules.		
7 Nano-tubes Carbon rods	Approximately half of the diameter of a DNA molecule. a needle of a record player comparing with a cell.	Confirming the change of DNA accompanying cancer.	Confirming mutation and its position in a molecular level using computer analysis.	Figure omitted.
8 Gd doped Nano-particles	MRI contrast medium. Iron oxide is also used for detection chemical of cerebral tumor.	MRI.	Iron oxide type contrast medium exists for liver and kidney. By protecting leakage of magnetic field, precise information can be obtained.	
9 adsorption of neuro-transmitter like histamine and antipruritic	Ion doped titanium oxide synthesized by microwave chilled plasma, silica nano particle (atmospheric pressure plasma etc.)	Presuming electron avalanche and molecular adsorption at the charged nano particle surface from neurotransmitter.	Equipment is unnecessary. Monitoring is needed (in the future electric potential is measured).	See Figure 27.

(Comments) The data between 1 and 8 are translated by ISI based on the information of the National Cancer Institute (NCI) in the US and literature by research institutes in the US and Europe. The data 9 was obtained by the author and Kogoma in Sophia University.

3.5.2. Economical Consideration for the Particle Design and Manufacturing by a Plasma

Economical consideration for the manufacturing of particles using a plasma is carried out here.

Plasma processing with a vacuum system often shows more significance in development of equipment and infra-cost than the product yield. However, atmospheric pressure plasma and microwave chilled plasma need much lower initial investment than a conventional ion injection method. In addition, since they do not contain vacuum systems (particle trapping device will be connected if it is necessary), it is advantageous that up-scaling is easily performed with low cost, and that sequential expansion is possible by multiple parallel lines. In fact, particles production using atmospheric pressure plasmas needs licensing of advanced techniques and basic techniques. In spite of this, some large scale projects are started relying on this technology. If recycling technique of used gas is established, there is a possibility in the significant improvement of economical situation, and the merit for power consumption may exceed other techniques. As for the labour cost, production processes is costly, if it can be engaged by only specific engineers including the operation of plasma and UV. Therefore it is thought that superiority to various production methods will not be retained in the future.

From these viewpoints, it is indicated that nano-structured particles can not be synthesized easily, and that gradual transition from low-volume of various productions to mass production with developing structural design and production methods is required. Therefore general large scale plant with no degree of freedom is unsuitable, and is not easily diverted. Sufficient preparation is necessary for investment.

In the case of ion doped titanium oxide whose synthesis is aimed in this chapter, the production devices consist of

[1] Maintenance of infrastructure of a general chemical laboratory,
[2] Vacuum impregnation equipment or an experimental equipment,
[3] Temperature controllable dryer, thermostat,
[4] High speed mixer and airless gun for liquid (spray device)
[5] Plasma system
[6] Vacuum ultraviolet irradiation system,

and so on.

The aimed particles shall be successfully synthesized by using them, accompanied with materials design and various software. In the future, together with these methods, the significance of plasma will be enhanced by particle design which synthesizes nano-structured porous materials in gas phase dissociation of metal-organic compounds like TEOS at atmospheric or middle low pressure, and ion injection or bio-molecule bonding.

4. Design of Nano-Coating Forming Particles and its Typical Model

Synthesis methods of thin film like silica on inorganic compound surfaces like a sol-gel method have been widely performed. However, as the size of the mother particles which will be coated is smaller, segregation tends to occur. When they are eventually compounded and used, secondary processing such as re-pulverization and re-dispersion are indispensable. In addition, in the sol-gel coating which is once segregated, hydrolysis will not occur perfectly. In many cases organic functional groups will remain there. These organic functional groups form large space in a coating by facing with each other by hydrogen bonding etc., and the coating will be looser. This is caused by the use of reaction catalysis. It is indeed inappropriate for engineers who are in charge of next step production process that thin film forming materials like metal-organic compounds are completed without passing uniform continuous thin film forming process by adding heating process regardless of acid or alkali.

For example, when a silica thin film is formed, tetrametoxysilane (TMOS) can be used if it is to be formed in a short time only, and tetraethoxysilane (TEOS) is not necessary to be used whose reaction speed is very low. For the synthesis of titanium oxide thin films, titanate-propoxide or isopropyl-titanate is often used.

Furthermore if a zirconia thin film is formed, several zirconate compounds are provided though they are extremely unstable.

Here, design of complete reactions during plasma polymerization process of coatings are performed by reaction prediction of these materials in a plasma and chemical reaction prediction and analysis of the precursor production process before film synthesis.

The properties and performance of the synthesized thin films are discussed below. When the thin films are characterized, not characterizing simply by dispersiveness of particles as a group or TEM images, but closed-up observation of the thin films using HREM can lead to clear verification of novel nano thin films. At the same time, additional characterization of the thin films is sometimes necessary by choosing among EDX, FTIR, XPS, UV-VIS, thermal analysis, shear wear test, compulsive mixing test in a mixture, Hoover Muller test (colouring of pigment), elution test of heavy metals or prohibited organic molecules, long range stability test, solvent affinity test, electric property, or electronic property measurement. Depending on the objectives, they are more.

In particular, when the sizes of mother particles are large, they are difficult to characterize using TEM or HREM. The authors have polymerized various thin films using particles whose mother particles' maximum sizes are approximately 10 plus several μm and several hundreds μm by atmospheric pressure plasma, microwave plasma or middle low pressure plasma [13]. There is a trend that minimum particle size of mother particles is no more than 10 plus several nm. In the future nano-clusters can be applied.

The achievement of particle design up to now and future works are listed below.

4.1. Model of Formation of Silica Thin Films and its Problems

TEOS monomer is expressed as $Si(OC_2H_5)_4$. Its structure is pyramidal, and its ends have the ethoxy group. Figure 10 shows the structure of SiO_4 synthesized using TEOS. In order to synthesize a thin film by continuing this crystalline unit, a plasma with too high energy or too long plasma treatment shows adverse result. When plasma engineers synthesize thin films, many of them possibly imagine a CVD method first of all. In order to deposit mother particles and thin film forming material all over the surface uniformly in a plasma space, long time and a system device are necessary. In fact many TEOS adsorb at a specific crystal plane on which they preferably deposit, and a thin film is unevenly synthesized. As for silica, long time plasma irradiation promotes crystalline transition of silica which will be shown in Figure 13.

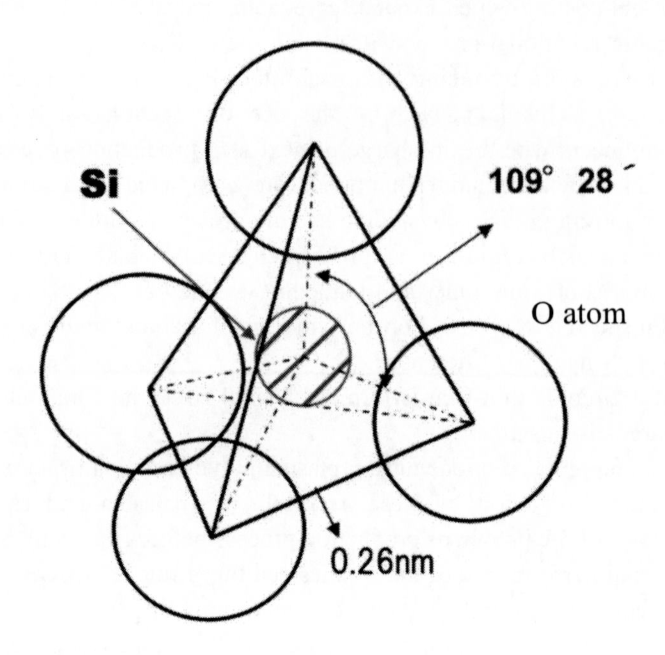

Figure 10. SiO_4 structure model.

The most stable processes for silica thin film synthesis using a plasma is as follows;

[1] 1st process: preparation of silica source solvent like TEOS

Study on the properties of mother particles which is deposited is performed (water solubility, solvent unsuitability etc.). Solubility of silica source (solvents are sometimes unused) is tested in advance.

[2] 2nd process: determination of mass of silica thin film

Using the surface area ratio of the mother particles, the approximate required amount of the silica source for a specific film thickness can be theoretically estimated. However, in a practical process, its yield and retention ratio differ significantly. Therefore the film thickness must be obtained basically by the direct measurement using HREM. Many source materials can relatively easily form amorphous silica nano thin films. However, some organic colouring matter, pigment, and inorganic materials show selective thick deposition on a specific crystal face, while opposite effect on others. It can be said that uniform thin film formation is obtained by a suitable pretreatment.

[3] 3rd process: Formation of precursor of silica thin film

After determining silica synthesis method, required mass of TEOS etc. is estimated. Since they are obtained by a very simple calculation, its explanation is omitted here.

However, reaction efficiency and loss of the source materials during the reactions must be measured experimentally. This process is normally carried out by a wet method or a semi-wet method using a high-speed mixer. It is desirable that a silica source is gasified and sprayed uniformly on particles. If it is unsuccessful, coalescence of particles occurs as shown in Figure 11.

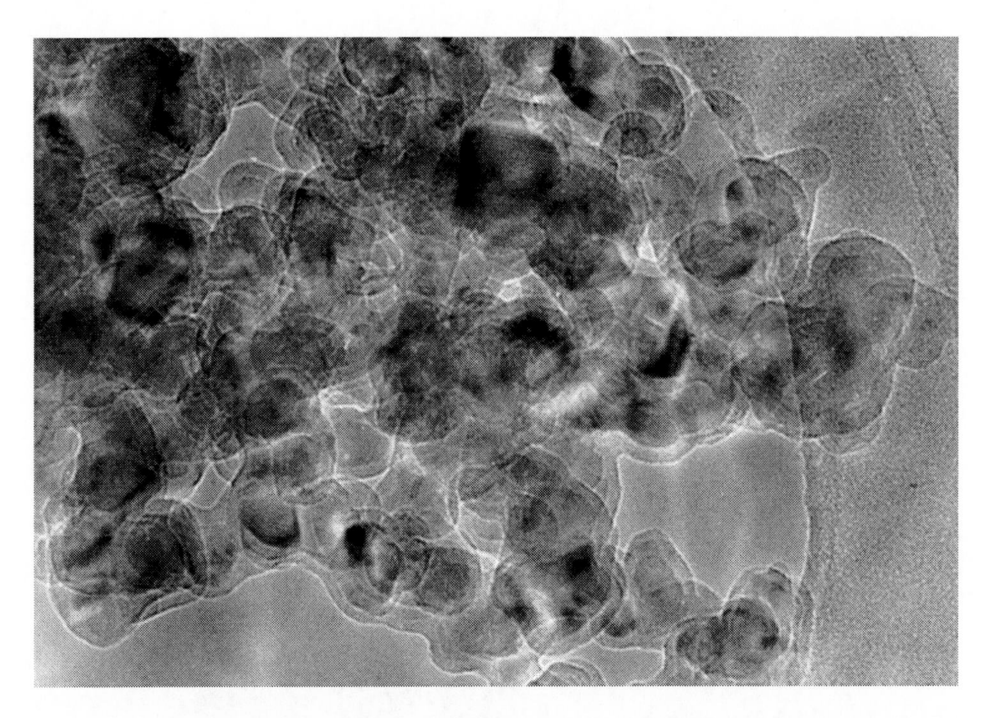

Figure 11. Coalescence due to the excess use of silica source. 20 nm TiO_2 particles are completely combined.

[4] 4th process: maturation and drying of precursor

At the maturation process, generation of the aimed silica precursor is promoted. The most important issue in this maturation process is to collect data using XPS or HREM. FTIR can also be used as a simplified method, but is not a definitive method.

[5] 5th process: plasma treatment

The dried powders obtained at the 4th process are exposed in a plasma. Here the energy of the plasma must be determined by the conditions of mother particles and silica thin films. For organic colouring matter of weak bond strength a soft atmospheric pressure glow discharge plasma should be chosen. If certain heating is acceptable, microwave child discharge plasma may also be applicable. However, even though it is a low temperature plasma, when particles directly pass the discharge field, silica may be crystallized, microcrystal may be randomly formed, and these become fragile films. Figures 12 and 13 show HREM images of a fine thin film and a fragile thin film after crystallization, respectively.

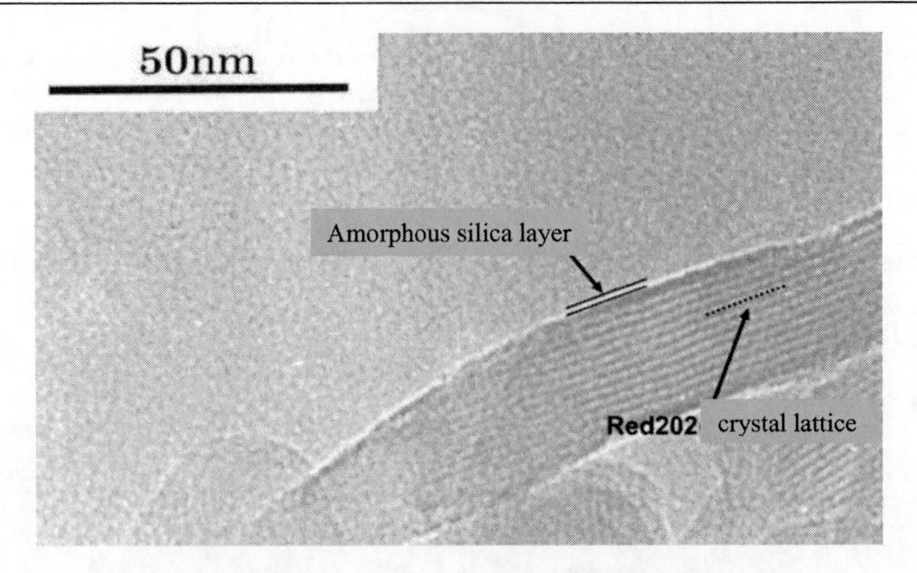

Figure 12. HREM image of silica coating at the red organic colouring matter.

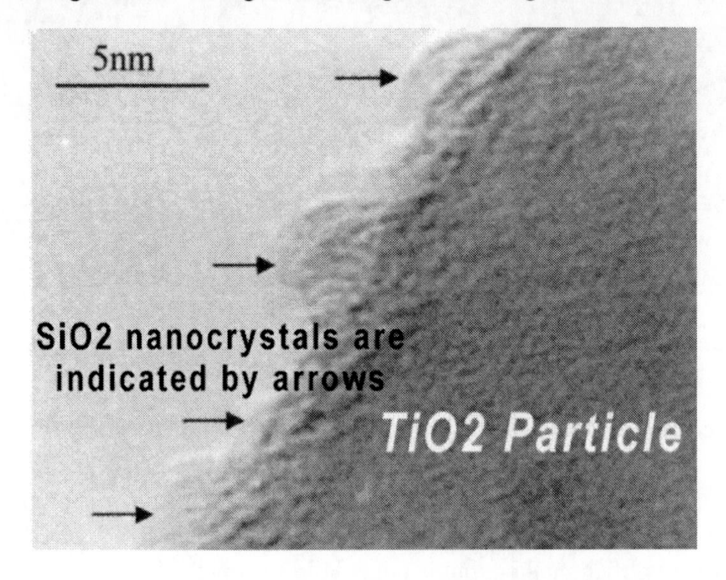

Figure 13. structure of silica coating at the titanium oxide particle surface which received excess plasma energy.

4.2. Design of Thin Film Formation Excepting Silica

There are many source materials which can be used for various kinds of thin film formation. However, if the ratio of metallic element in a metal-organic compound or that of MeOx (Me: metal) is small, a large amount of metal-organic compound is required, which disturbs thin film formation. In addition, if the compound is unstable in the ambient air or solvent, or can not remain stable during a certain duration, abnormal hydrolysis occurs during film formation, and uniform and continuous coating can not be obtained. In particular, care must be paid for some unstable compounds such as zirconate and titanate.

When the viscosity of the compound is high, the polymerization in the subsequent plasma reaction is not completed. Therefore preparation of source material mixture consisting of the starting material and its diluting medium is vitally important for the film synthesis of alumina and tin oxide etc. Based on the results obtained so far, very stable source materials for film synthesis include neo-alkoxy-zirconate coupling agent, neo-alkoxy-alminate coupling agent, radical dissociated metal-organic materials etc. In addition, although it takes time, it is possible to treat the surfaces of mother particles with silane coupling agent in advance, and to use them as the precursor of film synthesis.

4.3. Materials Design of Thin Films with Various Combinations

Multilayer deposition of nano-coatings is a completed technique in semiconductor industry. However, if the substrate is particles, it is very difficult. Nevertheless utilizing the above mentioned basic technique, multilayer deposition of nano-coatings on particles became possible. It will be a next question why such a technique is necessary.

In the case of monolayer film synthesis, if all the required properties of mother particles are not satisfied such as electric properties, chemical stability, optical scattering properties, dispersion property at the very surfaces, mechanical strength and electric mobility, hybrid film formation is important for the synthesis of highly functional particles.

For example, amorphous metallic soft magnetic material like a Ni-Co type which also demonstrates high frequency shielding dielectric property is easy to corrode, and the frequency dispersion of leakage of radio waves is high. In order to avoid them, coating with high dielectric constant and high resistance must be deposited on the surface.

However, in the composite production system which requires soft magnetic material at the end, when compounding with resin or other adhesives is designed, surface treatment of materials such as zirconia or alumina which may induce poor dispersion or bonding failure does not necessarily result in optimum production.

4.4. Design for Thin Film Formation for Uniform Dispersion of More Than One Component Particle Mixture, and Surface Modification of Thin Films

In order to use more than one component with different surface polarization for mixing and dispersing in one compound system, surface condition must be unified. If surface polarization is different, segregation occurs immediately regardless of the mixture between inorganic material particles, between organic material particles, or inorganic and organic material particles. One representative example for coordinating surface polarization is silica coating.

For example, in the case of colour filters for liquid crystal displays, representative example of difficult dispersiveness is blue in RGB. Its composition is mainly phthalocyanine blue compounded with a small amount of purple organic pigment. In the case of green, yellow Ni organic compound is mixed as an adjusting material in phthalocyanine, but the Ni organic compound is very difficult to disperse.

In order to solve the problem, these mixing phases are individually coated with silica, adding silanol at the surface so as to orient the polarization negatively with a uniform level.

Supplying electrons at the particle surfaces using atmospheric pressure plasma additionally plays an important role in improving dispersion of two kinds of colouring matter in the case of colour filters. These phenomena are shown in Figure 14. Figure 15 shows the coupling model at the silica coated particles. Plasma treatment enables radical cross-linking of coupling functional groups and anchoring. The kind of coupling must be chosen according to the aims, and the corresponding process must be carried out.

As mentioned above, if expensive treatment is terminated after adding polar functional groups only, it does not show high cost performance.

Therefore, for the most ideal particle design, formation of the interface layer at the particle surfaces must be performed which can achieve cross-linking and polymerization freely in a compound.

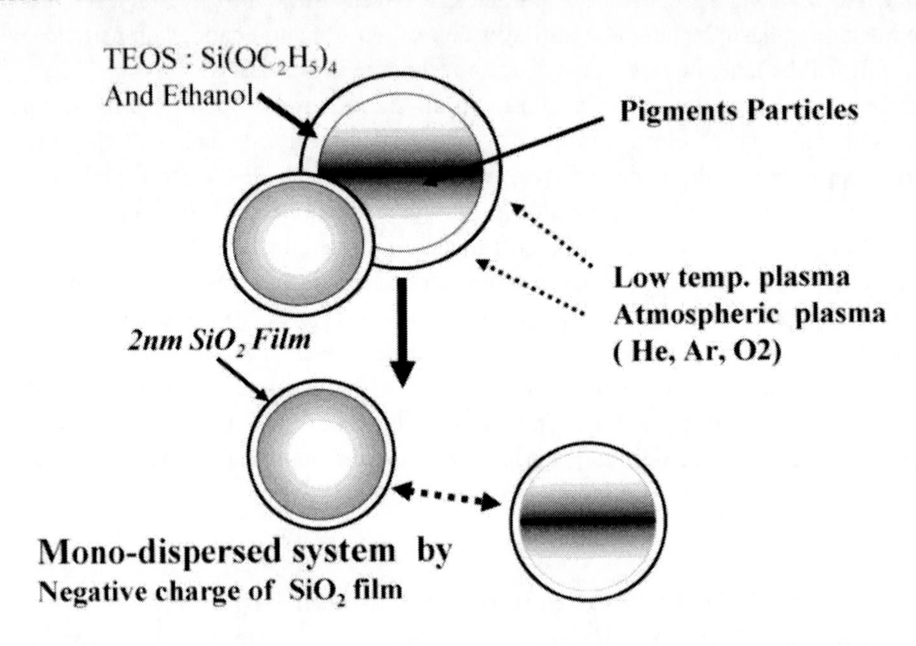

Figure 14. Uniform dispersion model of no less than two kinds of pigments, and plasma polymerized silica.

For the application of plasma, use of neo-alkoxy-titanate coupling agent, neo-alkoxy-zirconate coupling agent, and alminate coupling agents which can show radical bonding as well as conventionally used silane coupling agents compounding with various monomer, polymer and photo-resist materials can lead to design of novel materials. Figure 15 shows the interface of a representative titanate coupling agent. Here two acryl groups exist at the edges of hydrolytic functional groups of the coupling agent. The electron transfer with the radical opens the double bonding, which reacts with the particles that are surface-treated.

The particle surface which is excited by atmospheric pressure plasma and ultraviolet opens the allyl group and shows radical coupling. The terminal functional groups determine properties of dispersion, electron acceptance, heat resistance, corrosion resistance, polymerization, crosslinking, adhesion, and insulating.

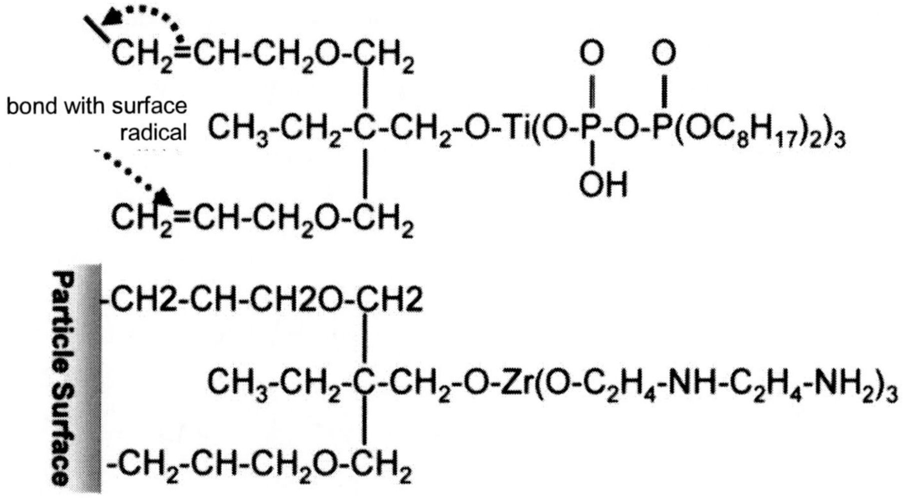

CH_2=CH-CH_2O-CH_2

bond with surface radical

CH_3-CH_2-C-CH_2-O-Ti(O-P(=O)(OH)-O-P(=O)(OC_8H_{17})$_2$)$_3$

CH_2=CH-CH_2O-CH_2

Particle Surface

-CH2-CH-CH2O-CH2

CH_3-CH_2-C-CH_2-O-Zr(O-C_2H_4-NH-C_2H_4-NH_2)$_3$

-CH_2-CH-CH_2O-CH_2

Figure 15. Coupling to the plasma treated particle surface, and its mode.

5. Highly Purified Model of Organic Superfine Particles

It is thought that the highest purity of organic compound during synthesis is approximately 95 % or slightly higher, which is different from inorganic compounds. It is impossibile to manufacture high purity organic compound of for example 11N (eleven nine) which is seen for inorganic silicon. Furthermore in organic compounds, many impurities exist such as unreacted components and source material as well as isomers. It is therefore very valuable to exclude them as many as possibile, and obtain high purity materials. In the case of polymer particles used for biomaterials hydrophilic treatment on the surfaces and stabilization by deposition of silica nano coatings are important future works.

For example, blue colouring matter which shows a sharp maximum absorption band is not produced yet. The organic colouring matter used for colour filter of LCD is very weak in temperature increase during high temperature baking or difficult to disperse. In particular although Ni compounds are necessary for formation of green colour filter, but its dispersive property is substantially poor.

In order to solve such problems, as a method to remove impurities [14, 15], untreated Carmin 6B (Lithol Rubine BCA. Legal colouring matter Red202) which is not treated by resin (pine resin) was coated with silica, which is the most representative coloring matter and was attempted to be highly purified as biomaterials.

5.1. Nanotechnology for High Functionality and High Quality of Organic Super Particle

[1] Objectives: Here pigment surfaces are hard landed with silica coating using TEOS, and simultaneously basic property of pigment is improved.

[2] Production method: Red202 which is not treated with the resin is emulsified with TEOS and water. Immediately after that, they are adsorbed at the surface, and then treated with a plasma so that silica is deposited without pinholes. The plasma used here was He/O_2 plasma method by atmospheric pressure glow discharge (C coupled) developed at Sophia University. The process is schematically illustrated in Figures 16 and 17.

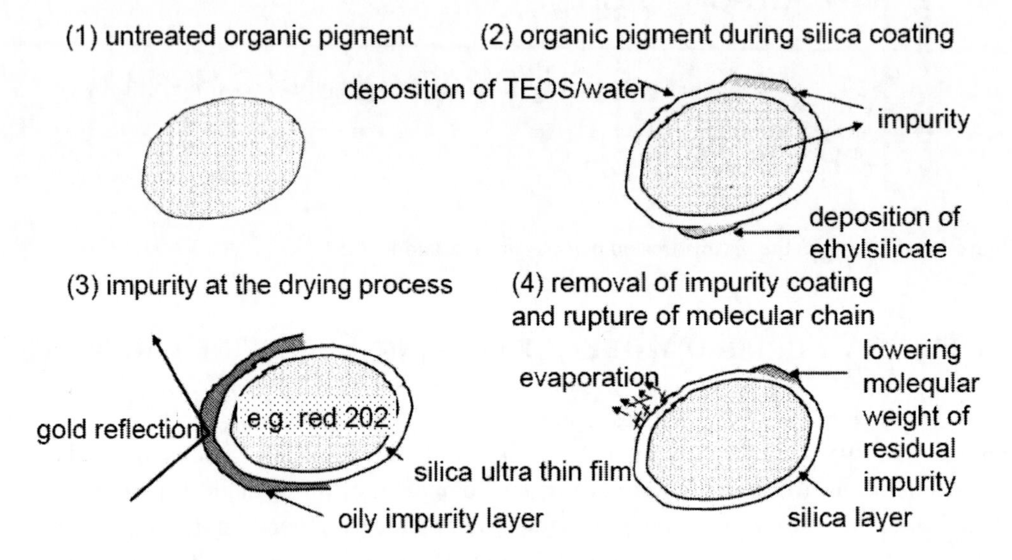

Figure 16. Model of silica coating and removal of impurity.

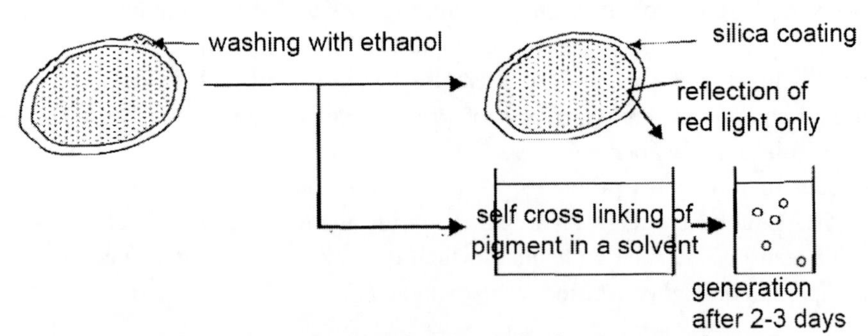

Figure 17. Removal of impurity at the silica coated pigment surface by plasma.

TEOS adsorbed at the particle surfaces will separate un-reacted materials and intermediates adsorbed at the particle surfaces outside of the TEOS coating. In the next drying process, these oily impurity coatings will surround a bottom part of the pigment. In this step TEOS coatings become a precursor of amorphous silica. However, at this condition, injected light into the impurity layer with a low reflective index shows total reflection at the pigment surface, resulting in appearance of golden total reflection. Therefore pure red would not be obtained.

Plasma treatment will attack such impurity layers and evaporate them. The photo-image of the plasma treated dispersive material is shown in Figure 18.

Figure 18. Dispersed substance after plasma treatment.

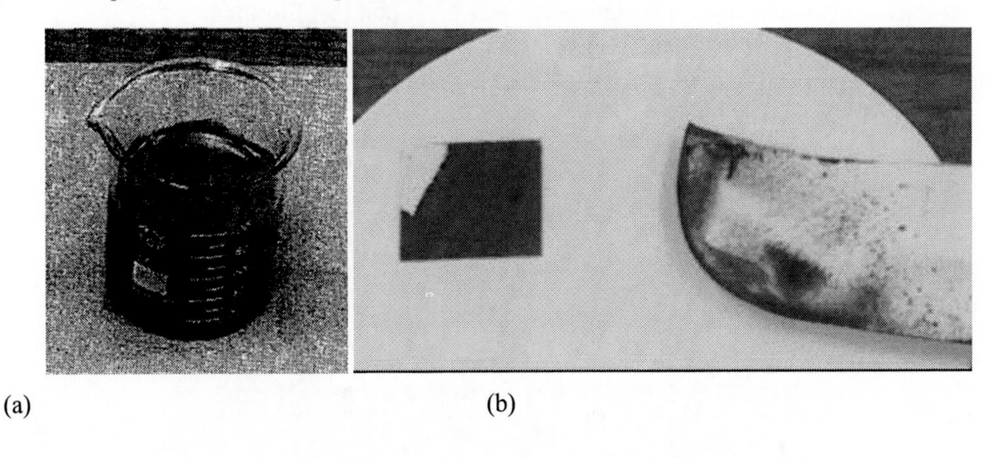

(a) (b)

Figure 19. (a) golden reflection light. (b) left and right are taken from the specimen in figure 18 at the liquid surface and the bottom, respectively. Rdéd202: Lithol Rubine BCA= Carmine 6B.

Figure 19 (a) and (b) shows golden reflective light by the impurities of untreated red pigment. In both cases of the dispersed liquid surface (a) or dispersive material cake (b) reflection light involves gold. Figure 20 shows the setup of plasma treatment. In the case of water dispersed material of silica-coated red pigment, in order to reduce the amount of impurity by long plasma treatment parallel plate type middle low pressure He plasma (100 m long) was used at Ritsumeikan University.

Figure 20. A helium discharge at 100 Torr, and the treatment of pigment. A glow discharge is seen at around 1 cm from the parallel plate capacitive anode. The organic colouring matter is placed 4 cm away from the discharge, and is treated for 10 minutes to synthesize silica using half-dissociated TEOS.

Figure 21 shows the filtrate and the pigment-dispersed material of ethanol-washed silica-coated pigment after sufficient plasma treatment by the above mentioned method. The filtrate shows light orange. The pigment-dispersed material hardly shows golden reflective light. It is therefore indicated that the impurities in the organic pigment were almost perfectly separated.

Furthermore, Figure 22 shows in the orange drainage the growth of red pigment by self cross-linking after two days. It is considered that N=N bond was formed.

In such a way, it is found that plasma can not only treat surfaces but also affect purification significantly in terms of dissociation reaction of ethyl-silicate.

Figure 21. Separated pigment which is not reacted after plasma treatment (left). When this light orange part overlaps with pure Red202 colouring matter which is the reaction intermediate of the source material, it shows golden reflection.

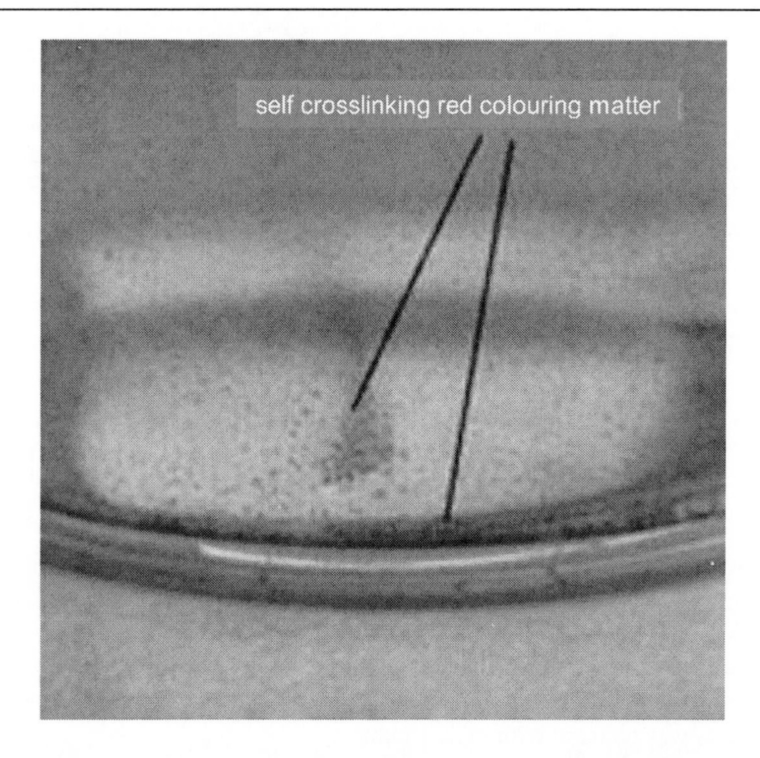

Figure 22. The red particles are produced by the formatting of N=N bonding due to the self radical polymerization of unreacted pigments (leaving the separated reaction intermediate for two days).

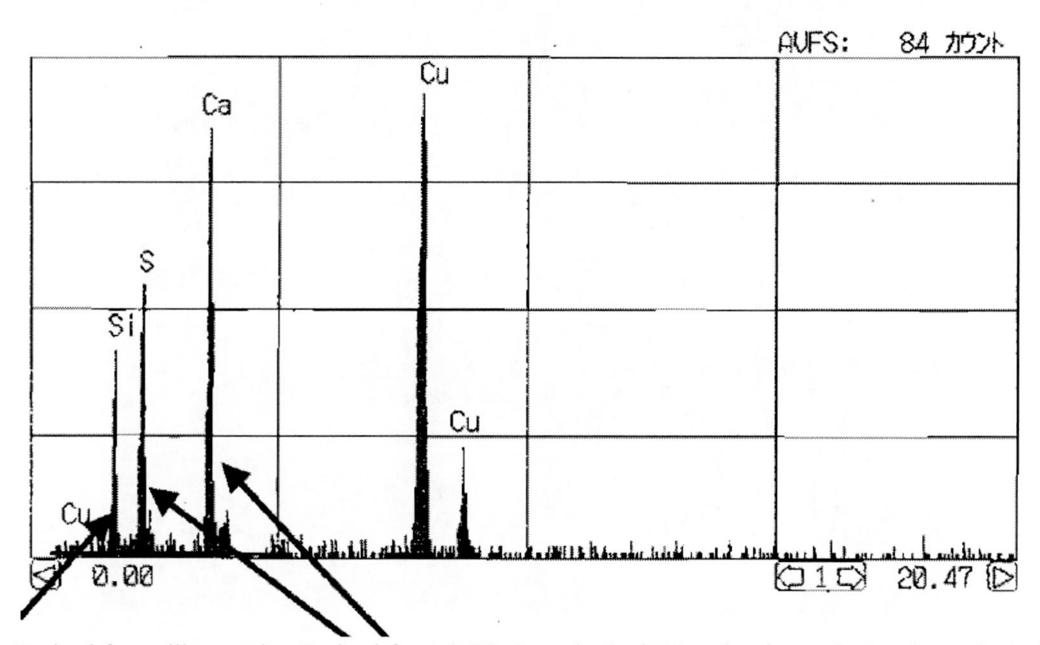

Derived from silica coating Derived from C-SO_3Ca at the Red202 molecule terminal at the surface of colouring matter.

Figure 23. EDX of Red202 of the silica coated particle on the copper mesh. Silicon which is not involved in the source is clearly detected.

Figure 23 shows the EDX data of silica-coated particles of the red pigment (Red202). The 100kV transmission electron microscopy was used for the confirmation of the existence of silica which is formed on Red202. S and Ca in the EDX analysis data indicate the original bonding states. Si is derived from the silica coating which is formed at the surface.

It is noted here that if TEOS is attempted to be treated simultaneously in a chamber using the CVD method, Red202 will be immediately damaged. SO_3Ca is cut from benzene ring and the colour changes from red to black.

Red202: Lithol Rubine

BCA= Carmine 6B $C_{16}H_{12}CaN_2O_6S$

6. Images of Silica-Coated Particles Captured by HREM

It is often believed that detailed information on organic colouring matter is not obtained using high resolution electron microscopy, since the colouring matter is evaporated by the electron beam. The author and Kaito in the Science and Engineering Department, Ritsumeikan University, broke-through this difficulty. The equipment used is H9000 transmission electron microscopy, and acceleration voltage was 300 kV. Figure 24 and 25 show the images of the particles captured at the limit of the system. Furthermore analysis of inorganic particles was realized with high accuracy.

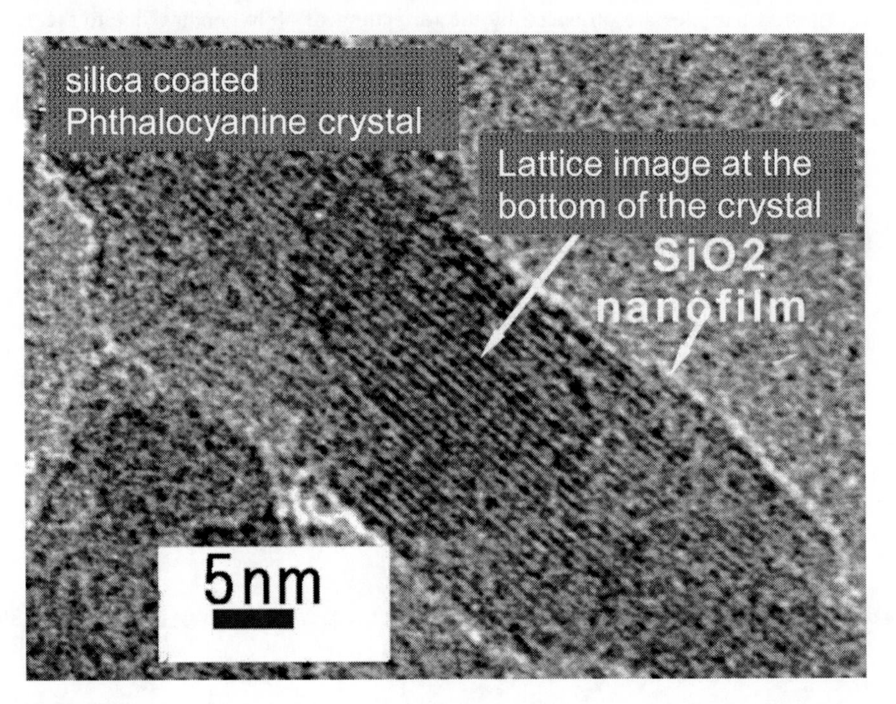

Figure 24. HREM image of silica coated copper phthalocyanine.

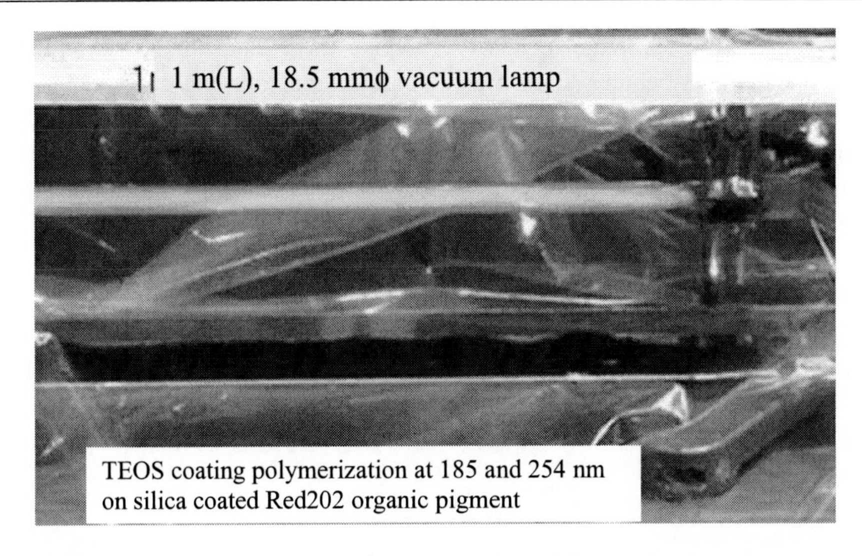

Figure 25. UV equipment which can polymerize SiO_2 coating (right).

The lattice spacing of nano-silica coated organic colouring matter is measurable. However, observation and imaging of crystal lattice image of untreated organic compound is significantly difficult. It is seen that they are dispersed o a significantly thin flake like single-crystalline from the particle contrast.

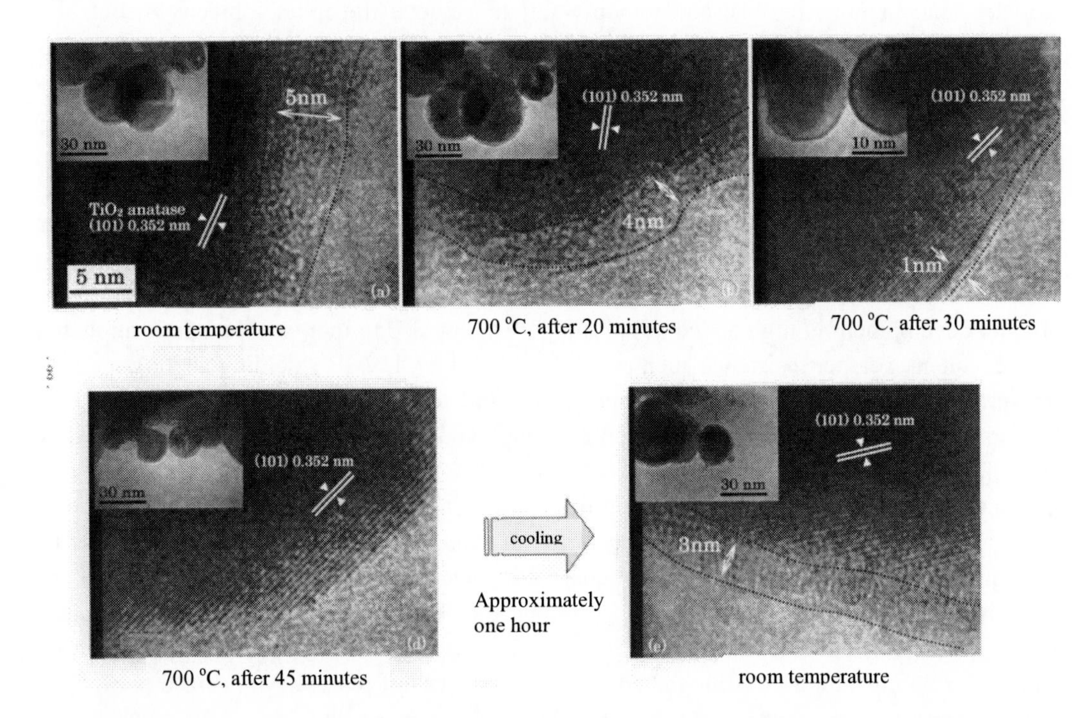

Figure 26. HREM image of nano-silica coated titanium oxide synthesized by plasma. In situ observation during heating and moving silica. It is in situ observation during heating in HREM of the specimens of titanium oxide on which silica single nano coating is deposited. It is seen that the silica coating is crystallized and absorbed in titanium oxide. Finally, as the cooling proceeds, silica coating is re-precipitated at the particle surface with reduced thickness of 3 nm. It is indicated that some are injected as SiO_4 into titanium oxide crystal.

It was said to be impossible to observe precise crystalline conditions and molecular conditions at high magnification, since organic colouring matter can immediately evaporate by the electron beam by using a normal 100 kV electron microscopy. The use of 200-300 kV HREM further enhances this phenomenon. However, organic particles deposited with 1 nm – several nm thick silica nano coating show high resistance against electron beams. Its internal crystalline structures were clarified. It is thought that this operation method would be useful for detailed observation of living issues, and in the future DNA or viruses.

HREM analysis is the core in the nano level silica coating technology. It is particularly important for the quality control. As mentioned before, the design of particles will not be completed by the simple plasma treatment only. However, plasma treatment will contribute a lot for inducing the new achievement and affect the design of the final products.

SUMMARY

Design of particles was performed which will significantly contribute to the next generation industry such as organic colouring matter, inorganic highly functional particles, phosphors, and biomedical nano particles aimed for medicine and pharmacy, considering the application of atmospheric pressure plasma and microwave chilled plasma in mind.

A plasma, which is one of the revolutionary techniques for the perfect dispersion of particles, can in turn, if misused, not only fail to achieve the initial objectives, but also accompany complete damage of the material.

By efficiently utilizing the research achievements that have been developed by many predecessors, conception and extension of more rational material designs have been developed. The outcomes have been provided to the markets in the areas of electric materials, chemical engineering, and cosmetics, with subsequent commercial productions. In the future, seeing worldwide, endless challenges are waiting in the areas where they can contribute to human being.

In this respect, organic compound particles are expected to be applied not only to the product development in high added-value categories, but also to more stable colouring matter with excellent weather resistance such as ink for the inkjet, UV-IR shielding coating material (or spray), gravure printing for the shrink films used for PET bottles, acryl emulsion high functional compounds, many other compounds, high added-value electric devices, FPD photo controlled films, and window materials for automobiles, buildings, ships, and aeroplanes.

In the case of inorganic compound particles, materials design is indispensable for particle materials which are applicable for various areas of global applications with already covering most areas and leading, cross-linking boundary areas such as electronic, imaging, medical and medicine, dairy life related materials, adsorption, magnetic properties, optical shielding, and scattering and its hybrid type interfaces. Figure 27 shows the model of the particle design for which plasma must be used. Charged condition of surfaces of plasma synthesized particles is a novel nanostructure that can adsorb and fix protein, amino acid, and neurotransmission materials. This property can lead to the future prospect of development.

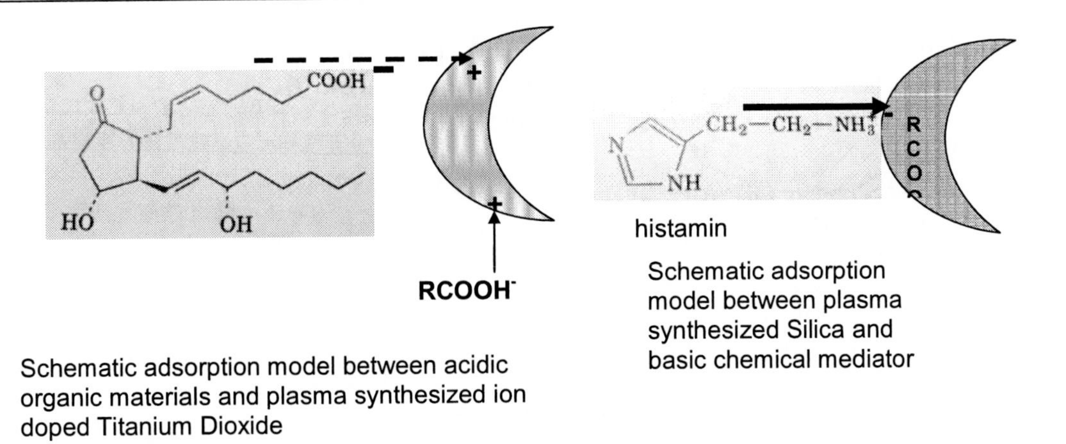

Figure 27. Interaction model at the plasma synthesized nano-particle surface in which synaptic transmission of itch is prevented by blocking H1 receptor of histamine.

ACKNOWLEDGMENTS

The author is grateful to have an opportunity to contribute to this review book. Professor Kogoma in Sophia University, Professor Kaito in Ritsumeikan University and Professor Shimada in Hokkaido University are highly appreciated for developing the techniques together with the author. The author is grateful for Dr M. Makino in the national Sanatorium Oku-Komyo-En for valuable advices in the view of medicine, and many people who have assisted the author. As for the basic techniques of cosmetics, ISI Ltd. and Pias Corp. have collaborated for development of basic thin film synthesis techniques.

REFERENCES

[1] Inoue, Y.; Takeda, A.; Kawasaki, K. *Abstract Jpn. Inst. Metals Autumn Meeting*, 1998, *289*, 355.
[2] *ISI-Chem Eng, Tiny particles*, 1994, *39*.
[3] Takeda, A.; Miyahara, M.; Sakatani, H.; Sato, H.; Sato, H. *Jpn. Patent.* 2002, JP2002309173.
[4] Takeda, A.; Kogoma, M.; Horiuchi, C. *Jpn. Patent.* 2002, JP2002308716.
[5] Takeda, A.; Sato, T.; Kaito, C.; Kaneko, S. *Thin Solid Films* 2003, *435*, 211-214.
[6] Takeda, A.; Sato, T.; Kaito, C. *Jpn. Ass. Crystal Growth* 2002, *29(2)*, 108.
[7] Shirakura, A.; Takeda, A.; Kaito, C. *Proc. World Pak. Conf.*, Michigan State Univ. 2002.
[8] Shimada, S.; Yoshimatsu, M. *Thin Solid Films* 2000, *370(1-2)*, 146-150.
[9] Kaneko, K.; Hayamizu, T. *Jpn. Patent.* 2006, JP2006253454.
[10] Frey, A.; Neutra, M.R.; Robey, F.A. *Bioconjugate Chem.* 1997, *8(3)*, 424-433.
[11] Bender, A.; Immelmann, A.; Kreuter, J.; Rubsamen-Waigmann, H.; von Briesen, H. *Int. Conf. AIDS.* 1996, Jul 7-12; 11:64 (abstract no. MoA.1056).

[12] Enormous number of articles related to nanotechnology and nanoparticle design in the comprehensive study in cancer at National Cancer Institute (NCI), USA.

[13] Mori, T.; Tanaka, K.; Inomata, T.; Takeda, A.; Kogoma, M. *Thin Solid Films* 1998, *316*, 89-92.

[14] Takeda, A.; Sato, T.; Kaito, C. *Jpn. Ass. Crystal Growth* 2002, *29(2)*, 167.

[15] Takeda, A.; Kaito, C. *Surf. Finishing* 2002, *55(12)*, 79-81.

In: Generation and Applications of Atmospheric Pressure Plasmas ISBN: 978-1-61209-717-6
Editors: M. Kogoma, M. Kusano and Y. Kusano

Chapter 17

ATMOSPHERIC PRESSURE PLASMA STERILIZATION AND ITS BIO-APPLICATIONS

Tetsuya Akitsu*[*1], Siti Khadijah Binti Za aba [1, 2], Hiroshi Ohkawa[1], Keiko Katayama-Hirayama[1], Masao Tsuji[3], Naohiro Shimizu[4] and Yuichirou Imanishi[4]

[1]Interdisciplinary Graduate School of Medicine and Engineering, University of Yamanashi, Yamanashi, Japan
[2]School of Mechanic Engineering, Universiti Malaysia Perlis (UniMAP), Pejabat Pos Besar Kangar, Perlis, Malaysia
[3]Yamanashi Industrial Technology Center, Yamanashi, Japan
[4]NGK Insulators LTD., Aichi, Japan

INTRODUCTION

Recent technological development enabled non-thermal plasma treatment over a large area and large volume at the normal atmospheric pressure. Along with industrial applications such as surface cleaning and fictionalizations, applied research in biotechnology is also carried out, including treatment of living tissues in medical applications, inactivation of pathogenic microorganisms, and improvement of bio-compatibility of implants used for extra-cellular matrix. In the low temperature sterilization of medical devices, the ethylene oxide (EtO) sterilization is the primary choice [1-6]. Because of the explosive properties of this gaseous agent, novel sterilization scheme has become strictly important for the low temperature sterilization of heat-sensitive materials, used in various medical devices. The most popular low temperature sterilization method uses gaseous agent such as ethylene oxide and liquid such as formaldehyde. Ethylene oxide sterilization is one of the central pillars supporting the sterility assurance of one-time use medical care materials. Some trials to reduce the flammable properties using the mixture of chlorofluorocarbon (CFC) need reconsideration from the view point of the environmental restriction. The method without

* Correspondence author: Interdisciplinary Graduate School of Medicine and Engineering, University of Yamanashi, Takeda 4-3-11, Kofu, Yamanashi. 400-8511 Japan, Tel: +81-55-220-8478, akitsu@yamanashi.ac.jp

CFC, namely 100% EtO sterilization has been developed; still many problems are left unresolved [7]. The examples include long term carcinogenic properties, for example, by the residual compound adsorbed by the surface of sterilized materials. At high density, a residual gaseous agent may cause acute poisoning. The chronic exposure to low density gaseous agent may cause internal disease, and cancer producing effect. In Japan, the use of EtO is controlled by a law of the control of emission of chemical compounds being effective since 2001. Thus, the gas sterilization is performed following rigorous protocols. The aeration cycle is significantly longer than the exposure needed for the sterilization that reduces the efficient rotation of expensive medical equipments [3, 8-10]. The liquid chemical agents have known shortcomings such as the skin damage by the residual compounds experienced by hospital or health care staffs, but the chemical sterilization is being performed in major aspects of the total disinfection system [6, 9-13]. As alternative methods to replace the EtO sterilization, hydrogen peroxide plasma and radiation sterilization using electron beam have been developed. Another attractive method, the gamma ray sterilization needs huge and isolated site for the safety operation. As all sterilization method does, the gamma ray sterilization destroys the surface of materials as well as the chemical bonding and the cross-link inside the material. Thus, the properties of polymer can be modified [14]. All of the weak points of the existing sterilization methods have been accelerating the development of alternative methods. The hydrogen peroxide plasma sterilization is approved by FDA in the United States and by the Ministry of Public Welfare in Japanese Government, and being introduced to medical facilities [6, 9, 10, 15, 16]. Sterilization scheme is also required in the disinfection control of the production of the disposable health-care equipments. The hydrogen peroxide plasma sterilization can have larger market share in Japan, for the control of the shelf-stock and the prevention of the medical malpractice, following the gradual scale down of EtO sterilization. It is needless to stress the importance of the sterility assurance of implants and recycled health-care equipments, such as an endoscope in medical facilities. The required properties of ideal sterilization equipment can be described in the following sentences. The processing time: a shorter processing time compared with the auto-claving or dry heat sterilization, order of one hour, matched to variety of medical or health care materials to be sterilized at low temperature around 60 degrees C as in the case of EtO or Hydrogen peroxide plasma. Safe, harmless and low damage process for operators, patients and objects are requested.

The basic scheme for the low temperature plasma sterilization is the destruction of the intrinsic structure of microbes by the radical and ultra-violet radiation excited by the discharge, oxidation by atomic oxygen or hydroxyl radicals, collision and intrinsic excitation by rare gas and nitrogen molecules, the destruction of DNA by the Vacuum-Ultraviolet radiation, etc. This sterilization process has safe and harmless mechanism, because the plasma radicals and UV light show sufficiently short life-time. After the discharge is switched off, the excited gas molecule and radicals remain for several micro or milli seconds. Thus, we can understand that no risk by residual plasma radicals may happen to operators [1, 17-19]. Process should exhibit sufficient sterilization efficiency to microorganisms, while the deterioration of exposed objects should be controlled at the minimum level. Until recently our knowledge on the sterilization and inactivation mechanism was limited. Only the empirical relation between the plasma conditions and the sterilization efficiency was learned from the low-pressure plasma sterilization experiment. The mechanism is quite different from the conventional method such as auto-claving. The role of plasma radical was distinguished and the inactivation mechanism was learned in the experiment using spore forming bacteria [20].

In this section, topics are organized as follows; review on the plasma sterilization experiment using rare-gas, oxygen and nitrogen mixture and other schemes; brief history of the first step of plasma sterilization, study on basic process and contribution to the inactivation mechanism. A problem of organic loading effect will be discussed after some experimental comparison and a study on the plasma sterilization at the normal atmospheric pressure.

1. Plasma Sterilization

Plasma is ionized state of gas excited by electric field of stationary, direct current, or alternative direction at high frequency. Some difficulties always happen when the interdisciplinary research field in collaboration with the medical science. The nomenclature "Plasma" can indicate completely different idea, blood plasma. Well, in this side, the plasma consists of negatively charged high energy electrons and positively charged low energy ions, and both have almost equal densities. Non-thermal plasmas are generated by an ionization process, in which the electric field accelerates electrons and complete ionization, and the energy transfer from electron to heavy particles before the termination of one cycle [20-27].

In plasma discharges at normal atmospheric pressure, neutral atoms and molecules are mixed with ionized particles. Each group shows different thermal equilibrium of different temperature. This state is called non-thermal plasma. Neutral particles have internal energy of vibration or electronic excitation. Relaxation from the excited state causes emission of light and radicals. Light is emitted by excited state in the vicinity of the object, and then distracts large molecules of microorganism. Otherwise, the collisional sputtering (Intrinsic excitation) and oxidation can reduce the structure of the microorganism to volatile compounds. Because the UV-C light emission in the vicinity is especially efficient, specific compound such as NO is selected in the sterilization process [17-19, 20, 22-27, 31, 32]. A technical term "Glow discharge" indicates a low pressure cold cathode discharge, but the same term is frequently used for a kind of discharge phenomena at the atmospheric pressure. For example, experimental conditions to generate low temperature and homogeneous "atmospheric pressure glow" is searched by number of researchers. Probably homogeneous glow discharge is realized in extremely rare experimental conditions, or at least needs pre-ionization by electron beam or soft-X ray. Actually, most of the experimental scheme generates quasi-homogeneous state by the accumulations of small electron avalanche, at the normal atmospheric pressure region, of the characteristic length of 100 μm. For the low temperature plasma processing, the application of spatial after glow is also important. Plasma sterilization can be realized in the exterior of discharge region, by UV-irradiation or the afterglow. Here afterglow contains long life radicals and relatively lower population of high energy particles compared to the discharge region. In the discharges of oxygen-nitrogen mixture, sterilization is realized by reactive species such as excited oxygen atom, super anion O_2-, and nitrogen oxide by oxidation process. Recent study showed termination by nitrogen atoms, azide N_3-, is also important process. In the afterglow process, excited particles must be transported within their short lifetimes. Sufficient flow rate of discharge gas is also important for the cooling of discharge regions or elimination of partially ionized trace of discharge. In most cases, for the excitation of homogeneous "glow" discharge, minimum gas flow rate is known to exist. On the other hand, cascade excitation is also important.

The benefit of the sterilization process in the afterglow region can be summarized as in the following part. In the discharge region, especially in the high density plasmas, equivalent gaseous temperature reaches to several hundred or several thousand °C. On the other hand, in the afterglow region, the temperature can be reduced to temperatures lower than around 50 °C which is required for the treatment of thermally sensitive materials. Directly exposed materials are decorated by ion bombardment that is sometimes inconvenient. In the afterglows region, the populations of charged particle are relatively insufficient for the sheath. Actually activated neutral particles play primary role in the sterilization in the afterglows, and discharge field is not necessary [33, 34]. Afterglows can be expanded in relatively larger volume, but sterilization can be completed within shorter time in the discharge field. In each mode, generation of sufficient amount of reactive species should be verified by chemical indicator or other physical method.

Generation scheme of low temperature plasmas for the sterilization, pre-ionization and temperature control needs special experimental maneuver. Plasma container can be designed without forced cooling just like fluorescent lamps, because positively charged ions and neutral particles have relatively low energy compared with electrons. From the comprehensive review by Eliasson and Kogelschatz, neutral-gas temperature can be reduced as low as room temperature in pulse discharge (corona and dielectric barrier discharge). Generally, the neutral gas temperature becomes higher when the neutral gas density increases, because the increase in pressure or particle densities causes the frequency of energy-transfer collisions between electrons and heavy particles, and positively charged ions and neutral atoms become higher [35]. Thus, pulse modulation of excitation is necessary for the control of the neutral gas temperature in the discharge region, at atmospheric pressure. Otherwise, the temperature easily reaches to several hundred °C. Directly exposed low temperature plasma sterilization can be realized in low current density corona discharge and pulsed discharge in DBD electrode structure excited by direct current, low frequency region (1 to 100 kHz), high frequency region (1 to 100 KHz, 13.56, 27.12 and 40.68 MHz at the harmonics frequency of 13.56 MHz) or micro wave region at 2.45 GHz [18, 19, 36-39].

2. Pulse Modulated High Frequency Plasma Sterilization at the Atmospheric Pressure

The author's first encounter with the atmospheric pressure homogeneous glow was a report on experiment presented at International Symposium on Plasma Chemistry, ISPC 8 in Tokyo, by a group of Sophia University, by Okazaki et al [40-42]. The development and the chemical applications of the atmospheric pressure discharge (APG) were carried out by specialists in chemistry as an extension of the ozone generation. In this section, we present the performance in the inactivation of microorganisms of pulse-modulated, high-frequency plasma among various choice of generation scheme at the atmospheric pressure. The activity was evaluated by using biological indicators of spore forming bacteria of *Bacillus* genera and vegetative state of several opportunistic infection bacteria. Figure 1 shows the experimental scheme.

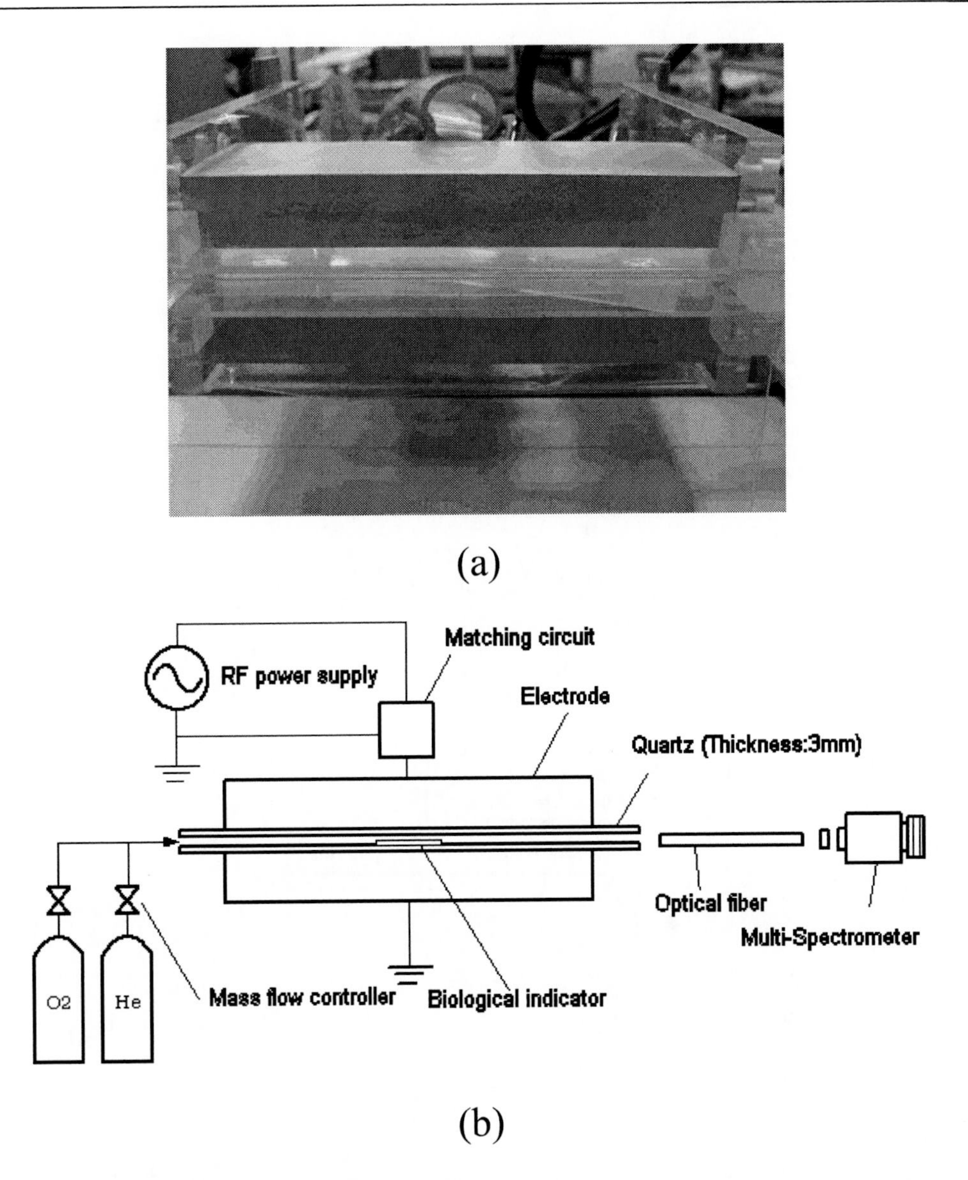

Figure 1 Pulse modulated high frequency discharge at atmospheric pressure (a) PWM-APG discharge and (b) experimental set-up.

This apparatus was energized by a pulse-modulated, high-frequency generator at 27.12 MHz. For the examination of the antibacterial effect we used mortality of biological indicators. For example, *Bacillus atrophaeus* inoculated on to various carriers was exposed to the discharge directly or indirectly through porous non-woven fiber sheet. Both sterilization time and neutral gas temperature varied depending on the incidental power and the pulse width of the modulated high frequency [43, 44]. Typical examples for the neutral gas temperature and the exposure time as a function of the pulse width are shown in Figure 2 (a), and exposure time and the sterilization as a function of the partial pressure of oxygen is shown in Figure 2(b). *Bacillus atrophaeus* is a kind of biological indicator used in the validation of EtO sterilization.

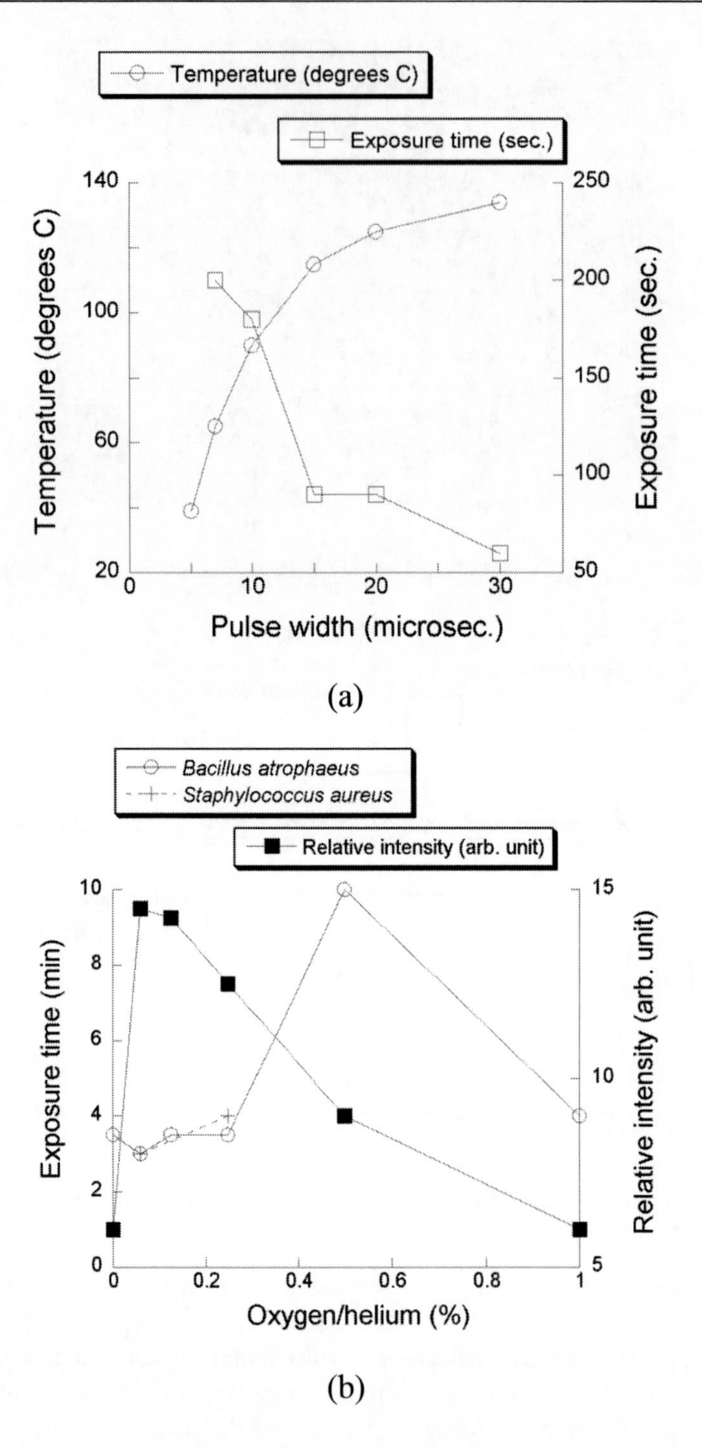

Figure 2. Dependence of disinfection efficiency on operational parameters. (a) Dependence of neutral gas temperature on pulse width and necessary exposure time to plasma discharge for biological indicator, *Geobacillus stearothermophilus*, He: 1.5 L/min; O_2: 1.0 mL/min and (b) dependence of relative intensity and exposure time on oxygen partial pressures for biological indicators: empty circle, *Bacillus atrophaeus* ATCC 9372 ($4.8x10^6$ CFU) and cross lines indicating *Staphylococcus aureus* ATCC 6538 ($1.3x10^8$CFU); Carrier: glass slide, RF power and frequency: 670 W, 27.12 MHz; Pulse width and interval: 10 μs; Helium: 1.5 L/min.

In the pulse-width modulated high frequency discharge at the normal atmospheric pressure, the surfaces of electrodes are covered with dielectric barrier of fused quartz. The volume of the discharge was 70 mm x 150 mm x 3 mm gap. The temperature of aluminum block, the electrode was stabilized at 20 °C by a forced circulation of coolant. Mixture of oxygen and helium was supplied from a multi-aperture nozzle attached to a lateral surface. When high frequency discharge was excited, the temperature of neutral gas rapidly increased to 200 - 300 °C which is higher than the survival limit of organisms. An experimental condition below 90 °C was realized with 50 percent duty and the pulse widths of 10 μs, incidental power from 670 to 700 W on average. The best sterilization was realized when the partial pressure of oxygen was 0.06 percent in the shortest disinfection time of 180 s. In this condition, the intensity of the observed spectral emission from excited oxygen atom was at the maximum. Thus, in the authors' speculation, the result suggests that the advanced oxidation was playing the main role in the sterilization [43-45]. In this experiment, temperature of neutral gas was measured using a florescent-type thermometer based on the relaxation time of fluorescence coated on an optical fiber probe (FL-2000, ANRITSU Measurement Co.). This optical-fiber thermometer enabled the measurement in high electric field. Emission spectrum was measured in the vicinity of the triplet spectrum at 777 nm, using a multi-channel spectrometer (Mono-Spec 18, Thermo-Vision Colorado Inc., USA) installed with a Pelletier-cooled CCD camera (ST-6V, SBIG Inc.). An optical-fiber was installed facing the extremity of the discharge gap. The multi-channel spectroscope has sensitivity from 200 to 800 nm.

Anti-biological effect was tested on the basis of the mortality of micro-organisms by measuring the population of non-cultivable state of microbes exposed to the discharge. Species of microbe were cultivated in soy-bean casein digest (SCD) fluid culture medium (Becton Dickinson Company Ltd.). 0.01ml was sampled and cultivated on SCD agar culture medium, as shown in the later section. Microbe sample was collected using a platinum wire from colonies on the agar dish and dissolved in distilled water. The density was adjusted to approximately 10^7 CFU/mL, and then 0.1 mL of the suspension was sampled and inoculated on sterilized glass with a size of 18 mm x 18 mm square and 150 μm thick. The inoculated carrier was left dry in a biological clean bench and sealed in a *Tyvek* pouch of a non-woven polyethylene fabric, 0.15 mm in thickness. A preferred solution using the plasma process is post-sterilization of objects through sterile package. Selected species of microbe are *Bacillus atrophaeus* ATCC 9372. When the initial population was $1.1x10^7$ CFU, propagation was observed after 20 minutes of exposure. Thus sterilization of sterility assurance level 6 (SAL6) was failed when the density was too high. On the other hand, samples of $5.4x10^5$ CFU was sterilized in 20 minutes and $2.7x10^4$ CFU in 5 minutes. Thus, sterilization was realized inside the sterile package in lower population. Similar experiment was repeated using *Geobacillus stearothermophilus* ATCC 7953, and samples of $6.6x10^5$ CFU was sterilized in 20 minutes and $3.3x10^4$ CFU in 5 minutes. This result is summarized in Table 2 in the later section. *Geobacillus stearothermophilus* is thermophilic spore forming bacterium that grows best at higher than normal temperatures used in the validation of the autoclave sterilization system.

Figure 3 shows the preparation of the biological indicators used in this experiment. Wide variety of anti-bacterial effect was tested using a variety of spore forming bacteria of the *Bacillus* genera and a part of microbes that cause opportunistic infections on the basis of the mortality or non-cultivable state of microbes. The selected species of microorganisms: Gram stain positive: *Staphylococcus aureus* ATCC 6538; Gram stain negative: *Escherichia coli* ATCC 8739, *Salmonella enteritidis*, yeast and mold: for example, *Candida albicans* ATCC 10231, and spore forming bacteria: *Bacillus atrophaeus* ATCC 9372, *Bacillus subtilis* ATCC 6633, *Geobacillus stearothermophilus* ATCC 7953, and Bacillus pumilis ATCC 27142 were all from Raven Biological Laboratories Company [6, 46-52], where ATCC stands for the American Type Culture Collection and CFU stands for Colony Forming Unit, the density of cultivable end-spore. In the incubation and sterility judgments, we followed protocols for sterility testing in the Japanese Pharmacopoeia, the JP 14th edition [53].

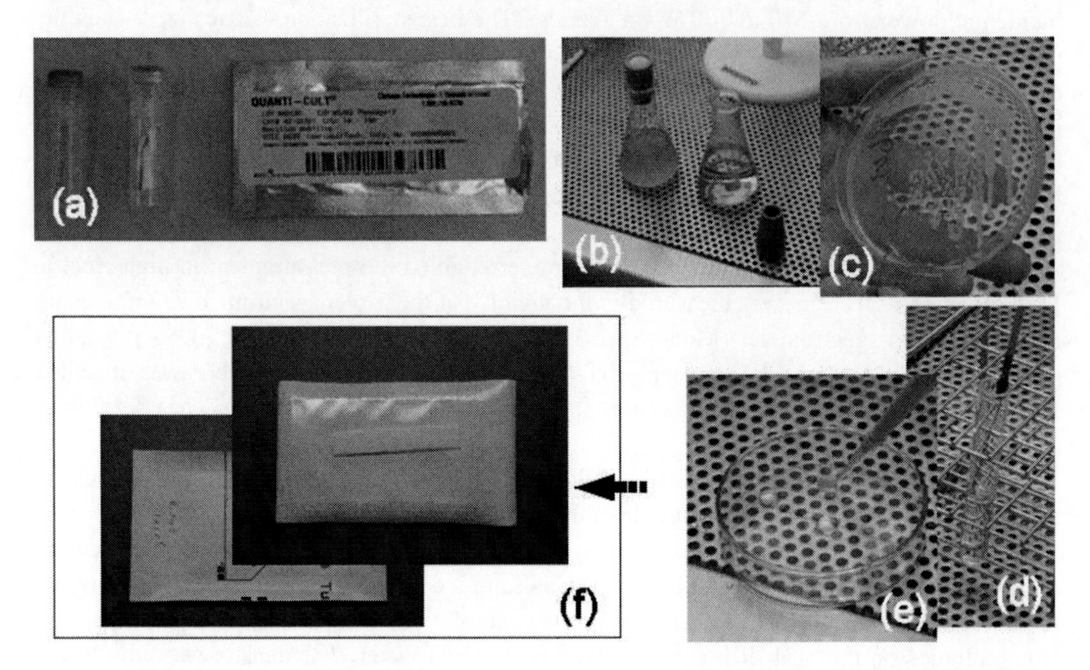

Figure 3. Preparation of biological indicators. (a) - (c) Incubation of biological sample, (d) adjustment of density in a suspension, (e) inoculation on to carrier, and (f) biological indicator carrier sealed in a sterile package, *Tyvek* ®.

After the plasma treatment, the cover glass carriers were removed from the *Tyvek* package and were incubated in 100 mL of SCD liquid culture medium. *Geobacillus stearothermophilus* was incubated at 57 degrees C, *Candida albicans* at 25 °C, and other microorganisms were incubated at appropriate temperatures. After 7 days of incubation, sterility testing was judged based on turbidity of the culture medium. If $6\log_{10}$ grade sterilization was performed successfully, culture medium remained clear.

Commercially available biological indicators: Attest ™ type 1262 and type 1264, from 3M Co. were used for screening of experimental conditions. One vial of Attest ™ type 1262 contained *Geobacillus stearothermophilus* ATCC 7953, 8.6×10^5 CFU and type 1264 contained *Bacillus atrophaeus* ATCC 9372, 4.8×10^6 CFU. Incubation was carried out in a specially designed incubator at appropriate temperatures, 57 degrees C for *Geobacillus stearothermophilus* and 35 degrees C for *Bacillus atrophaeus*. After the incubation for 48 hours in vials, the sterility was judged based on the turbidity, or a change of the incubation medium according to pH indicator.

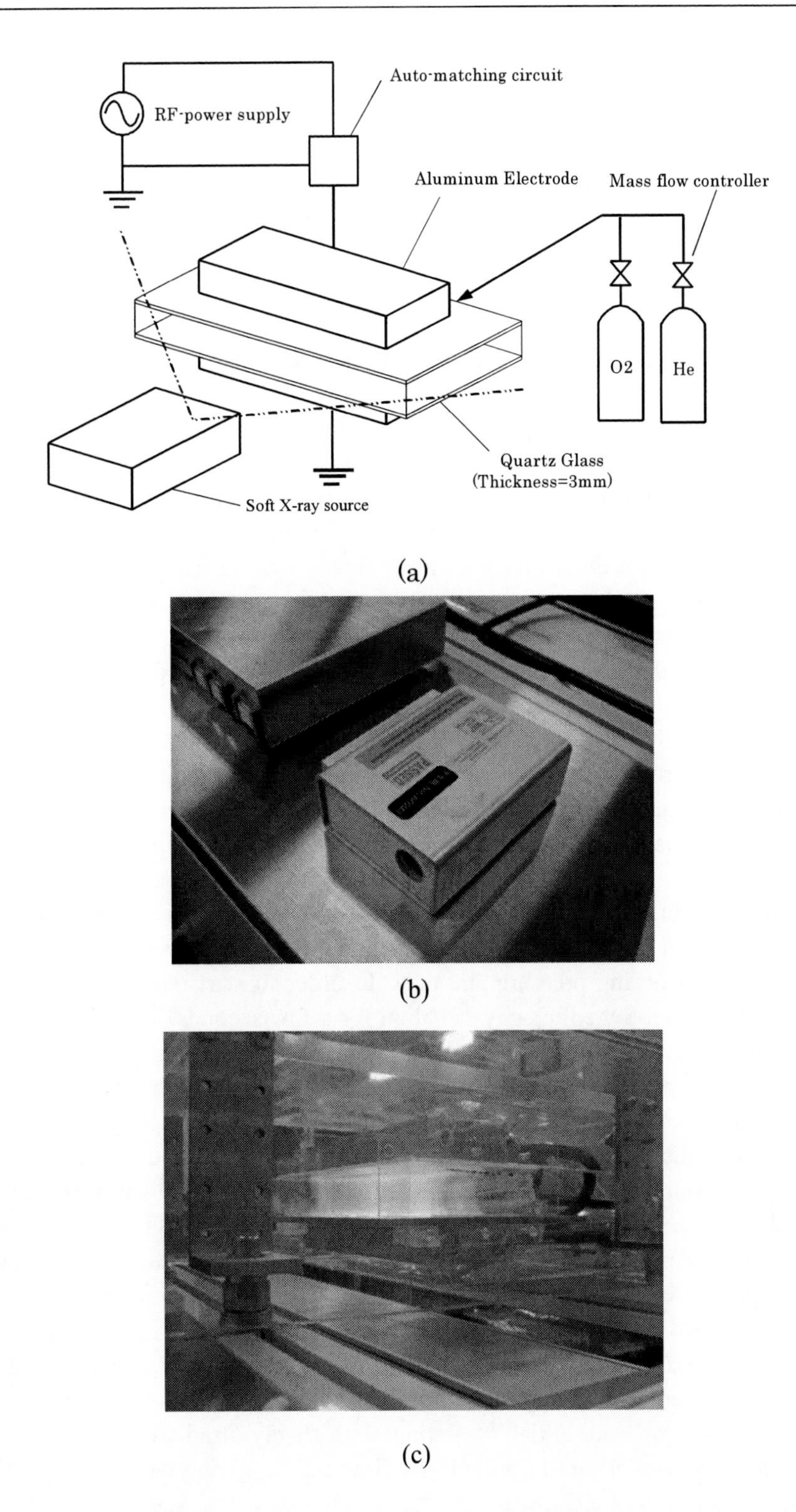

Figure 4. Soft-X ray pre-ionization of wide-gap radio frequency (RF) glow at atmospheric pressure. (a) A bird-view figure showing the layout of optical ionizer and discharge electrode, excited by 13.56 MHz, 20 mm discharge gap and (b) experimental setup of the soft X-ray source and (c) fully matured wide-gap high-frequency discharge at the normal atmospheric pressure, gap 28.5 mm in He flow.

For disinfection of 3-D object or development of hydrophilic property, a wider discharge gap is needed. It is, for example, 15 to 20 mm for small syringe, sample vials and needles. Even in helium based discharge medium, some 10 - 20 kV is needed for appropriate breakdown. The initial breakdown voltage decreases significantly when the discharge volume was pre-ionized by sufficient amount of photo emission from auxiliary electrode radiated by strong UV, electron beam or X-ray. This method is extremely useful for the triggering of wide gap discharge at normal atmospheric pressures. Figure 4 shows an experimental result using soft X-ray pre-ionization.

The dimensions of the electrode were 50 mm x 150 mm in width and length, respectively. Fused quartz plates of 3 mm thickness covered the surface. The dimension of this quartz plate was 30 mm larger than the electrode to prevent direct electrical discharge. The apparatus consists of a linear actuator for programmed transportation of samples of microorganism. The volume of the discharge region was surrounded by heat resistant glass and a removable shutter made of synthetic rubber. An array of aperture of 1 mm diameter was set along the lateral surface of the discharge volume and mixture of helium/oxygen was supplied. The flow rate of the working gas was controlled with mass-flow controllers; in a range from 1 mL/min to 5 L/min. Helium gas has a principal effect to homogeneously excite plasma at high pressures, because of its low breakdown voltage. The discharge gap spacing was varied from 1.5 mm to 30 mm. Wider clearance of the discharge gap is desirable for inline treatment of objects. However, higher voltage is needed to initiate the breakdown when wider discharge gap is employed. Applications of impulse voltage, for example, can initiate discharges in a wide gap. The electromagnetic interference in the pulsed high voltage might cause undesirable footprints on the surface and damage electronic components. Pre-ionization using soft X-ray irradiation can initiate electric discharge between wide gaps without EM damage. The soft X-ray source, Type L-6941, Hamamatsu Photonics Co. Japan, generates soft X-ray with a wave length centered at 0.2 nm; equivalent to a photon energy between 3 and 9.5 keV; with an average dose of 15 mSv/h, and it enables initial ionization over a wide range within 1 m in radius at wide working pressure and area. In order to start the discharge; a thin-wire electrode was illuminated by soft X-ray radiation for a few seconds. The localized discharge develops to fully spread a wide-gap discharge with increasing discharge power.

In the following part we present the result of initial breakdown characteristics; high-frequency power necessary to start discharge. Figures 5 and 6 show typical experimental results for 1.5 to 10 mm and 1.5 to 28.5 mm of gap distances, in helium, helium/oxygen and helium/inert gas mixture at the atmospheric pressure. For the case of a gap spacing of 10 mm, one can find that the required power was 200 W for the initial breakdown, comprising approximately 1.4 kV peak voltages. The volumetric pre-ionization effect alone is insufficient to reduce the power requirement. To resolve this problem, a fine copper wire of 0.15 mm diameter was inserted perpendicularly to the center of the edge. In the experiment, the discharge medium was helium, helium/argon mixture or helium/oxygen mixture. The initial powers were measured under the following conditions: (1) gas flow without initial pre-ionization, (2) gas flow and initial triggering with X-ray irradiation, (3) gas flow and triggering with a fine wire electrode, and (4) gas flow and triggering with a combination of X-ray irradiation on a fine wire electrode.

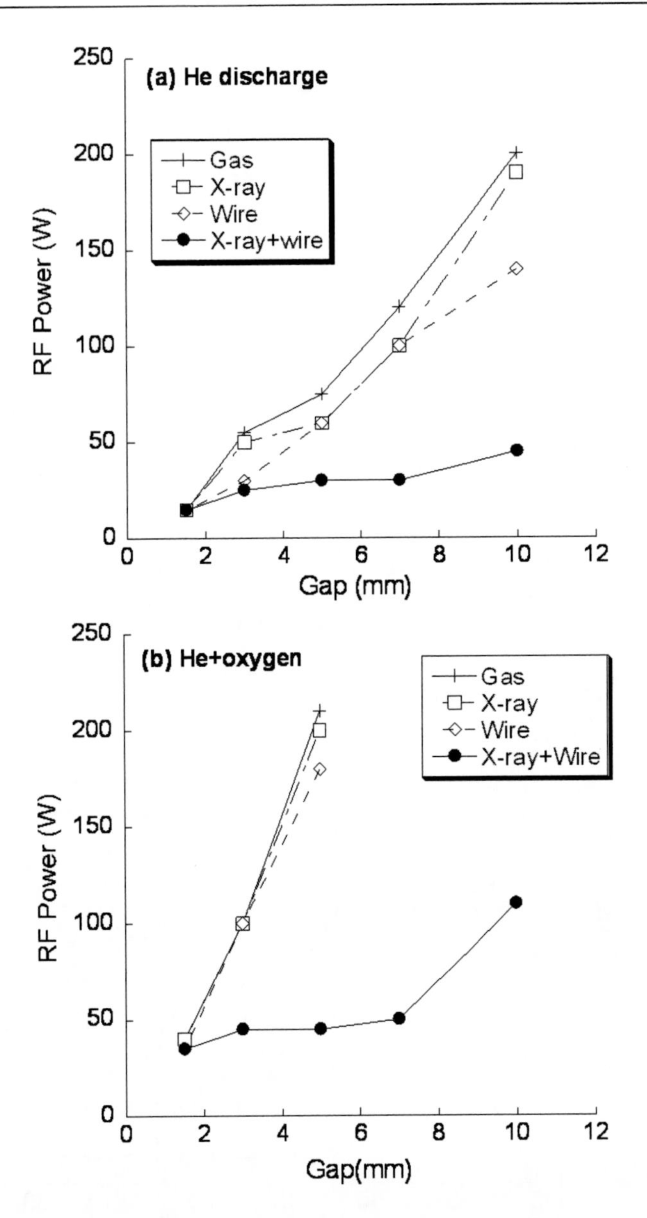

Figure 5. Initial breakdown characteristics. Required RF power for the initiation of APG (a) in He and (b) in He-diluted oxygen (2.5%). Gap: 1.5 - 10 mm.

Figure 5 shows that the required radio-frequency (RF) power was significantly reduced when soft X-ray irradiation was combined with insertion of a fine wire with or without the reactive component of mixture. The required HF power to initiate the discharge decreased approximately 10 W, and when the fine wire electrode was inserted simultaneously, a decrease by 40 W. Figure 5 (a) shows that the power requirement increased to 200 W at a gap distance of 5 mm, when the discharge medium was mixture of electro negative gas oxygen and helium. With soft X-ray ionization on a fine wire electrode, the required RF power decreased to as low as 50 W.

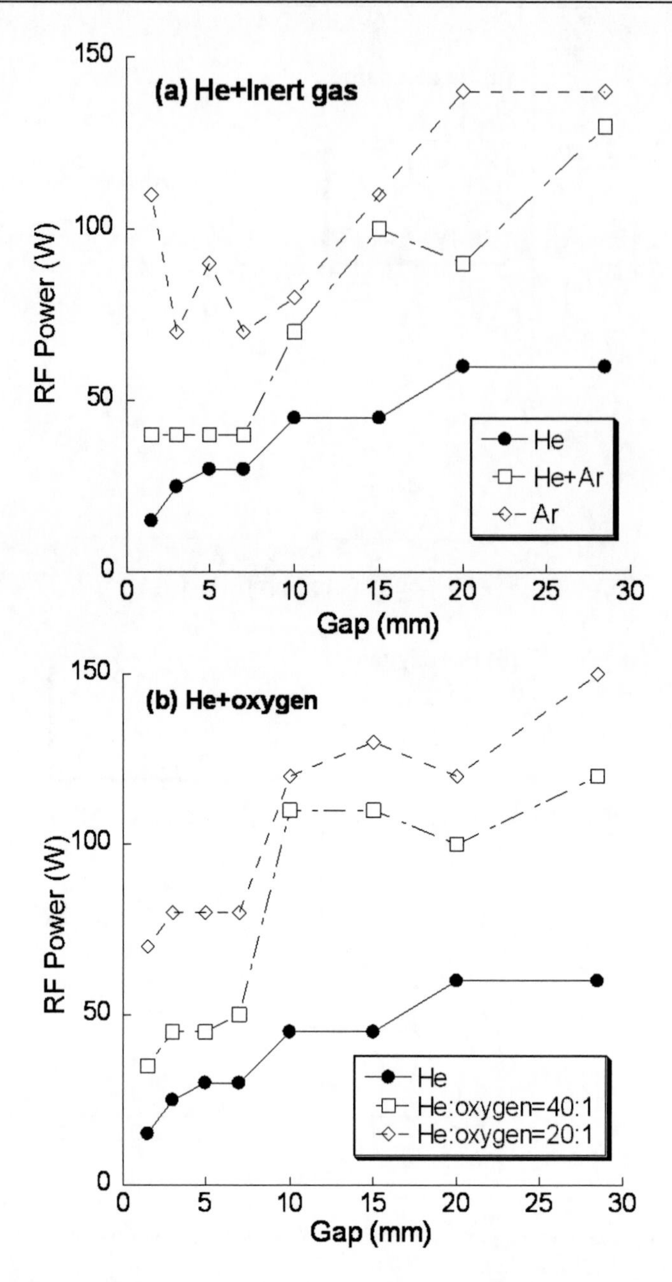

Figure 6. Initial breakdown characteristics. Required RF power for the initiation of APG (a) in He-diluted Ar and (b) in He-diluted oxygen (2.5% and 5%). Gap: 1.5 - 28.5 mm.

Figure 6 shows the dependence of RF power requirement on the discharge gap up to 28.5 mm. When the discharge medium was helium, the initial power is approximately 70 W. Mixture of molecular gas such as oxygen or even argon increases the initial power requirements. Generally, voltage limiting circuit is installed in power supplies, for the prevention of internal breakdown by reflected high frequency signal. Hence, in wider gaps, the initiation of discharge was inconvenient without soft X-ray pre-ionization on an inserted fine wire electrode.

Pre-ionization by a variety of experimental schemes: pulsed spark discharge by an auxiliary electrode; irradiation by laser and X-ray can be solutions of the initial breakdown of a large volume. Another, somewhat more difficult problem arising is temperature control of neutral gas [44].

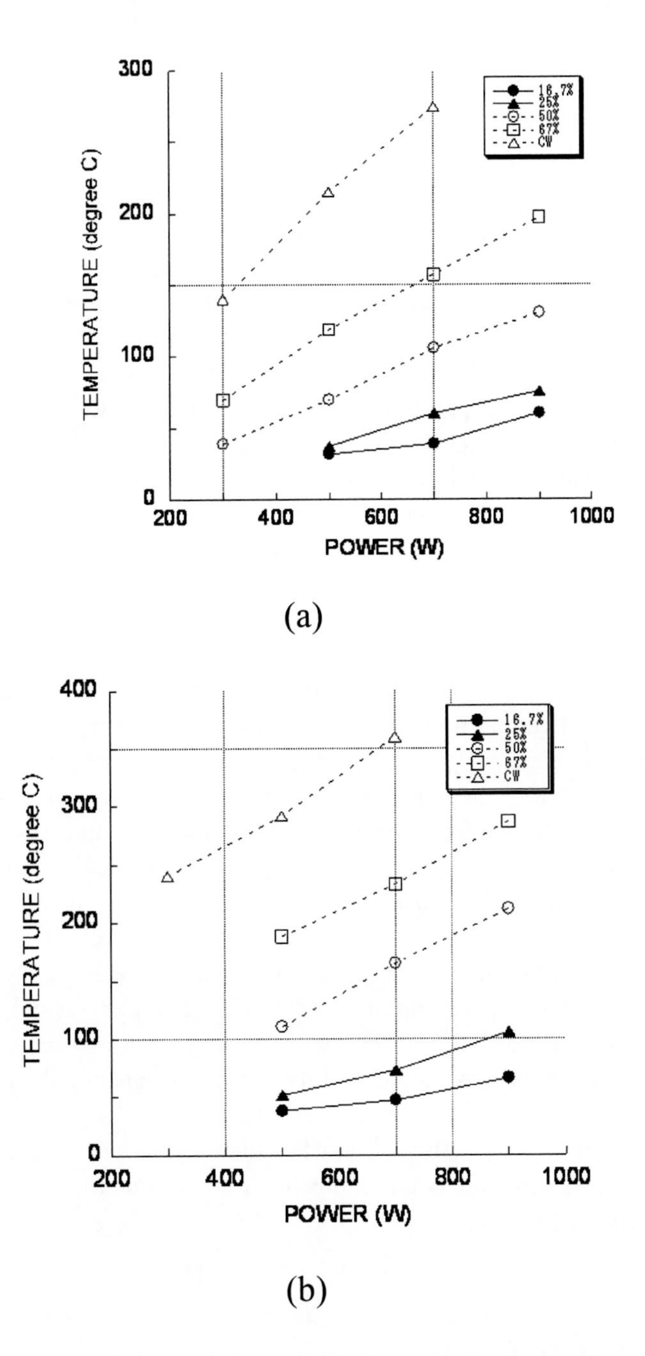

Figure 7. Dependence of neutral gas temperature on the incidental power and the pulse width in (a) 3 mm and (b) 5 mm discharge gap. Duty ratio: 16.7% stands for 10 μs width and 60 μs intervals; 25%, 10 μs width and 30 μs intervals; 50%, 10 μs width and 10 μs intervals; 67%, 20 μs width and 10 μs intervals, at frequency 27.12 MHz, He flow rate 1.5 L/min.

Figure 7 shows the dependence of neural gas temperature as functions of the incidental power and the pulse width of the modulation. An increase in temperature was observed when the high frequency power was increased. At a fixed power, a decrease in temperature was observed when the pulse duration become shorter. The coverage of the discharge becomes a serious problem, when the plasma source was operated at relatively lower power. In this experiment, the temperature of discharge region was measured with a fiber probe. In the case showing gap distance of 3 mm, we can find an appropriate experimental condition for low temperature plasma sterilization (T<90 degrees C). Narrow allowance of experimental conditions around, the average power is 670 to 700 W, pulse width of 10 μs and the same interval. In the case of Fig 7 (b) with 5 mm gap, no appropriate conditions were found. Neutral gas temperature decreases when the duty of pulse decreases, however, antibacterial effect also decreases.

Choosing the experimental condition to maintain the average temperature at 90 °C, the partial pressure of oxygen was varied to find out the minimum sterilization time for a biological indicator: *Geobacillus stearothermophilus*. The intensity of oxygen atom was at the maximum, when the partial pressure of oxygen was 0.06%, whereas the oxygen and He flow rates were 0.001 L/min and 1.5 L/min, respectively. The authors' speculation of the basic process of the oxidation sterilization is as follows:

$$He + electron \rightarrow He^{*}\left(2^{3}S, 2^{1}S\right) + electron$$

$$He^{*} + O_{2} \rightarrow He + O^{*} + O$$

Higher disinfection efficiency can be expected for a higher concentration of atomic oxygen excited via the Penning process. Increase in the oxygen flow, however, results in an increase in the energy relaxation process, and a decrease in the population of helium in the excited state. The discharge medium contains low concentration of oxygen and nitrogen originated in the back stream from the environment, because this system is open to the surrounding atmosphere. Thus, the highest efficiency was observed near the lower end of the controllable concentration. The disinfection time decreases again at higher oxygen concentrations, at this end generated ozone was detected. In Table 1, we summarize the result of disinfection experiment of microorganisms of vegetative state and in Table 2, the result of sterilization experiment of spore forming bacteria of *Bacillus genera: Bacillus subtilis*, *Bacillus atrophaeus*, *Geobacillus stearothermophilus* and *Bacillus pumilis*. In the incubation and sterility judgments, we followed the sterility test in the Japanese Pharmacopoeia, the JP 14th edition.

The spore of these microorganisms was inoculated on glass carrier at different spore densities. In Table 2, we summarize the disinfection time for *Bacillus atrophaeus* measured at various mixture ratio of oxygen.

Table 1 Antibacterial effect of plasma processing

A. Yeast and other bacteria (Tyvek ® package)

Species and density	Plasma treatment (min)					Dry heat* (min)				
(CFU)	1	3	5	10	15	1	3	5	10	15
Escherichia coli ATCC 8739 ($1.6x10^7$)	-	-	-	-	-	+	-	-	-	-
Salmonella enteritidis ($3.5x10^7$)	-	-	-	-	-	+	+	+	-	-
Staphylococcus aureus ATCC 6538 ($4.7x10^7$)	+	+	-	-	-	+	+	+	+	+
*Candida albicans A*TCC 10231 ($5.1x10^6$)	-	-	-	-	-	+	-	-	-	-

*Dry heat at 90 °C

B. Fungus and other bacteria (direct exposure)
PWM-APG at 27.12MHz

Species and density	Plasma treatment (min)				Dry heat* (min)		
(CFU)	1	2	3	5	1	3	5
Escherichia coli ATCC 8739 ($4.4x10^7$)	-	-	-	-	-	-	-
Pseudomonas aeroginosa ATCC 9027 ($5.0x10^7$)	-	-	-	-	-	-	-
Salmonella enteritidis ($4.0x10^7$)	+	-	-	-		+	+
Enterobactor aerogenes JCM1235 ($6.4x10^7$)	+	-	-	-	+	+	+
Lactobacillus plantarum JCM 1057 ($4.5x10^6$)	+	+	-	-		+	+
Staphylococcus aureus ATCC 6538 ($6.8x10^7$)	-	-	-	-	+	-	-
Aspergillus niger ATCC 16040 ($5.5x10^5$)	+	+	+	+	+	+	+

*Dry heat at 90 °C

C. Spore-forming bacteria: High density (direct exposure)
PWM-APG at 27.12MHz

Species and density (CFU)	Plasma treatment (min)			Dry heat* (min)
	5	10	20	20
Bacillus subtilis ATCC 6633 ($5.8x10^6$)	+	-	-	+
Bacillus atrophaeus ATCC 9372 ($5.1x10^6$)	+	+	+	+
Geobacillus stearothermophilus ATCC 7953 ($3.8x10^5$)	-	-	-	+
Bacillus pumilis ATCC27142 ($6.6x10^6$)	+	+	+	+

*Dry heat at 90 °C. PWM-RF power and frequency: 670 W, 27.12 MHz; Pulse width and interval: 10 μs; Gas: He, 1.5 L/min; O_2, 1.0 mL/min, glass carrier.

D. Antibacterial effect of plasma processing (direct exposure)
CW-APG at 13.56 MHz

Species and density	Plasma treatment (sec)					
(CFU)	5	10	20	30	40	60
Escherichia coli ATCC 8739 ($4.4x10^7$)	+	-	-	-	-	-
Pseudomonas aeroginosa ATCC 9027 ($5.0x10^7$)	-	-	-	-	-	-
Salmonella enteritidis ($4.0x10^7$)	+	-	-	-	-	-
Enterobactor aerogenes JCM1235 ($6.4x10^7$)	+	+	-	-	-	-
Lactobacillus plantarum JCM1057 ($4.5x10^6$)	-	-	-	-	-	
Staphylococcus aureus ATCC 6538 ($6.8x10^7$)	+	+	-	-	-	-
Candida albicans ATCC 10231 ($5.1x10^{6)}$	+	-	-	-	-	
Aspergillus niger ATCC16404 ($5.5x10^5$)		-	-	-	-	

CW-HF APG, 13.56 MHz, 670 W, Gap: 10 mm, He 5 L/min

Table 2. Antibacterial effects through Tyvek package, PWM-APG)
A. Test of antibacterial effect
BI: *Geobacillus Stearothermophilus*. Carrier: glass.

Density (CFU)	Plasma treatment (min)			Dry heat* (min)
	5	10	20	20
6.6×10^5	+	+	-	+
3.3×10^4	-	-	-	+
1.6×10^3	-	-	-	+
8.2×10^1	-	-	-	+

B. Test of antibacterial effects
BI: *Bacillus atrophaeus.* Carrier: glass.

Density (CFU)	Plasma treatment (min)			Dry heat* (min)
	5	10	20	20
1.1×10^7	+	+	+	+
5.4×10^5	+	+	-	+
2.7×10^4	-	-	-	+
1.4×10^3	-	-	-	+
6.8×10	-	-	-	+

*Dry heat at 90 °C. PWM RF power and frequency: 670 W, 27.12 MHz; Pulse width and interval: 10 μs; Gas: O_2/He (0.06%; He: 1.5 L/min; O_2: 1.0 mL/min).

C. Spore-forming bacteria: High density (direct exposure, HF-APG)

Species and density (CFU)	Direct exposure (sec)			
	10	20	30	40
Bacillus subtilis ATCC 6633 (5.8×10^6)	+	+	-	-
Bacillus atrophaeus ATCC 9372 (5.1×10^6)	+	+	-	-
Geobacillus stearothermophilus ATCC 7953 (3.8×10^5)	-	-	-	-
Bacillus pumilis ATCC27142 (6.6×10^6)	+	+	-	-

RF APG at 13.56MHz, 670W, Gap: 10mm, He 5L/min.

The maximum antibacterial effect was observed at the minimum disinfection time at the oxygen partial pressure from 0.06 to 0.09 %. Disinfection time of *Staphilococcus aureus* was measured at various flow rate of oxygen, at 0.06% and 0.25% around the optimum condition. *Staphilococcus aureus* was sterilized in three minutes for the 0.06% group, and in four minutes for 0.25% group. For the 0% group, no sterilized case was observed. This result also reinforces the close relationship between generation of oxygen radicals and disinfection efficiency.

When biological indicator was sealed in sterile packages, longer disinfection time is needed compared with the direct exposure, because part of plasma radicals is consumed by porous non-woven fiber. In Table 2, one can find that the group of 10^7 CFU remained alive after twenty minutes of plasma treatment. This indicator can be sterilized by direct exposure in several minutes. The group of 10^4 CFU was sterilized in twenty minutes, and group of lower spore densities, $2.7x10^4$, $6.8x10^1$ were sterilized in five minutes. Thus, necessary plasma treatment was shorter for lower density groups. Similar result was observed for *Geobacillus stearothermophilus*. Group of $6.6x10^5$ CFU was sterilized in twenty minutes, and group of $3.3x10^4$ and $8.2x10^1$ CFU were sterilized in five minutes. Thus, the required plasma treatment was shorter for lower spore density groups. Every group prepared at different spore densities remained alive after dry-heat treatment at 90 degrees C for twenty minutes. At lower densities such as natural pollution, the disinfection will not occur inside the sterile package, but this pollution can be sterilized from the outside of the package.

Table 3 Antibacterial effect

A. Case of PWM-RF APG at 27.12 MHz

Classification of microorganism	
Gram-stain positive *Staphilococcus aureus,* Disinfection confirmed	Gram-stain negative *Echerichia coli* *Salmonella enteritidis* *Pseudomonas aeruginosa* Confirmed
Yeast / Fungus *Candida albicans* *Aspergillus niger* Confirmed	Spore-forming bacteria *Bacillus atrophaeus/ Bacillus subtilis* *Geobacillus Stearothermophilus* Confirmed, low CFU case

PWM HF: 670W, 27.12 MHz; pulse-width 10 μs; He: 1.5 L/min.; O_2: 1 mL/min., Gap: 3 mm

B. Case of CW-APG at 13.56 MHz

Classification of microorganism	
Gram-stain positive *Staphylococcus aureus* *Lactobacillus plantarum* *Enterobactor aerogenes* Disinfection confirmed	Gram-stain negative *Escherichia coli* *Salmonela enteritidis* *Pseudomonas aeruginosa* Confirmed
Yeast / Fungus *Candida albicans* *Aspergillus niger* Confirmed	Spore-forming bacteria *Bacillus atrophaeus/ Bacillus subtilis* *Geobacillus stearothermophilus* *Bacillus pumilis* Confirmed

CW RF power: 670 W, 13.56 MHz, CW; He: 5 L/min.; Gap: 10 mm.

In Fig. 8, we show a typical result of the survival curve in high frequency atmospheric pressure discharge, RF APG, sustained by 13.56 MHz, 250 W in He 1.5 L/min and O_2 5 mL/min. Electrodes: 50×150mm in width and length, gap 2 mm.

Instead of the sterility judgment on the base of the turbidity and pH indicator, we estimated the time dependent decreasing constant of the population of the survived microorganism. The biological indicator was *Geobacillus stearothermophilus* inoculated at 10^6 CFU on glass carrier and polypropylene carrier.

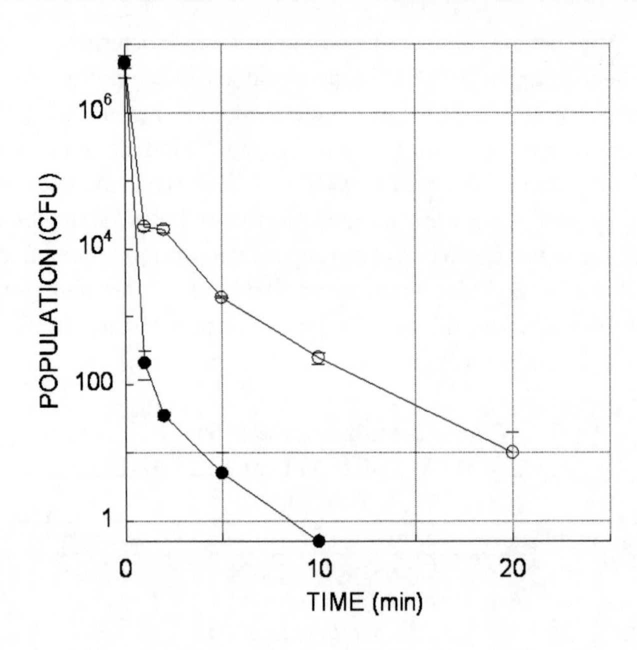

Figure 8. Survival curves for CW-APG discharge, in He 1.5 L/min. O_2 5 mL/min. Average power: 250W, Plain dielectric barrier electrode, gap 2 mm. Empty symbols show polypropylene carrier and filled symbols show glass carrier, $1.0 \pm 0.1 \times 10^6$ CFU *Geobacillus stearothermophilus*.

The experimental result indicates an interesting dependence on the material. Because this plasma discharge in helium-diluted oxygen did not emit UV-C radiations, the primary effect in the disinfection is surface-oxidation by oxygen radicals. Comparing two disinfection curves, one can find that the empty symbols, polypropylene carrier shows a typical retardation of the disinfection effect after the early stage. This result indicates an organic loading effect. Among several possible mechanisms, this retardation can originate in hydrogen covering the polymer surface. This result shows that the organic loading effect depends on the material of carriers, although we paid reasonable attention if the surface is "clean" and no organic binder is used in the inoculation step.

In the later part, we are going to discuss some review papers in which the multiple-step formation is explained as synergistically UV-C radiation and oxidation by oxygen radicals. Simpler experiment as in the present case can show similar retardation in multiple steps. At the final step of the disinfection, the number densities of cultivable spores become low. The possibility of detecting cultivable spores in this sampling process of such rare samples shows statistical behaviors in Poisson distributions. It should be noted that for the assurances of the sterility result should be proven by independent sampling without serial dilution phase.

An advantage of oxidation disinfection is that excited oxygen cleans of organic compounds such as proteins by changing to volatile compounds. Figure 9 shows optical microscope images of *Candida albicans* and *Staphylococcus aureus* recorded by inverted microscope system installed with polarization optics and digital image enhancement (Olympus Co. Ltd. Japan).

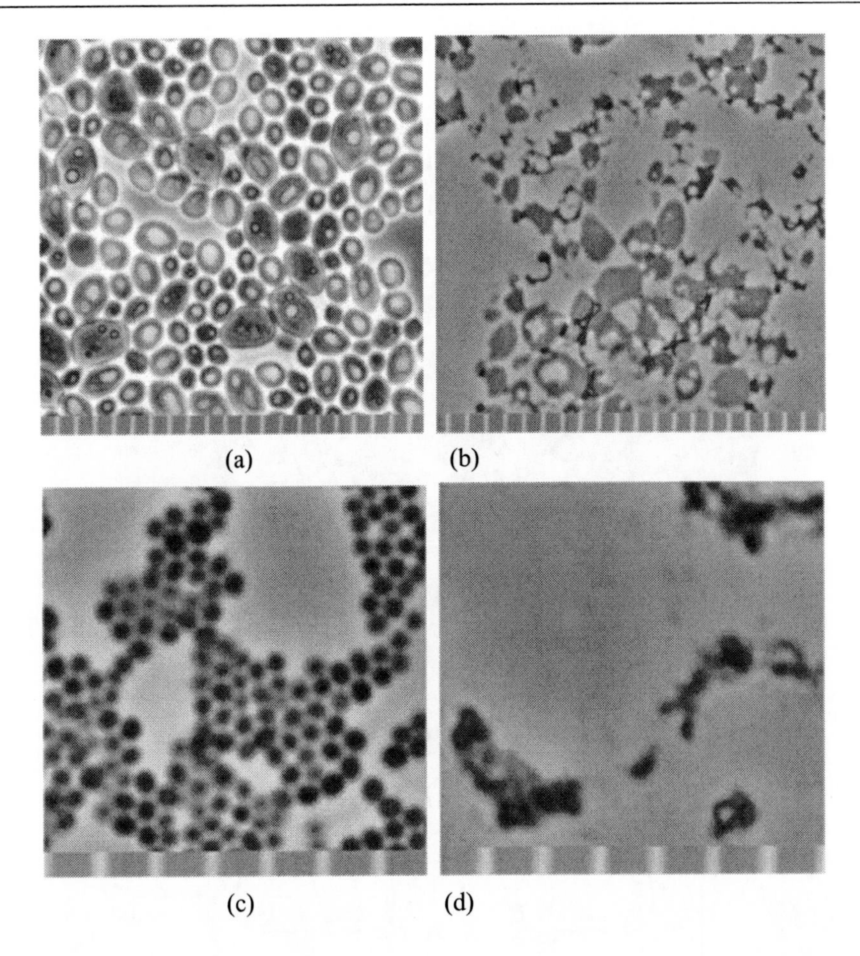

Figure 9. Sterilization and post-cleaning of *Candida albicans* before (a) and (b) after the plasma treatment, *Staphylococcus aureus* before (c) and (d) after the exposure. Observation of digital-enhanced optical microscope image. Pulse-modulated high frequency APG, 670 W, 27.12 MHz; width and interval: 10 μs; exposure time, 5 minutes. Scale: 5 μm/division.

Figure 9 (a) shows an image of living cells of *Candida albicans*, typically 5 μm in diameter. After the plasma treatment for five minutes, the spherical cellular structure was destroyed and residue of cell leakage was cleaned in part by the oxidization as shown in Fig. 9 (b). Figure 9 (c) shows a similar result for *Staphilococcus aureus* showing a round shape approximately 1 μm in diameter and (d) shows an image of a specimen exposed for five minutes. The process destroyed round shape of cells and the leakages were cleaned in part after oxidization.

Next, we describe a development of disinfection process using nitrogen and helium/nitrogen mixture. Atmospheric glow discharge in helium is attracting great attention, but consume of inert gas without recovery can result in higher processing cost. For the uniformity of the early breakdown phase, high pressure discharges can be generated by fast rising high-voltage pulse with very short pulse duration and the electrical discharge must be terminated before the arc transition. Static-induction thyristor and inductive energy storage system can provide pulse generation scheme. Figure 10 shows a temporal evolution of the voltage and current wave and a schematic drawing for a streamer discharge device. The center electrode structure consists of Re/ W wire of 0.6 mm in diameter. The wire electrode was surrounded symmetrically by a dielectric barrier; quartz tube attached with perforated aluminum ground electrode, 20 mm in diameter. The voltage increase at 10^{11} V/s and homogeneous breakdown develops along the high voltage wire. This asymmetric structure of the dielectric barrier discharge was necessary to keep the sustain phase.

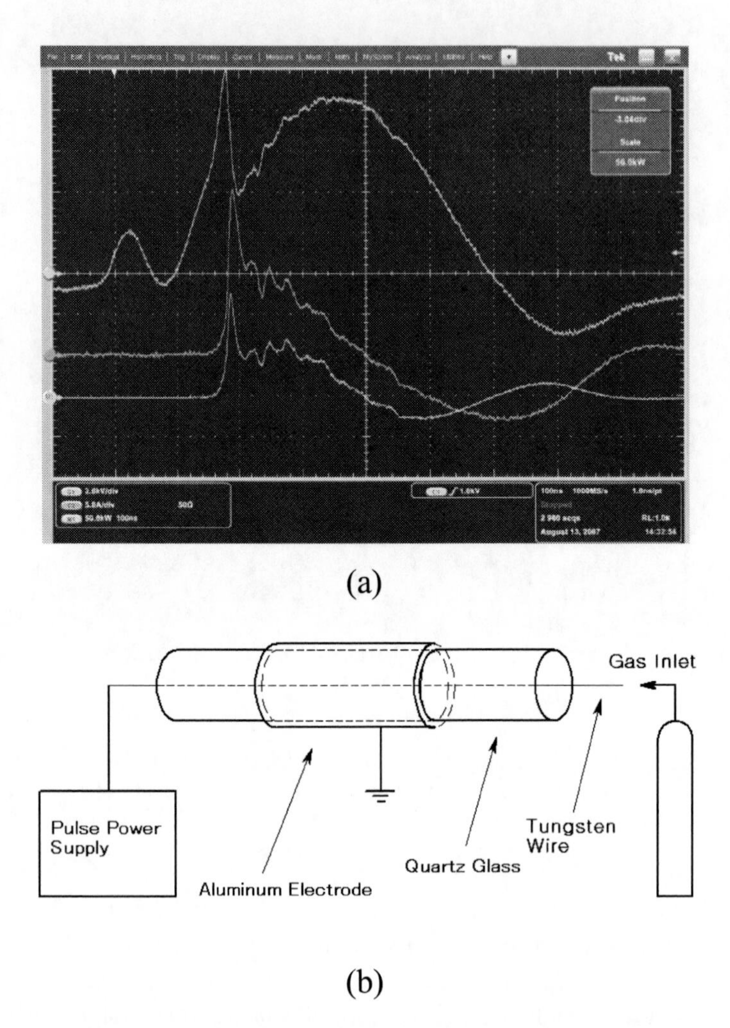

(a)

(b)

Figure 10. Inactivation of spore forming bacteria by pulsed streamer-discharges. (a)Temporal evolution of voltage pulse: breakdown, sustain and reversed voltage phase. Upper trace is voltage (5kV/div, yellow) and middle trace is current (5A/div, blue).2k pulse/s, gap distance 10 mm. The instantaneous power generated by mathematical operation (50kW/div, red). (b) Experimental setup of a cylindrical configuration: wire electrode surrounded by a coaxial, dielectric barrier electrode, diameter 20 mm, length 100 mm, at ground potential.

Survival test of selected microorganism by atmospheric pressure plasma was tested using spore forming bacteria, *Geobacillus stearothermophilus* ATCC 7953. Sterilization experiment in relatively inactive inert gas, helium and nitrogen shows antibacterial effect comparable to oxidizing process and smaller dependence on the carrier material as a result of the non-thermal excitation by the inductive energy storage pulse-power supply. For the estimation of the initial population and the number of survivors, serial dilution was carried out in the following way. 10 mL and 9 mL sequence of sterilized water was measured with a pipette. After the exposure, the spore on each carrier was expanded in 10 mL of sterile water by a super-sonic mixer, and then diluted in sterile water to 10, 10^2, 10^3, 10^4 times dilutions. 1mL was sampled from appropriate dilution and incubated. Same maneuver were repeated for untreated control indicators to determine the initial population of spores.

In the experiment, dried biological indicator was extracted from vial and set in a sterilized Petri dish. At the same time, control indicator was extracted from a vial and expanded in 10 mL of sterilized water in supersonic bath. 1 mL of the suspension was sampled with measuring pipette and added to new sterilized water of 9 ml and mixed. Similarly 1 mL was sampled and serial dilution was formed. 1 mL of dilution was sampled from appropriate stage and mixed with autoclaved non-selective meat-broth agar. This recovery medium was heated again in a boiling water bath and cooled down to 40 to 50 °C, prior to the mixing. 1 mL of suspension was mixed with this medium inside sterilized Petri dish and kept still until the medium forms solid gel.

Solidified dish was shielded with paraffin tape (NOVIX), and then incubated at 57 degrees C, for 48 hours. The antibacterial effect was evaluated using spore forming bacteria, *Geobacillus stearothermophilus.* This specie is frequently used in the evaluation of the plasma sterilization, because *Geobacillus stearothermophilus* survive at high temperature, 115 °C, and form easily identifiable colony on non-selective agar. The number of survivors was estimated from the count of colony multiplied with the dilution ratio. The survival curve shows two stages of the sterilization process. In the initial stage, the population of microorganism decreases faster than the second stage where the rate saturates. This result shows the organic loading effect in the cases of polymer materials. In the vicinity of polymer materials, the sterilization biological indicator is retarded. Pulsed discharges were excited between a sufficiently wide gap, 10 mm in He and 4.5 mm in N_2 below the required level < 60 °C, by natural convection and gas flow without forced circulation of coolant. Sterilization using nitrogen mixture shows minor dependence on the carrier materials: polypropylene and glass plate. Pulsed excitation of nitrogen discharge shows strong antibacterial effect.

Figure 11 shows a typical example for the survival curve for pulse discharge in He diluted oxygen. Average power: 30 W, in the optimized condition of He diluted N_2 discharges. Empty symbols: polypropylene carrier BI and filled symbols show glass slide cover carrier BI. In this experiment, the spore on the carrier was expanded in sterile water, 10 mL by an ultrasonic cleaner for 30 min., and then diluted in sterile water to 5 stages: 10, 10^2, 10^3, 10^4 times dilutions. Appropriate dilution stages were sampled and incubated on non-selective nutrition agar dish for 48 hours at 57 °C. In conditions near the endpoint, less dilution stage was needed.

One can find two stages of the survival curve. In the initial stage, the population of survived spores of microorganism rapidly decreases, then the rate saturates in the second stage. In the cases of polymer materials, the second stage needs long time to be clean up. This result indicates that organic material in the vicinity of the biological indicator can retard the completion of the sterilization. The result of the next experiment indicates similar problem for sterile packaged objects. Figure 11 survival curves for disinfection process of streamer discharge in helium diluted nitrogen mixture. Comparing with the result for helium/oxygen mixture, helium/nitrogen mixture discharge shows less retardation of the disinfection observed the case of polymer carriers.

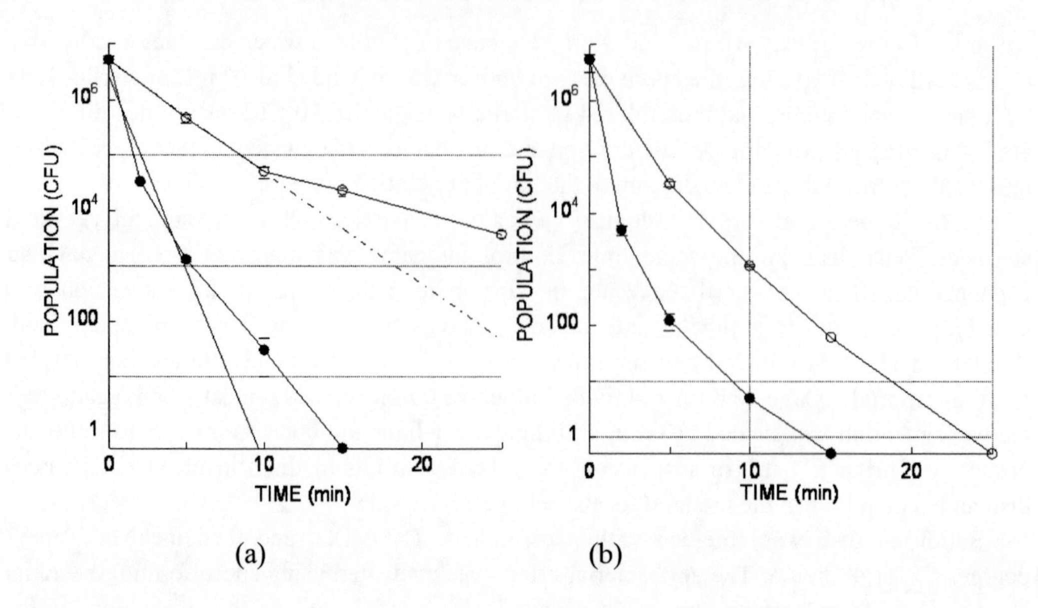

Figure 11 Comparison of pulse-excited discharge (a) helium flow rate 1 L/min. O_2 3 % and (b) N_2 1 %. Average power: 30 W. Cylindrical configuration. Empty symbols shows polypropylene carrier and filled symbols glass carrier. Spore density of control was $1.0 \pm 0.1 \times 10^6$ CFU, *Geobacillus stearothermophilus.*

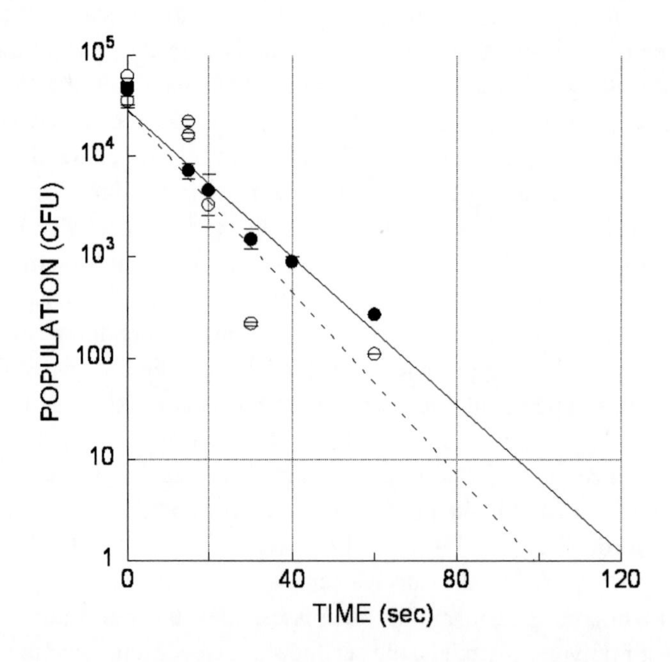

Figure 12. Survival curves for streamer discharge. N_2 60 °C, pulse width 150 ns, peak voltage 16 kV, Parallel wire configuration, three wire electrodes, 600 μm in diameter, gap 4.5 mm separated from a dielectric barrier electrode at ground level, control: $4.5 \pm 1 \times 10^4$ CFU, Disinfection time 120 s. Biological indicator: *Geobacillus stearothermophilus*.

Figure 12 shows a typical example for the survival curve for non-helium nitrogen discharge in parallel wire configuration. The gas temperature was 60 degrees C, pulse width 150 ns, peak voltage 16kV. High voltage electrodes consist of three wire electrodes, 600 μm in diameter, gap 4.5 mm separated from a dielectric barrier electrode at ground level, control: $4.5\pm1\times10^4$ CFU. The result shows that the disinfection can be completed within 120 s. This experiment was repeated changing the gas temperature and the decreasing time constant D was as small as 5 - 10 s, when the gas temperature was optimum (>60 °C). This sterilization based on the strong radiation of UV lines by N_2 and NO. A few case of unsatisfactory sterilization was caused by the presence of cold spot, probably a shadowed region in the biological indicators.

The disinfection process strongly depends on the state of the spore. Wide antibacterial spectrum was confirmed for the atmospheric pressure plasma based on helium-oxygen mixture. The spore forming bacteria shows strong resistance against the plasma treatment.

Commercially available biological indicators of spore-forming bacteria: *Attest*™ 1262 and *Attest*™ 1264, from 3M Co., were also used for screening test. *Attest*™ type 1262 vial contained *Geobacillus stearothermophilus* ATCC 7953 at 8.6x 10^5 to 1.4x10^6 CFU, type 1264 *Bacillus atrophaeus* ATCC 9372 at 5.0x10^5 to 4.8x10^6 CFU.

Experimental results shown in Tables were taken from collaboration with Dr. Hiroshi Ohkawa, one of the authors. Most of data was taken by a clump-less biological indicators developed by Yamanashi Industrial Technology Center under the care by Dr. Masao Tsuji, in collaboration. These data cover the result from PWM-APG and HF-APG in He diluted O_2, oxidation sterilization and nitrogen sterilization in helium free process. Similar effort should be dedicated towards the study on the antibacterial spectrum of the nitrogen streamer discharges using several kinds of spore forming bacteria and some pathogenic bacteria of opportunistic infection.

The compatibility with other medical care materials and research on wider antibacterial spectrum will be necessary to establish the state of this antibacterial treatment in the sterilization in the production of medical and health-care materials and precaution sterilization. Following issue is planned to complete the present study.

- ✓ Temperature dependence found in the test of antibacterial effect using *Bacillus atrophaeus* ATCC 9372 in nitrogen streamer discharges.
- ✓ Control of spore cohesion due to the water repellency of the carrier or flocculation in the suspension.
- ✓ The experimental result using *Aspergillus niger* ATCC16404 is case sensitive. The reason for this conflict should be attributed to the spore cohesion due to the water repellency.

3. The First Step of Plasma Sterilization

From the best knowledge of the authors, the early experiment on the plasma sterilization at normal atmospheric pressure was described in 1996, IEEE Trans action on Plasma Science, by Laroussi. We are curious how and when the schemes of the variations were established. In 1996, also sterilization using microwave excited plasma was reported by Chou *et al.* We found numbers of major group carrying this research area described in Boudam *et al.* (2006), Kuzmichev *et al.* (2000), Laroussi (2000), Moissan (1999), Massines (2003), and Roth (2001) [17, 20, 22-17, 55-60]. Sterilization using remote plasma and micro-plasma can be found in

Refs. [4, 61-63]. Ozone disinfection is one of remote plasma schemes, so including the invention of ozone generator, the long history from early 1890's should be included [35, 64]. Patent claims the first report of plasma sterilization by Menashi in 1968 [28].

Another patent by Ashman in 1972 describes a plasma sterilizer consisting of atmospheric pressure argon discharge excited by pulsed high frequency applied to an antenna surrounding a bottle of glass so that the inner side of the bottle contacts the plasma and is sterilized. Menashi reported in this experiment that inside of the glass bottle containing 10^6 spores was sterilized within 1 s. The sterilization time in the early stage reports of plasma sterilization is often very short. If current technique of remote plasma is used, and if it is reexamined caring about the temperature control of neutral gas, it turns out that required sterilization time is as long as 10-60 minutes. It looks as if plasma sterilization using the newest system is less efficient. However, it is thought that in the early stage experiments, sterilization was established because of short term excess heating due to the contact of the plasma onto the surface of the inside bottle. This phenomenon was named micro-incineration by Peeples and Anderson (1985) [38, 39]. If a very short time sterilization method is proposed independent of the kinds of microorganisms, one should suspect if it would be due to thermal effect. Ashman and Menashi (1972), Bithell (1982), Boucher (1980) etc. also mentioned the relation between a discharge in a gas and sterilization technique [2, 65-67].

As for experiment in the range of microwaves, Boucher (1980) realized sterilization with microwave discharge at the frequency of 2.45 GHz [66, 67]. After that an improved microwave plasma system was developed. In order to sterilize inside the glass bottle, microwave is applied from the outside, and laser pulse was used to ignite a discharge. Most of these early-stage experiments used He or Ar as an inert gas. In order to enhance the sterilization efficiency, use of halogen such as chlorine or brome (Ashman and Menashi 1972) and seeding in aldehyde as a carrier gas (Boucher 1980) were reported. Jacobs and Lin (1987) used hydrogen peroxide solution (sterilization agent) as a carrier gas instead of rare gas [2, 66-69]. This experiment consists of two step processes. An object is placed in a chamber, in which air is pumped out and H_2O_2 is introduced. It is diffused and the sterilant is condensed at the object surface, followed by the dissociation process by an RF discharge. The plasma dissociation process ensures no toxic residuals on the sterilized object, while the sterilization effect is realized by the condensation of the diffused sterilant at the surface. This method is in principle same as the hydrogen peroxide low temperature plasma sterilization method which rapidly increases the share recently [15, 68-75].

Boucher (1985) emphasized that some gases (e.g. CO_2) are effective for deactivating spores of bacteria [67]. If the spores are immersed in water prier to plasma treatment, they are more easily sterilized, reported by Ratner et al. [76]. These experimental studies showed that plasma sterilization can be realized by using various types of discharges in most discharge gases (O_2, N_2, air, H_2, halogens, N_2O, H_2O, H_2O_2, CO_2, SO_2, SF_6, aldehyde (organic acid)) [4, 61, 65, 77-79]. Boucher (1980) described the effect of UV radiation on to microorganisms [66]. The UV light of the wavelength between 250-280 nm which corresponds to 3.3-6.2 eV and the maximum absorption wavelength for deoxyribo (DNA) and other nucleic acids results in strong sterilization. However, this effect is limited by the opt-chemical reaction at the coating, and thus the depth of the UV effect is limited up to approximately 1 μm thick layer only. If bacteria, which does not form spores and is covered with thin coating, is sterilized, the UV effect for sterilization is superior to oxygen radicals. If spores with high resistance against sterilization are treated, UV influences coating on the protein. In order to treat DNA etc. at the core, these barriers must be removed by surface etching using diffusion of free radical atoms or excited molecules. Spores of *Bacillus subtilis*, which is used for the test of sterilization effect as a biological indicator, may further become deposited assemblies. Boucher (1985) concluded that plasma sterilization effect on spores was not affected by the existence or nonexistence of UV. It is however interpreted that the micro-incineration by the contact with a plasma played the major role in the sterilization as was often the case for the early stage sterilization research.

The next generation plasma sterilization experiment approached to the recent research level. Nelson and Berger (1989) showed that oxygen plasma was very efficient for sterilization of *B. subilis* and *Clostoridium sporogenes*. For this experiment surface treatment was carried out by high energy ions for which an RF (13.56 MHz) discharge is generated between two parallel plate electrodes in a reactive ion etching (RIE) system. The gas pressure was as low as 0.02-0.2 Torr, and the discharge power was 200 W. The density of *B. subtilis* decreased log 10 exceeding 3.5 after 5 minutes exposure. Fraser *et al.* (1976) also shows inactivation of *B. subtilis* by O_2 plasma, where 15-minute exposure for 300 W input power was required [80].

It is noted that at this stage the microorganisms which should be used for the evaluation of sterilization effect were chosen. The test must be performed with using the most resistant microorganisms. Since other microorganisms show lower resistance, guarantee of sterilized level can be performed by the test using one kind of standard microorganism. In the early stage, as mentioned by Boucher (1985) the American environmental protection agency demanded to test spores of both aerobic and anaerobic spore-forming bacteria (*B. subtilis* and *Clostoridium sporogenes*, respectively) in order to show the sterilized condition. The cultivation evaluation of the anaerobic bacteria requires special cultivation equipment. The current standard is the evaluation of inactivation using aerobic cultivation of *B. subtilis var. niger* (*Bacillus atrophaeus*) and *Geobacillus stearothermophilus* in order to check the guarantee of the sterilized level of autoclave and sterilizing equipment. Boucher (1980) reported that *G. stearothermophilus* was slightly more resistive than *B. subtilis* against plasma sterilization (low pressure at 1.6-1.1 Torr, air plasma). Similar result was reported by Kelley-Wintenberg *et al.* (1992) using atmospheric pressure plasma [48]. On the contrary Hury *et al.* (1998) found that *G. stearothermophilus* was easier to inactivate than *B. subtilis* using low pressure O_2 plasma [78]. Krebs *et al.* (1998) found that *G. stearothermophilus* was more resistive than *B. subtilis* in a system based on hydrogen peroxide plasma [72]. Baier *et al.* (1992) compared *G. stearothermophilus* and *E. coli*, and claimed that Gram negative bacteria was more resistive due to their excess protein and Lipopolysaccharide walls using RF (35 MHz) Ar plasma at the middle pressure range [4]. In such a way, the test technique for testing plasma sterilization system was established using vegetative propagation condition of the Gram negative such as *B. subtilis*, *G. stearothermophilus* and Escherichia coli (*E. coli*).

Kelley-Wintenberg *et al.* (1998) used atmospheric pressure discharge in air in order to inactivate microorganizms including bacteria and endospores. Plasma with relatively low ion and electron densities (10^{10} cm^{-3}) was generated at a gas temperature near the ambient temperature. The DBD was generated between series-coupled plate electrodes driven at the low frequency between 1-8 kHz. A sample was fixed at the insulator surface which covers the bottom electrode, exposed directly or via holes of a package to the plasma. The gas flow rate was established as an important parameter to promote the uniformity of a plasma. The required time for the inactivation is very short. Typically 10^6CFU *B. subtilis* spores showed the decrease of more than 5log10 in 5 minutes. As the mechanism, sterilization by reactive species such as O_2-, hydroxyl (OH), NO, hydrogen peroxide, and radicals like ozone were assumed. The survival graph obtained from the experiment shows two phases consisting of the first slow and the second fast steps. The first step has the maximum D1 (decimal reduction time). Once the concentration of the reactive species becomes high enough for fatality, the second step proceeds very promptly derived to irreversible damage of a cell and dissipation. This is the reason why the very short D2 value was obtained. In the experiment by Kelley-Wintenberg *et al.* (1998) *S. aureus* and *E. coli* ($5.0x10^4$) were coated on a propylene carrier, and the survival graphs of *S. aureus* and *E. coli* directly exposed in the atmospheric pressure DBD in air and exposed in the semi-permeability bag were compared.

The contrasting experiment was described by Khomich *et al.* (1997) [33]. In the report on the UV sterilization of low pressure DC discharge at the pressure between 0.05-0.2 Torr they showed that optical emission from a plasma did not play an essential role in the inactivation of microorganisms. The energy of photons emitted from the dc discharge will not exceed a few eV. On the contrary, Lisovskiy *et al.* (2000) showed that the ion bombardment to the microorganism surface played a major role in the inactivation as the ion energy was as high as 100-200 eV using an RF capacitively coupled parallel plate electrodes [81]. Khomich *et al.* (1998) and Soloshenko *et al.* (1999) reported that the effects of UV irradiation and neutral particles for sterilization could be separated, and concluded that the inactivation by the neutral particles took two times longer than the sterilization process in the oxygen discharge, and that 5-6 times longer in the air discharge [34, 82]. The excited neutrals are able to pass the shield plate while the effect of UV is limited due to the shadowing. Therefore sterilization of complicated shape of the object must rely on reactive neutrals rather than UV. Soloshenko *et al.* (1999, 2000) further showed that the most efficient precursor for sterilization was O_2, followed by CO_2, H_2, Ar and N_2 [82, 83]. Soloshenko *et al.* (1999) separate the effects of UV radiation and reactive neutrals, generated a DC discharge at 0.2 Torr, measured the survival evolution of *B. subtilis*, and discussed the result. The result by Soloshenko (1999) indicates that the survival curve is not a single line but contains two or three linear segments. The first step is the fastest, and thus has the smallest D value (decimal reduction time), while the second step is the slowest, and thus has the maximum D value. Soloshenko *et al.* (1999) pointed out the similarity of the survival curves among this result, sterilization using a mercury lamp with 0.1 mW/cm^2, and that of air plasma with the power density of 1.5 mW/cm^2. The inactivation by Hg lamp is a two-step process, and the first step shows the smallest D value. Warriner *et al.* (2000) obtained similar survival curves using the UV (248 nm) excimer laser [84]. Moreau *et al.* (2000) performed sterilization in an afterglow from microwave discharge sustained in pure Ar, a gas mixture of O_2 and Ar, and that of N_2 and O_2. The gas pressure in the chamber was in the range between 2-7 Torr, and the gas flow rate was in the range between 0.5-2 SLM. The test spore was *B. subtilis* which was coated on a 10 mm diameter carrier placed at the bottom of the stainless chamber. The total inactivation time for initial spore population 10^6 was approximately 40 minutes, which is much longer than the required time for the direct contact with a plasma in 2 % O_2 and 98 % N_2. Moreau *et al.* (2000) showed that the survival curves of *B. subtilis* spore in the afterglow tended to show three-step sterilization process. By adding Ar and O_2, the spores were perfectly inactivated in 40 minutes [60]. In addition similar three-step sterilization process was recognized for the survival curve of *B. subtilis* spores exposed in the after glow of O_2/N_2 mixture. In this experiment, the maximum O concentration was obtained at 15 % O_2 content in the gas mixture, while the maximum UV emission intensity was seen at 2 % O_2 in the gas mixture. It is shown that there exist three-step survival processes in the Ar afterglow sterilization mixed with O_2, by comparing the sterilization effect of Ar afterglow with and without 5 % O_2. It is thus indicated that inactivation of endospores in the Ar afterglow in this experiment derived from the effect of the UV photons. The third step which is specific for the O_2/Ar gas mixtures relates to the existence of O atoms, and addition of O_2 in Ar reduced approximately 25 % of D2 which represents the step of the most stagnated of sterilization. Ricard *et al.* (2001) determined the O_2 content in N_2 based gas mixture for producing maximum oxygen atom concentration [O] in the optimized sterilization process using titration. [O] initially increased as the increase of the O_2 content, approached to the maximum, and saturated at the O_2 content of no less than 12 %. On the contrary, the UV emission intensity which relates to the excited NO molecules produced in the afterglow became the maximum at the O_2 content of no more than 2 % [85]. In the case when the UV emission intensity is the maximum, it was shown that the efficiency of inactivation was the maximum. Concentrations of NO_α and NO_β showed different dependencies on the O_2 content in the gas. As the UV is absorbed by oxygen molecules, the excess partial pressure of O_2 in the gas mixture reduces the effect of the UV. However, in order to establish the cleaning of spores deposited at the second step, the existence of the oxidative agents like O atoms is indispensable. Moisan *et al.* (2001) pointed out synergy effect between UV photons and O atoms [20, 22-27, 30].

Lerouge *et al.* (2000) used the process of direct exposure of *B. subtilis* endospores in the microwave discharges (2.45 GHz) in various gases such as O_2, CO_2, O_2/Ar, O_2/H_2, O_2/H_2/Ar and O_2/CF_4. In order to retain low gas temperature, pulse modulated microwave power of 30 ms rest and 30 ms pulsed operation was performed. It was the proposal of plasma sterilization using the pulsed module method noticing generation of low temperature plasma [17-19, 57]. Except the case of O_2/CF_4 gas mixture in which only the first and second phases exist, the inactivation showed three-phase survival curves. In this experiment, it seems that they would like to show the analogy of the plasma sterilization to the etching of polymers and silicon type semiconductors. The plasma sterilization which mainly relies on UV light has less effect on etching, and the shows minor deformation of the spore shapes. Comparing with the O_2/CF_4 gas mixture process, it is thought that the damage to polymeric material acting as carriers is low.

4. Reactive Species Involved in Plasma Sterilization Process

Summarizing the review articles by Barbeau *et al.* (2006) etc., it is necessary to assume the following three elementary processes for the interpretation of survival curves;

[1] direct destruction of genetic material of microorganisms by UV radiation,
[2] erosion of atoms by photo-desorption: desorption induced by UV. Cutting chemical bonding, derived from photo induced dissociation reactions connecting to chemical reactions forming volatile components from atoms intrinsic of microorganisms. Byproducts are volatile molecules such as CO and CH_x.

[3] erosion of atoms by etching; he etching is due to the formation of volatile components by reactive species from a plasma or an afterglow onto the surface of microorganism. The reactive species are atomic or molecular radicals such as O, OH. Under the thermal equilibrium condition, products of the oxidation process are small molecules such as CO_2 and H_2O.

In some cases, etching is enhanced by synergy effect with reactive species induced by UV photons. These volatile products are removed by evacuation or gas exchange. A plasma discharge field, or reactive species and UV radiation highly depends on gas pressure. For example, ozone is easily generated at atmospheric pressure, but at low pressure, oxygen atoms dissociated from O_2 are transported to a surface. On the other hand, at around atmospheric pressure, UV photons are strongly reabsorbed in a plasma. This effect prevents UV from arriving at a sample surface to be sterilized. It is particularly significant for vacuum UV (VUV, whose wavelength is no more than 180 nm) photons. The UV photons generated in a low pressure discharge plasma is less affected by the re-absorption, and it is expected that the sterilization treatment by UV would be more efficient. The efficiency of UV photons in the sterilization process relates to fundamental mechanisms of inactivation of genetic material and erosion of composing atoms of microorganisms by the UV radiation. These processes depend on the intensity of UV radiation. Reactive species like metastable singlet O_2 as well as oxygen atoms and ozones are found in an O_2 based discharge gas. Photons can not be radiated by electronic dipolar transition through very high speed de-excitation process (10^8 s), and thus these excited species have long lifetimes (several s). Metastable species induce chemical reactions which activate other particles, gases, or particles on exposed sample surface, transport energy, and de-excitation without radiating photons. The metastable species thus play a role of storing energy for plasma chemical reactions. In order to explain the case when the three survival curves are obtained, the following mechanisms are assumed, in which both reactive species and UV photons exist, and spores will be inactivated by the UV photon radiation on the DNA material.

In the first step, UV directly radiates at the top layer of the dissociated spores and deposited spores and the destruction of DNA occurs. This process defines the fastest reducing time corresponding to the smallest D value. However, this process can not result in the complete sterilization as the surface can be covered with debris, or the penetration depth of UV photons is limited for the spores locating beneath deposited spores. The second step consists of the slowest mechanism corresponding to the maximum D value. It reflects the necessary time for eroding inactivated spores and the debris on living spores. In order to inactivate all genetic materials by direct UV radiation, the covering debris must be sufficiently removed by etching. When this step is completed, the third step starts and the sterilization rapidly completes. This scenario explains the reason why the third step is observed before completing sterilization. In practice in spite of the fact that complicated processes such as plasma exposure, and cascade dilution are experienced, the third step starts within relatively small fluctuation of the delay times using the indicators whose number of fungus is adjusted to equal.

In case when neutral reactive species like O atoms exist and the UV radiation intensity is the maximum, this delay becomes the shortest (Moreau *et al.* 2000, Philip *et al.* 2001). The second step inactivation mechanism is restricted by the erosion process, and thus still larger D value is observed. For the experiment using a Hg lamp or a UV laser in which photo-desorption is the only mechanism for erosion, erosion proceeds faster as etching starts to act in addition to photo-desorption by addition of oxygen. As the substrate temperature increases, the D2 time similarly becomes shorter. Laroussi *et al.* (2000) and Lerouge *et al.* (2000) observed reduction of the size of spores, directly indicating the occurrence of erosion. Since only a part of initial density of spores is inactivated, this process starts. When fluxes of oxygen atoms and UV photons exist, the final third phase starts relatively promptly. Otherwise, it can delay several hours [31, 32]. The D3 time constant is close to D1, and is the fast process. It is indicated that direct inactivation of DNA by UV radiation may occur during the third inactivation process. Sometimes D3 shows much larger value than D1, especially when sterilization treatment is performed with UV photons only using a Hg lamp or a UV laser.

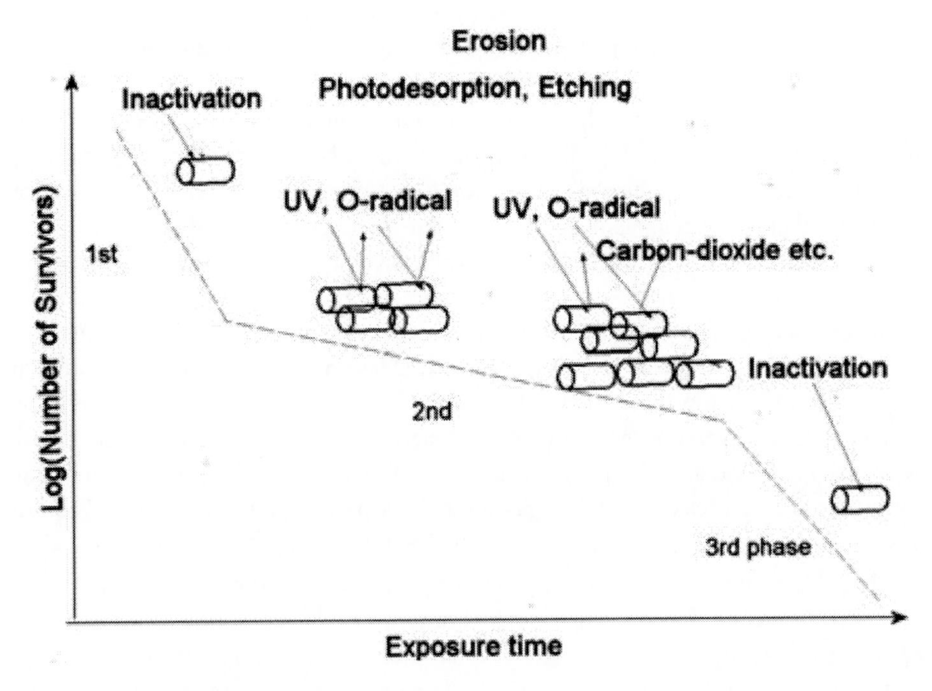

Figure 13. A survival curve for plasma sterilization.

Figure 13 shows a model of a survival curve. By measuring the number of the regenerative spores which are not inactivated as a function of time, necessary time to be reduced down to 10 % is determined. Because of the load effect of organic materials of deposited spores the second phase shows the largest D value.

The recent trend of atmospheric pressure plasma research is to replace expensive helium as a discharge medium to relatively inexpensive N_2 [54, 58, 86-88]. Boudam *et al.* (2006) and Massines *et al.* (2003, 2005) confirmed that highly uniform atmospheric pressure plasma could be generated in a small gap of approximately 500 μm between dielectric-barrier-discharge electrodes even in N_2 mixed with approximately 500 ppm O_2 or oxidative agent like N_2O (Townsend discharge mode). These components are intrinsically involved in the ambient air as main components. Unless they are activated by the discharge electric field, they will not demonstrate the effects of destructing lives. Only when the discharge is on, the reactive species exist demonstrating corresponding sterilization effect, while once the discharge is off, the radicals will disappear within as long as several ms. It therefore indicates that gas exchange time is unnecessary as is the case for EtO sterilization. Inactivation by N_2 plasma shows specific plasma sterilization, characterized by the existence of two or three identical phases in the survival curve of spores. Furthermore, because of the existence of UV radiators characteristics of N_2 plasma with small amount of O_2 like NOγ, the inactivation process dominated by UV. In the case of middle or low pressure discharge process, the inactivation by N_2 plasma is well-understood. Here, the inactivation mechanisms are inactivation of DNA by specific UV radiation, erosion of microorganisms by excited N_2 molecules and oxygen atoms, and dehydration and etching of composing atoms. It deductively is expected that the technique would also be applicable to inactivation for membrane toxin of *Escherichia coli* like endotoxin and abnormal prion by the erosion of these proteins. Since they are just proteins without genetic material, it is believed that they can not be destructed by a conventional high pressure steam sterilization process. In this case, the etching by excited molecules is the fastest inactivation process.

The experimental results introduced here are the evaluation of sterilization property of atmospheric pressure plasma in O_2 diluted with helium. The observation of optical emission spectra using multi-spectra spectroscopy did not detect intense emission in the UV-C region. The major sterilization mechanism is therefore the oxidation. Even though the process is simpler than synergetic effect with UV, survival curve shows complicated time evolution. Since each microorganism shows different sterilization time by carriers and aseptic packing, discussion considering the complicated conditions is necessary for the understanding.

As is indicated by Boudam *et al.* (2006) and Massines *et al.* (2005) for atmospheric pressure plasma in N_2, the discussion for the roles of UV photons and reactive species is already solved if it is for low pressure plasma sterilization. The erosion effect due to the synergetic effect between UV photons and O atoms is required for minimizing the sterilization time. In the case of N_2/O_2 gas mixture, a partial pressure of O_2 is required for maximizing the UV radiation intensity from NOγ band and relative density of O atoms. In practice, the highest intensity of NOγ occurs at an O_2 partial pressure lower than that when O atoms are maximum, and thus by experimental compromise the condition near the optimum is realized. The sterilization methods using a UV lamp or a UV laser only instead of plasma are particularly inefficient when objects are thickly deposited with each other. On the contrary to the sterilization by the exposure in a plasma, UV photons can be disturbed by the charging and shielding effects of organic materials, and thus requires longer time of inactivation of genetic material. The sterilization process by direct contact to a plasma can realize shorter time for sterilization than by the contact to the afterglow. The important disadvantage of the plasma sterilization relying on UV is the dependence on the actual thickness of microorganism to be inactivated which disturbs reaching the UV photons at the genetic material. This result contains the organic material loading effect for aseptic packaging and polymer carrier material. It is necessary that the erosion action works on any materials covering microorganism, suggesting that the required time for establishing sterilization would be longer.

Reports on a plasma sterilization must specify not only a plasma generation method but also initial number of bacteria and concentration of spores, preparation method of suspension, the condition of coverage by aseptic packaging, the surface condition of a carrier material. In addition the information on organic load by albumin et al. used for fixation of spores and cleanliness of carrier must be described. When low frequency sinusoidal inverter or RF power supply is used, homogeneous atmospheric pressure plasma (Townsend mode) in N_2 instead of helium is applicable only for a small gap of electrodes as small as approximately 500 μm. Since it is gas molecule which has less thermal conductivity, further advanced atmospheric pressure research requires development of new technologies for pulsed power excitation which shows less Joule heating.

ACKNOWLEDGMENT

The evaluation of microorganisms was performed by Dr Masao Tsuji at Yamanashi Industrial Technology Center. Valuable guidance for the experiment was obtained from Dr Masuhiro Kogoma at Sophia University. Many experiments and research achievements were realized by significant efforts of the postgraduate students at the Interdisciplinary Graduate School of Medicine and Engineering, and Graduates from the Electrical and Electronic System Engineering, Faculty of Engineering, University of Yamanashi.

REFERENCES

[1] Anderson, D.C. *Navy Med.* 1989, *Sept-Oct*, 9-10.
[2] Ashman, L.E.; Menashi, W.P.; *US patent*, 1972, 3 701 628.
[3] Agency for Toxic Substances and Disease Registry (ATSDR) 2003, CAS 75-21-8, UN#1040.
[4] Baier, R.E.; Carter, J.M.; Sorensen, S.E.; Meyer, A.E.; McGowan, B.D.; Kasprzak, S.A.; *J. Oral Implantol.* 1992, *18*, 236-242.
[5] Block S.S. In *Disinfection, Sterilization, and Preservation*; Block, S.S.; Ed.; 4th Ed; Lea and Febiger: London, UK, 1991; Ch 9.
[6] Kohn, W.G.; Collins, A.S.; Cleaveland, J.L.; Harte, J.A.; Eklund, K.J.; Malvitz, D.M. *Morbidity and Mortality Weekly Report (MMWR)* 2003, *RR-17*, 52.
[7] Ernest, A. *Adv. Steriliz.* 1995, *1*, 1-3.
[8] Alfa, M.J.; DeGagne, P.; Olson, N.; Puchalski, T. *Infect. Cont. Hosp. Epidemiol.* 1996, *17*, 92-100.
[9] Steelman, V.M. *Aorn. J.* 1992, *55*, 773-787.
[10] Steelman, V.M. *Urol. Nurs.* 1992, *12*, 123-127.
[11] Ehrenberg, L.; Hiesche, K.D.; Ostermqan-Golkar, S.; Wennberg, I. *Mutat. Res.* 1974, *24*, 83-103.
[12] Holyoak, G.R.; Wang, S.; Liu, Y.; Bunch, T.D. *Toxicology*, 1996, *108*, 33-38.
[13] Zhang Y.Z.; Bjursten, L.M.; Freij-Larson, C.; Kober, M.; Wesslén, B. *Biomater.* 1996, *17*, 2265-2272.
[14] Henn, G.G.; Birkinshaw, C.; Buggy, M.; Jones, E. *J. Mater. Sci. Mater. Med.* 1996, *7*, 591-595.
[15] Kyi, M.S.; Holton, J.; Ridgway, G.L. *J. Hosp. Infect.* 1995, *31*, 275-284.
[16] Rutala, W.A.; Green, M.F.; Weber, D.J. *Infect. Control. Hosp. Epidemiol.* 1999, *20*, 514-516. and *Am. J. Infect. Control.* 1998, *26*, 393-398.

[17] Lerouge, S.; Guignot, C.; Tabrizian, M.; Ferrier, D.; Yagoubi, N.; Yahia, L.H. *J. Biomed. Mater. Res.* 2000, *52*, 774-782.

[18] Lerouge, S.; Fozza, A.C.; Wertheimer, M.R.; Marchand, R.; Yahia, L.H. *Plasmas Polym.* 2000, *5,* 31-46.

[19] Lerouge, S.; Wertheimer, M.R.; Marchand, R.; Tabrizian, M.; Yahia, L.H. *J Biomed. Mater. Res.* 2000, *51*, 128-135.

[20] Moisan, M.; Barbeau, J.S.; Moreau, S.; Pelletier, J.M.; Tabrizian, M.; Yahia, L.H. *Int. J. Pharmaceutics.* 2001, *226*, 1-21.

[21] Delocroix, J.L; Bers, A. *Physique des Plasmas*; InterEditions/CNRS Edition : Paris, FR, 1994 ; Ch 1.

[22] Moisan, M.; Moreau, S.; Tabrizian, M.; Pelletier, J.; Barbeau, J.; Yahia, L.H. PCT WO 00/72889.

[23] Moisan, M.; Barbeau, J.; Pelletier, J. *Sci. Techn. Appl.* 2001, *229*, 15-28.

[24] Moisan, M.; Barbeau, J.; Pelletier, J. ; Philip, N. In *Proc. 13th Int. Colloquium Plasma Proc. (CIP2001) in Le VideSci. Techn. Appl. (Numero Special : Actes de Colloque)* 2001, 12-18.

[25] Moisan, M.; Saoudi, B.; Crevier, M.C.; Philip, N.; Fafard, E.; Barbeau, J.; Pelletier, J. In *Proc. 5th Int. Workshop Microwave Discharges* (Zinnowitz, Greifswald, DE). 2003.

[26] Moisan, M.; Moisan, M.; Barbeau, J.; Crevier, M.C.; Pelletier, J.; Philip, N.; Saoudi, B. *IUPAC Pure Appl. Chem*. 2002, *74*, 349-358.

[27] Moisan, M.; Philip, N.; Saoudi, B. *PCT WO* 2004/011039 A2, 2003.

[28] Menashi, W.P. *US Patent* 3 383 163, 1968.

[29] Moats, W.A. *J. Bacteriol.* 1971, *105*, 165-171.

[30] Moisan, M.; Pelletier J. *Microwave Excited Plasmas*; Elsevier: Amsterdam, NL, 1999.

[31] Philip, R.E.; Saoudi, B.; Barbeau, J.; Moisan, M.; Pelletier, J. In Proc. *13th Int. Colloquim Plasma Proc. (CIP2001) In Le Vide : Sci. Techn. Appl. (Numéro Spécial : Actes de Colleque)* 2001, 245-247.

[32] Philip, R.E.; Saoudi, B.; Crevier, M.C.; Moisan, M.; Barbeau, J.; Pelletier, J. *IEEE Trans. Plasma Sci.* 2002, *30*, 1429-1436.

[33] Khomich, V.A.; Soloshenko, I.A.; Tsiolko, V.V.; Mikhno, I.L. In *Proc. 12th Int. Conf. Gas Dischage Appl.* Greifswald, 1997, *2*, 740-744.

[34] Khomich, V.A.; Soloshenko, I.A.; Tsiolko, V.V.; Mikhno, I.L. In *Proc. Int. Cong. Plasma Phys.* Prague, 1998, 2745-2748.

[35] Eliasson, B.; Kogelschatz, U. *IEEE Trans. Plasma Sci.* 1991, *19*, 1063-1077.

[36] Chau, T-T.; Kao, K-C.; Blank, G.; Madrid, F. *Biomater.* 1996, *17*, 1273-1277.

[37] Griffiths, C.N.; Raybone, D. *PCT (Patent)* WO 92/15336, 1992.

[38] Peeples, R.E.; Anderson, N.R. *J. Parenter. Sci. Technol.* 1985, *39*, 2-8.

[39] Peeles, R.E.; Anderson, N.R. *J. Parenter. Sci. Technol.* 1985, *39*, 9-14.

[40] Kanazawa, S.; Kogoma, M.; Moriwaki, T.; Okazaki, S. In *Proc. 8th Int. Symp. Plasma Chem. (ISPC-8)*, Tokyo, Japan, 3, 1839-1844.

[41] Okazaki, S.; Kogoma, M. In *Proc.* 2nd *Int. Symp. High Pressure Low Temp Plasma Chem.* Kazimirz, PL, 1989, 13.

[42] Yokoyama, T.; Kogoma, M.; Moriwaki, T.; Okazaki, S. *J. Phys. D Appl. Phys.* 1990, *23*, 1125-1128.

[43] Ohkawa, H.; Akitsu, T.; Tsuji, M.; KImura, H.; Fukushima, K. *Plasma Proc. Polym.* 2005, *2*, 120-126.

[44] Ohkawa, H.; Akitsu, T.; Tsuji, M.; Kimura, H.; Kogoma, M.; Fukushima, K. *Surf. Coat. Technol.* 2006, *200*, 6829-6835.

[45] Akitsu, T.; Ohkawa, H.; Tsuji, M. *Surf. Coat. Technol.* 2005, *193/1-3*, 29-34.

[46] Deng, X.T.; Shi, J.J.; Shama, G.; Kong, M.G. *Appl. Phys. Lett.* 2005, *87*, 153901.

[47] Holt, J.G. In *Berge's Manual of Systematic Bacteriology*; Williams & Wilkins: Baltimore, US, 1986; 2, pp 1130-1135.

[48] Kelly-Wintenberg, K.; Montie, T.C.; Brickman, C.; Roth, J.R.; Carr, A.K.; Sorge, K.; Wadsworth, L.C.; Tsai, P.P.Y. *J. Indust. Microbiol. Biotechnol.* 1998, *20*, 69-74.

[49] Levinson, W.; Jawetz, E. *Medical Microbiology & Immunology: Examination & Board Review*; 7th Ed Lange Medical Book, McGraw-Hill/Appleton & Lange: New York, US, 2002.

[50] Nadine, T.N.; Tourova, T.P.; Polaris, A.B.; Novice, E.V.; Grégorian, A.A.; Ivanova, A.E.; Lusaka, A.M.; Petunia, V.V.; Asimov, G.A.; Belayed, S.S.; Ivan, M.V. *Int. J. Sys. Evolutionary Microbiol.* 2001, *51*, 433-446.

[51] Nelson, C.L.; Berger, T.J.; *Curr. Microbiol.* 1989, *18*, 275-276.

[52] Nicholson, W.L.; Munakata, N.; Horneck, G.; Melosh, H.J.; Setlow, P. *Microbiol. Mol. Rev.* 2000, *64*, 548-572.

[53] The Japanese Pharmacopoeia JP14th, Tokyo, JP, B-628, 2001, *1*, 87-89 (in Japanese). Eng. ver. available at http://jpdb.nihs.go.jp/jp14e/.

[54] Boucher, (Gut). R.M. *Med. Device Diagnost Indust.* 1985, *7*, 51-56.

[55] Kuzmichev, A.I.; Soloshenko, I.A.; Tsiolko, V.V.; Kryzhanovsky, V.I.; Bazhenov, V.Yu; Mikhno, I.L. In *Proc. Int. Symp. High Pressure Low Temp. Plasma Chem. (HAKONE VII)*, Greifswald, 2000, *2*, 402-406.

[56] Laroussi, M. *IEEE Trans. Plasma Sci.* 1996, *24*, 1188-1191.

[57] Lerouge S.; Wertheimer, M.R.; Yahia, L.H. *Plasma Polym.* 2001, *6*, 175-188.

[58] Massines, F.P.; Segur, P.N.; Gherardi, N.C.; Khamphan, C.; Ricard, A. *Surf. Coat. Technol.* 2003, *174-175*, 8-14.

[59] Montie, T.C.; Kelly-Wintenberg, K.; Roth, J.R. *IEEE Trans. Plasma Sci.* 2000, 28, 41-50.

[60] Moreau, S.; Moisan, M.; Tabrizian, M.; Barbeau, J.; Pelletier, J.; Ricard, A. *J. Appl. Phys.* 2000, *88*, 1166-1174.

[61] Bankupalli, S.; Dhali, S.; Madigan, M. In *Proc. 17th Int. Symp. Plasma Chem.*; Mostaghimi, J.; Coyle, T.W.; Pershin, V.A.; Salimi Jazi, H.R.; Ed.; Toronto, CA, ***2005***, pp 1072.

[62] Becker, K.; Koutsospyros, A.; Yin, S.M. ; Christodoulatos, C. ; Abramzon, N. ; Jaoquin, J.C. ; Brelles-Marino, G. *Plasma Phys. Control Fusion.* 2005, *47*, B513-B523.

[63] Birmingham, J.G. *IEEE Trans. Plasma Sci.* 2004, *32*, 1526-1531.

[64] Kogelschatz, U. *Plasma Chem. Plasma Proc.* 2003, *23*, 1-46.

[65] Bithell, R.M. *US Patent* 4 321 232, 1982.

[66] Borneff-Lipp, M.; Okpara, J.; Bodendorf, M.; Sonntag, H.G. *Hyg. Mikrobiol.* 1997, *3*, 21-28.

[67] Boucher, (Gut). R.M. *US Patent* 4 207 286, 1980.

[68] Jacobs, P.T.; Lin, S.M. *US Patent* 4 643 876, 1987.

[69] Jacobs, P.; Kowatsch, R. *Endosc. Surg.* 1993, *1*, 57-58.

[70] Brandenburg, R.; Maiorov, V.A.; Golubovskii, Yu.B.; Wagner, H-E.; Behnke, J.; Behnke, J.F. *J. Phys. D Appl. Phys.* 2005, *38*, 2187-2197.

[71] Cariou-Travers, S.; Darbord, J.C. *Sci. Tech. Appl.* 2001, *299*, 34-46.

[72] Krebs, M.C.; Bécasse, P.; Verjat, D.; Darbord, J.C. *Int. J. Pharmac.* 1998, *160*, 75-81.

[73] Marcos-Martin, M.A.; Bardat, A.; Schmitthaeusler, R.; Beysens, D. *Pharm. Technol. Eur.* 1996, *8*, 24-32.

[74] Vassal, S.; Favennec, L.; Ballet, J.J.; Brasseur, P. *Am. J. Infect. Cont.* 1998, *26*, 136-138.

[75] Vickery, K.; Deva, A.K.; Zou, J.; Kumaradeva, P.; Bissett, L.; Cossart, Y.E. *J. Host. Infect.* 1999, *41*, 317-322.

[76] Ratner, B.D.; Chilkoti, A.; Lopez, G.P. *Plasma Deposition, Treatment, and Etching of Polymers*; Academic Press: Boston, US, 1990.

[77] Birmingham, J.G.; Hammerstrom, D.J. *IEEE Trans. Plasma Sci.* 2004, *28*, 51-55.

[78] Hury, S.; Vidal, D.R.; Desor, F.; Pelletier, J.; Lagarde, T. *Lett. Appl. Microbiol.* 1988, *26*, 417-421.

[79] Pons, M.; Pelletier, J.; Joubert, O. *J. Appl. Phys.* 1994, *75*, 4709-4715.

[80] Fraser, S.J.; Gillette, R.B.; Olsen, R.L. *US Patent* 3 948 601, 1976.

[81] Lisonsky, V.A.; Yakovin, S.D.; Yegorenkov, V.D.; Terent'eva, A.G. *Probl. Atomic Sci. Technol.* 2000, *1*, 77-81.

[82] Soloshenko, I.O.; Khomich, V.A.; Tsiolko, V.V.; Mikhno, I.L.; Shchedrin, A.I.; Ryabtsev, A.V.; Bazhenov, V.Yu. in *Proc. 14th Int. Symp. Plasma Chem.* Prague, 1999, 2551-2556.

[83] Soloshenko, I.O.; Tsiolko, V.V.; Khomich, V.A.; Shchedrin, A.I.; Ryabtsev, A.V.; Bazhenov, V.Yu.; Mikhno, I.L. *Plasma Phys. Rep.* 2000, *26*, 792-800.

[84] Ricard, A.; Moisan, M.; Moreau, S. *J. Phys. D Appl. Phys.* 2001, *34*, 1203-1212.

[85] Warriner, K.; Rysstad, G.; Murden, A.; Rumsby, P.; Thomas, D.; Waites, W.M. *J. Appl. Microbiol.* 2000, *88*, 678-685.

[86] Massines, F.; Gherardi, N.; Fornelli, A.; Martin, S. *Surf. Coat. Technol.* 2005, *200,* 1855-1861.

[87] Massines, F.; Gherardi, N.; Naud, N.; Ségur, P. *Plasma Phys. Control Fusion* 2005, *47*, 1-12.

[88] Massines, F.; Gherardi, N.; Naudé, N.; Ségur, P. *Plasma Phys. Control Fusion* 2005, *47*, B577-B588.

INDEX

D

E

F

G

H

I

K

L

M

N

O

P

Q

R

T

U

V

W

X

Y

Z